CW01072460

"Tim Mann's *Defending Legal Freedoms in Indonesia* is an exciting new history and analysis of cause lawyering in Indonesia that sheds new light on the challenges to and successes of the major cause lawyering movement there. It provides a framework for thinking about the rocky course of cause lawyering and public interest law more broadly in Asia and beyond, and will become a crucial text for academics and lawyers as they think about these important areas."

Mark Sidel, Doyle-Bascom Professor of Law and Public Affairs, University of Wisconsin-Madison

"In this outstanding study, Tim Mann brings alive the many dilemmas and obstacles members of Indonesia's Legal Aid Institute have encountered when using the legal system to extend and defend democratic rights. As well as providing a definitive account of Indonesia's most distinguished human rights organisation, *Defending Legal Freedoms in Indonesia* is packed with valuable insights for scholars and activists anywhere interested in democratic decline and how to resist it."

Edward Aspinall, Professor of Politics, Department of Political & Social Change, Coral Bell School of Asia Pacific Affairs

Defending Legal Freedoms in Indonesia

Defending Legal Freedoms in Indonesia provides fresh insights into how cause lawyers navigate political and institutional change, by presenting and analysing the Indonesian Legal Aid Foundation (YLBHI), the oldest and most influential legal and human rights organisation in Indonesia.

Based on rich ethnographic research, this book charts the developments of the organisation since its founding in 1970, its contribution to the ending of the authoritarian, military-backed New Order (1966–98), its relative decline in the years following Indonesia's democratisation and its revival in recent years as Indonesian democracy and human rights have come under threat. The author examines the tactics the organisation has used, including show trials and cooperation with grassroots communities, organising them and educating them about their rights. It highlights how this organisation flourished more under an authoritarian regime than under democracy and how its present, prominent, adversarial-political version of cause lawyering is playing a leading role in civil society's resistance to further erosion of democracy and human rights. The book addresses recent democratic erosion under President Joko Widodo, and documents pivotal moments in Indonesia's contemporary history, such as the 'Reform Corrupted' mass demonstrations in 2019, illuminating how democracy shrinks, and how lawyers push back.

The first book on Indonesia's crucially important cause lawyering, activist lawyers' group, Defending Legal Freedoms in Indonesia will be of interest to researchers in Asian Law and Indonesian Studies. It is also an essential point of reference for future research in public lawyering in Asia.

Tim Mann works in the development sector. He is an associate of the Centre for Indonesian Law, Islam and Society (CILIS) at Melbourne Law School, University of Melbourne.

Routledge Law in Asia

Series editors:
Randall Peerenboom
Pip Nicholson
Michael Ng

16 Constitutional Interpretation in Singapore
Theory and Practice
Jaclyn L. Neo

17 Politics and Constitutions in Southeast Asia
Marco Bünte and Björn Dressel

18 Law and Society in Malaysia
Pluralism, Religion and Ethnicity
Edited by Andrew Harding and Dian Shah

19 Judicial Reform in Taiwan
Democratization and the Diffusion of War
Edited by Neil Chisholm

20 Constitutional Change in Singapore
Reforming the Elected Presidency
Edited by Jaclyn L. Neo and Swati S. Jhaveri

21 Crime and Punishment in Indonesia
Edited by Tim Lindsey and Helen Pausacker

22 Traditional Communities in Indonesia
Law, Identity, and Recognition
Lilis Mulyani

23 Defending Legal Freedoms in Indonesia
The Indonesian Legal Aid Foundation and Cause Lawyering in an Age of Democratic Decline
Tim Mann

Defending Legal Freedoms in Indonesia

The Indonesian Legal Aid Foundation and Cause Lawyering in an Age of Democratic Decline

Tim Mann

LONDON AND NEW YORK

First published 2025
by Routledge
4 Park Square, Milton Park, Abingdon, Oxon OX14 4RN

and by Routledge
605 Third Avenue, New York, NY 10158

Routledge is an imprint of the Taylor & Francis Group, an informa business

British Library Cataloguing in Publication Data
A catalogue record for this book is available from the British Library

ISBN: 978-1-032-78245-4 (hbk)
ISBN: 978-1-032-78246-1 (pbk)
ISBN: 978-1-003-48697-8 (ebk)

DOI: 10.4324/9781003486978

Typeset in Times New Roman
by Taylor & Francis Books

Contents

Illustrations

Figures

Table

Preface and acknowledgements

This book focuses on Indonesian cause lawyers and the strategies they use to promote social change. I owe a tremendous debt to the many Indonesian lawyers and activists who have dedicated their lives and careers to defending poor and marginalised Indonesians, and tirelessly fighting for Indonesian democracy, under incredibly challenging and sometimes dangerous circumstances.

Although this book is largely concerned with how cause lawyers push back against democratic decline, I began thinking about this project about a decade ago with a very different question in mind: what happens to activist lawyers after they get what they want? Under Soeharto's authoritarian New Order (1966–98), the Indonesian Legal Aid Foundation (*Yayasan Lembaga Bantuan Hukum Indonesian,* YLBHI) had developed a reputation as a hub of resistance to authoritarian rule, an outspoken champion of democratic values and human rights in the face of significant state repression. When the Soeharto regime finally collapsed in 1998, Indonesia began a sweeping process of democratic reform, instituting many of the reforms that lawyers had been demanding. Despite this, after the democratic transition, there was a perception that YLBHI had failed to meet the promise it had demonstrated during the authoritarian New Order years. In the early 2010s, I worked with the Jakarta office of the international nongovernmental organisation The Asia Foundation. During this period, I managed grants to YLBHI and its regional Legal Aid Institutes (*Lembaga Bantuan Hukum,* LBH), and came to know many current and former staff members. I occasionally heard disparaging comments from other donor organisations, and members of Indonesian civil society, suggesting that YLBHI had struggled to adapt to the changing times. Some believed that the democratic transition had provided new opportunities to contribute to policy reform and YLBHI had yet to capitalise on them.[1] Donors sometimes complained that YLBHI continued to deploy blunt direct action strategies, rather than undertaking "nuanced political and stakeholder analysis" that would allow for the development of "sophisticated and tailored" approaches to reform (Antlöv, Brinkerhoff, and Rapp 2010, 428).[2]

This was a view to which I was sympathetic – partially, at least. It is significant that most people expressing this viewpoint were, like me, based at international donor organisations, which tend to prioritise governance programming and formal policy change. I have no doubt that my own views were also coloured by my experiences working for a donor organisation with a strong governance focus. I was initially interested in exploring whether – and why – YLBHI had found it difficult to adapt to the democratic transition. As this book will make clear, however, my views quickly changed, especially as the extent of Indonesia's democratic decline became clear.

Before I begin, I must thank the people who helped this book come into being. It would not have been possible without the support of YLBHI. In particular, I would like to thank Asfinawati (Asfin), Muhamad Isnur, Siti Rakhma Mary Herwati (Rakhma), Febi Yonesta (Mayong), Arip Yogiawan (Yogi), Era Purnama Sari, Fanti Yusnita and April Pattiselanno Putri, along with the rest of the YLBHI staff. Thank you for your wisdom and enthusiasm – you were exceedingly accommodating and generous with your time. It was inspiring to come to Diponegoro 74 every day and learn from you all.

I would also like to thank Syahrul from LBH Banda Aceh, Ismail Lubis from LBH Medan, Wendra Rona Putra from LBH Padang, Arif Maulana, Pratiwi Febry and Muhammad Rasyid Ridha from LBH Jakarta, Willy Hanafi, Lasma Natalia and Abdul Muit Pelu from LBH Bandung, Abdul Wachid Habibullah, Mohamad Soleh, Habibus Shalihin and Sahura from LBH Surabaya, Haswandy Andy Mas, Rezky Pratiwi, Andi Muhammad Fajar Akbar, Edy Kurniawan Wahid, Salman Azis, Abdul Aziz Dumpa and Ady Anugrah Pratama from LBH Makassar, Zainal Arifin, Eti Oktaviani and Herdin Pardjoangan from LBH Semarang, Emanuel Gobay from LBH Papua, Barita Lumbanbatu and the many other LBH staff, including finance and support staff, who I spoke to in more casual conversations. Thank you, too, to Abdul Fatah, Alghiffari Aqsa, Carolina S. Martha, Dadang Trisasongko, Erasmus AT Napitupulu, Hasbi Abdullah, Hendardi, Herlambang P. Wiratraman, Indro Sugianto, Iqbal Felisiano, Luhut Pangaribuan, Meila Nurul Fajriah, Nurkholis Hidayat, Nursyahbani Katjasungkana, Patra M. Zen, Poengky Indarti, Ricky Gunawan, Robertus Robet, Taufik Basari, Todung Mulya Lubis, Yasmin Purba, Yunita, Patrick Burgess, Craig Ewers, Mohamad Doddy Kusadrianto, Renata Arianingtyas, Ajeng Wahyuni and Windu Kisworo.

This book started out as a PhD project, and I remain extraordinarily grateful to my supervisors, Tim Lindsey and Jennifer Beard, at Melbourne Law School, University of Melbourne. Tim, your enthusiastic encouragement, mentorship and friendship over several years has meant a great deal to me. You have been unwaveringly supportive of this project ever since I first began speaking about it with you years ago. Likewise, Jenny, I would not have completed my thesis without your thoughtful and incisive questions and calm encouragement. I could not have asked for better or more complementary supervisors. I am truly thankful to you both.

My thesis depended on the funding and scholarship I received from the University of Melbourne, and the assistance of the staff of the Research Office of the Melbourne Law School. I would also like to thank current and former staff at the Centre for Indonesian Law, Islam and Society (CILIS), Debbie Yu,

Ade Suharto, and Kathryn Taylor, for their support. I am indebted to Bivitri (Bibip) Susanti and Jentera Indonesia School of Law for facilitating and supporting my research process in Indonesia. I must also express my gratitude to other scholars and students in Melbourne, particularly Helen Pausacker, Annisa Beta, Tessa Toumbourou, Ken Setiawan and Laura (Lette) Nevendorff, whose guidance, advice and friendship helped me get through my PhD.

Additional research for this book was completed at the Nordic Institute of Asian Studies (NIAS) at the University of Copenhagen. It was a sad time to be at this important institution during its last year of operation. The closure of NIAS is an incredible loss for Asian studies in the Nordic region. I am grateful for the institutional support I received during my time there, and the opportunity to work with impressive and inspiring colleagues like Birgit Bräuchler, Terese Gagnon, Myunghee Lee, Cécile Medail, Rubkwan Thammaboosadee, Mai Van Tran, and Dechun Zhang. Thanks, too, to Inga-Lill Blomkvist, Frode Hübbe, Gerald Jackson, Ida Nicolaisen and Fanny Töpper for your warmth and companionship during the strange final months of NIAS.

I would also like to thank my friends and colleagues in Indonesia, from whom I have learned so much over the years. I would especially like to thank Sandra Hamid, Laurel MacLaren, Budhy Munawar Rachman, Lies Marcoes, Hana Satriyo, the late Leopold Sudaryono, Kharisma Nugroho, Fitriana Nur, Ahsan Jamet Hamidi, Yurifa and many more current and former staff at The Asia Foundation in Jakarta. I am equally grateful to current and former staff at Yayasan Humanis dan Inovasi Sosial, including Tunggal Pawestri, Jonta Saragih, Siska Dewi Noya (Chika) and Tazia Teresa Darryanto. You all shaped my understanding of Indonesia and Indonesian civil society in important ways. Thank you for your generosity in sharing your knowledge and experience and, above all, your friendship.

Finally, love and thanks to my parents, Susanne and David, for your constant support. This book is dedicated to them.

Notes

1 See Davis (2015, 4) for an example of this critique.

2 Antlöv, Brinkerhoff, and Rapp make this observation about Indonesian civil society in general terms, but their phrasing here mirrors the kind of criticism that I often heard about YLBHI.

References

Antlöv, Hans, Derick W. Brinkerhoff, and Elke Rapp. 2010. "Civil Society Capacity Building for Democratic Reform: Experience and Lessons from Indonesia." *VOLUNTAS: International Journal of Voluntary and Nonprofit Organizations 21 (3)*: 417–39. https://doi.org/10.1007/s11266-010-9140-x.

Davis, Ben. 2015. "*Financial Sustainability and Funding Diversification: The Challenge for Indonesian NGOs.*" Cardno, prepared for the Department of Foreign Affairs and Trade.

List of abbreviations

AJI (Aliansi Jurnalis Independen)	Alliance of Independent Journalists
amdal (analisis mengenai dampak lingkungan, or analisis dampak lingkungan)	environmental impact assessment
Bakin (Badan Koordinasi Intelijen Negara)	State Intelligence Coordination Agency
Bappenas, or officially, Kementerian PPN (Badan Perencanaan Pembangunan Nasional, or Kementerian Perencanaan Pembangunan Nasional/Badan Perencanaan Pembangunan Nasional)	National Development Planning Agency, or the Ministry of National Development Planning/National Development Planning Agency
BEM (Badan Eksektutif Mahasiswa)	Student Executive Body
BIN (Badan Intelijen Negara)	State Intelligence Agency
BMUA (Baku Mutu Udara Ambien)	Ambient Air Quality Standards
BNN (Badan Narkotika Nasional)	National Narcotics Agency
BPHN (Badan Pembinaan Hukum Nasional)	National Law Development Agency
BPN (Badan Pertanahan Nasional)	National Land Agency
CLS	Critical Legal Studies
CSO	civil society organisation
Densus 88 (Detasemen Khusus 88)	Special Detachment 88, the Indonesian National Police's counter-terrorism unit
DFAT	Australian Department of Foreign Affairs and Trade
DPD (Dewan Perwakilan Daerah)	Regional Representative Council
DPR (Dewan Perwakilan Rakyat)	People's Representative Council, the national legislature

DPRD (Dewan Perwakilan Rakyat Daerah)	Regional People's Representative Council
Elsam (Lembaga Studi dan Advokasi Masyarakat)	Institute for Policy Research and Advocacy
Fitra (Forum Indonesia untuk Transparansi Anggaran)	Indonesian Forum for Budget Transparency
FPI (Front Pembela Islam)	Islamic Defenders Front
FPKR (Forum Pemurnian Kedaulatan Rakyat)	People's Sovereignty Purification Forum
Gafatar (Gerakan Fajar Nusantara)	Archipelagic Dawn Movement
Golkar (Golongan Karya)	Functional Groups, the political party of former President Soeharto
HAM (hak asasi manusia)	human rights
ICCPR	International Covenant on Civil and Political Rights
ICEL	Indonesian Centre for Environmental Law
IDLO	International Development Law Organisation
ICJR	Institute for Criminal Justice Reform
ICW	Indonesia Corruption Watch
Ikadin (Ikatan Advokat Indonesia)	Indonesian Advocates League
ISSI (Institut Sejarah Sosial Indonesia)	Indonesian Institute of Social History
ITB (Institut Teknologi Bandung)	Bandung Institute of Technology
IUP (Izin Usaha Penambangan. May be specified further, as in IUP Eksplorasi, IUP Operasi Produksi)	Mining Business Licence. Mining Business Exploration Licence, or Mining Business Production Licence.
JMPPK (Jaringan Masyarakat Peduli Pegunungan Kendeng)	Community Network for the Kendeng Mountains
Kalabahu (Karya Latihan Bantuan Hukum)	Legal Aid Training, or Legal Aid Workshop
KIPP (Komisi Independen Pemantau Pemilu)	Independent Election Monitoring Committee
KLHS (Kajian Lingkungan Hidup Strategis)	Strategic Environmental Study
Komar (Komunitas Milah Abaraham)	Abrahamic Religion Community
KontraS (Komisi untuk Orang Hilang dan Korban Tindak Kekerasan)	The Commission for Disappeared and Victims of Violence
Komjak (Komisi Kejaksaan)	Prosecutors Commission

Komnas HAM (Komisi Nasional Hak Asasi Manusia)	National Commission on Human Rights
Komnas Perempuan (Komisi Nasional Anti Kekerasan Terhadap Perempuan)	National Commission on Violence Against Women
Kompolnas (Komisi Kepolisian Nasional)	National Police Commission
Kopkamtib (Komando Pemulihan Keamanan dan Ketertiban)	Command for the Restoration of Security and Order
Kostrad (Komando Strategis Angkatan Darat)	Army Strategic Reserve Command
KPA (Konsorsium Pembaruan Agraria)	Consortium for Agrarian Reform
KPK (Komisi Pemberantasan Korupsi)	Corruption Eradication Commission
KRHN (Konsorsium Reformasi Hukum Nasional)	National Consortium for Law Reform
KSUM (Komite Solidaritas untuk Marsinah)	Solidarity Committee for Marsinah
KUHP (Kitab Undang-Undang Hukum Pidana)	Criminal Code
KUHAP (Kitab Undang-Undang Hukum Acara Pidana)	Code of Criminal Procedure
KY (Komisi Yudisial)	Judicial Commission
LBH (Lembaga Bantuan Hukum)	Legal Aid Institute
LBH Apik (Lembaga Bantuan Hukum Asosiasi Perempuan Indonesia untuk Keadilan)	Legal Aid Institute of the Indonesian Women's Association for Justice (not related to the YLBHI network)
LBH Masyarakat, often styled as LBHM (Lembaga Bantuan Hukum Masyarakat)	Community Legal Aid Institute (not related to the YLBHI network)
LBH Pers (Lembaga Bantuan Hukum Pers)	Legal Aid Institute for the Press (not related to the YLBHI network)
LPH YAPHI (Lembaga Pengabdian Hukum Yekti Angudi Piadeging Hukum Indonesia)	YAPHI Legal Services Agency

LeIP (Lembaga Kajian dan Advokasi untuk Independensi Peradilan)	Indonesian Institute for an Independent Judiciary
LPSK (Lembaga Perlindungan Saksi dan Korban)	Witness and Victim Protection Agency
Malari (Malapetaka Lima Belas Januari)	The Fifteenth of January Disaster
MaPPI, sometimes MaPPI FHUI (Masyarakat Pemantau Peradilan Indonesia, Fakultas Hukum Universitas Indonesia)	Indonesian Court Monitoring Society (of the Faculty of Law, University of Indonesia)
MKMK (Majelis Kehormatan Mahkamah Konstitusi)	Honorary Council of the Constitutional Court
Monas (Monumen Nasional)	National Monument
MPR (Majelis Permusyawaratan Rakyat)	People's Consultative Assembly
MUI (Majelis Ulama Indonesia)	Indonesian Council of Ulama
Nasdem	National Democrat Party
NKK (Normalisasi Kehidupan Kampus)	'Normalisation' of Campus Life
NKTSAN (Negeri Karunia Tuan Semesta Alam Nusantara)	Archipelagic Nation of the Grace of the Lord of the Universe
Opsus (Operasi Khusus)	Special Operations
PBHI (Perhimpunan Bantuan Hukum Indonesia)	Indonesian Legal Aid Association
PDI (Partai Demokrasi Indonesia)	Indonesian Democratic Party
PDI-P (Partai Demokrasi Indonesia Perjuangan)	Indonesian Democratic Party of Struggle
Peradin (Persatuan Advokat Indonesia)	Indonesian Advocates Union
perda (peraturan daerah)	regional regulation
Perkap (Peraturan Kepala Kepolisian Republik Indonesia)	Police Chief Regulation
Perludem (Perkumpulan untuk Pemilu dan Demokrasi)	Association for Elections and Democracy
Perppu, or Perpu (Peraturan Pemerintah Pengganti Undang-Undang)	Government Regulation in Lieu of Law, or interim emergency law
PI (Perhimpoenan Indonesia)	Indonesia Association
PK (peninjauan kembali)	reconsideration, or reopening of a case by the Supreme Court

PKI (Partai Komunis Indonesia)	Indonesian Communist Party
PKS (Partai Keadilan Sejahtera)	Prosperous Justice Party
Polri (Kepolisian Negara Republik Indonesia)	Indonesian National Police
PPP (Partai Persatuan Pembangunan)	United Development Party
PRD (Partai Rakyat Demokratik)	Democratic People's Party
PSHK (Pusat Studi Hukum dan Kebijakan Indonesia)	Indonesian Centre for Law and Policy Studies
PTUN (Pengadilan Tata Usaha Negara)	State Administrative Court
PUDI (Partai Uni Demokrasi Indonesia)	Indonesian United Democracy Party
RKUHP/RUU KUHP (Rancangan Undang-Undang Kitab Undang-Undang Hukum Pidana)	Draft revised Criminal Code, or Criminal Code bill
RUU (Rancangan Undang-Undang)	draft law or bill
RUU Minerba (Rancangan Undang-Undang Pertambangan Mineral dan Batu Bara)	bill on mining of minerals and coal
RUU Pertanahan (Rancangan Undang-Undang Pertanahan)	bill on land (tenure)
RUU TPKS (Rancangan Undang-Undang Tentang Tindak Pidana Kekerasan Seksual)	bill on the crime of sexual violence
SAFEnet	Southeast Asian Freedom of Expression Network
SBSI (Serikat Buruh Sejahtera Indonesia)	Indonesian Workers Welfare Union
SDP (Sistem Database Pemasyarakatan)	Corrections Database System
Sespimti, or Sespimti Sespim Lemdiklat Polri (Sekolah Staf dan Pimpinan Tinggi Lembaga Pendidikan dan Pelatihan Polisi Republik Indonesia)	Senior Staff and Leadership College of the Indonesian National Police

SIUPP (Surat Izin Usaha Penerbitan Pers)	Press Publication Business Licence
SIUPP (Solidaritas Indonesia untuk Pembebasan Pers)	Indonesian Solidarity for Press Freedom
Supersemar (Surat Perintah Sebelas Maret)	11 March Instruction
Surpres (surat presiden)	Presidential Letter
TMII (Taman Mini Indonesia Indah)	Beautiful Indonesia in Miniature Park
TNI (Tentara Nasional Indonesia)	Indonesian military
TPHKI (Tim Pembela Hukum dan Keadilan Indonesia)	Team of Defenders of Law and Justice in Indonesia
UGM (Universitas Gadjah Mada)	Gadjah Mada Univeristy
USAID	United States Agency for International Development
Walhi (Wahana Lingkungan Hidup Indonesia)	Indonesian Forum for the Environment
Wantimpres (Dewan Pertimbangan Presiden)	Presidential Advisory Council
WIUP (Wilayah Izin Usaha Pertambangan)	Mining Business Licence Area
YLBHI (Yayasan Lembaga Bantuan Hukum Indonesia)	Commonly translated as Indonesian Legal Aid Foundation, a direct translation is Indonesian Foundation of Legal Aid Institutes

Glossary

akta pendirian	foundation document, deed of establishment
Aliansi Nasional Reformasi KUHP	National Alliance for Reform of the Criminal Code (KUHP)
aliran kepercayaan, aliran kebatinan	traditional belief
anggaran dasar	articles of association, or organisational constitution
anggaran rumah tangga	organisational bylaws
Ashura	A holy day observed by Shi'a Muslims
asisten pengabdi bantuan hukum	assistant legal aid lawyer
bantuan hukum	legal aid
beres	settled, sorted out, in order
buta hukum	ignorant of the law
cacat formil	procedurally flawed
celaka	disaster, harm, accident
Cipta Lapangan Kerja, or Cipta Kerja	Job Creation
Dewan Pembina	Advisory Board
Dewan Pengawas	Supervisory Board
Dewan Pengurus	Board of Directors
dwifungsi	dual function (of the military), the political and social role of the military
Fraksi Rakyat Indonesia	Indonesian People's Faction
Gejayan Memanggil	Gejayan Calling, referencing the name of a street in Yogyakarta that was the site of a major student protest in 1998
halal	permitted under Islam
Indonesia Vrij	Indonesia Free
Indonesia Menggugat	Indonesia Accuses
Insiden Monas, Tragedi Monas	Monas Incident, or Monas Tragedy, an attack on supporters of religious

	freedom by members of the Islamic Defenders Front (FPI)
Kamisan	Thursdays
kasar	rough, coarse, rude, crude
keberpihakan	literally partisanship, alternatively 'concern for…'
keterbukaan	openness
Koalisi Ornop untuk Konstitusi Baru	Civil Society Coalition for a New Constitution
Komite Masyarakat Sipil Untuk Pembaharuan KUHAP	Civil Society Committee for Criminal Procedure Code Reform
Komite Pembelaan Rakyat	People's Defence Committee
lembaran putih	white paper
mafia hukum or mafia peradilan	judicial mafia
Maulid Nabi	The commemoration of the birthday of the Islamic prophet Muhammad
menembus batas	breaking boundaries or crossing the boundary
Milah Abraham	'Religion of Abraham', a new religious movement connected to Gafatar
modernisasi hukum	legal modernisation
Mosi Tidak Percaya	Motion of No Confidence
naskah akademik	the academic study or academic paper that accompanies a new bill, sometimes translated as 'policy paper'
negara hukum	Rule of law. Literally, 'law state'
Ombudsman Republik Indonesia	Indonesian Ombudsman
Pancasila	The Five Principles, Indonesia's state ideology
pedagang kaki lima	informal street traders
Pedoman Pokok Nilai-Nilai Perjuangan YLBHI dan Kode Etik Pengabdi Bantuan Hukum	Guide to YLBHI Struggle Values and Code of Ethics for Legal Aid Lawyers
pembangunan hukum	legal development
pendidikan hukum kritis	critical legal education
pengabdi bantuan hukum	LBH legal aid lawyer, literally 'legal aid servant'
pengacara	lawyer
penodaan agama	blasphemy
penyadaran	consciousness raising
peraturan	regulation

perbuatan melawan hukum	unlawful act, act in contravention of the law
Peringatan	Warning
Peristiwa 27 Juli	27 July 1996 Incident
Rechtsstaat	state based on law, rule of law
reformasi	Reform, the reform era following the fall of Soeharto
Reformasi Dikorupsi	Reform Corrupted
Risalah Sidang	Summary of Court Proceedings
rumah demokrasi	home of democracy
Sidang Rakyat	People's Forum, or People's Hearing
somasi	summons
surat edaran	circular
surat keputusan	decision, sometimes translated as decree
ulama	Islamic scholar
undang-undang	statutes
yayasan	foundation

Part One

Cause lawyering and the birth of the legal aid movement in Indonesia

1 Introduction

Cause lawyering and democratic change

Reform Corrupted

In September 2019, Indonesia witnessed the largest student and civil society protests in the country since the fall of President Soeharto's authoritarian New Order regime two decades earlier. The protests were sparked by the national legislature (*Dewan Perwakilan Rakyat*, DPR) rushing through reforms to the 2002 Law on the Corruption Eradication Commission (*Komisi Pemberantasan Korupsi*, KPK) that were "disastrous for the KPK and Indonesia's anti-corruption drive" (Butt 2019). Political elites had long sought to weaken the KPK (Mietzner 2021, 162). This time, they finally got what they wanted.

Seemingly buoyed by their efforts to destroy the KPK, legislators insisted they would pass a revised Criminal Code (*Kitab Undang-Undang Hukum Pidana*, KUHP) by the end of the month, a job that lawmakers have grappled with for decades (Maharani 2019; Butt and Lindsey 2018, 200). The draft revised criminal code, known as the RKUHP or RUU KUHP,[1] would have dismantled key democratic reforms and was roundly criticised for introducing new and expanded criminal offences (Akbari 2019). Among its most controversial provisions, the draft code criminalised co-habitation of unmarried couples and all consensual sex outside marriage, and reintroduced offences for insulting the president and the government that had previously been struck down by the Constitutional Court (Ristianto 2019). In addition to the RKUHP, lawmakers also said that, before their terms expired on 30 September, they planned to pass a new land law and revisions to laws on mining, corrections and labour that were criticised for diluting environmental protections, threatening citizen rights and further weakening anti-corruption efforts (Putsanra 2019). At the same time, legislators infuriated activists by showing none of the same urgency for passing a long discussed bill on the eradication of sexual violence (Cahya 2019).[2]

The constellation of these multiple threats culminated in these landmark protests, which were dubbed *#ReformasiDikorupsi* (#ReformCorrupted) to capture the anger over the weakening of the KPK, and the sense that the achievements of the reform era (known as *reformasi*, and considered to have

DOI: 10.4324/9781003486978-2

begun in 1998 when Soeharto fell) were slipping away.[3] Student protests began outside the national legislative complex in Jakarta on 19 September, after the passage of the revised KPK Law. They continued over several days in multiple cities across Indonesia, with the largest protests on 24 September. Thousands of students in Jakarta blocked one of the city's main roads, displaying creative and often humorous banners that were shared widely across social media. According to *Tempo* magazine, about 50,000 demonstrators protested across the country on 24 September (Dongoran 2019).

The police response to the protests was excessive and brutal. The atmosphere in Jakarta at the time was on a knife edge. I spoke to activist and journalist friends who spent hours hiding in a university building after being pursued by police. One night during the protests, the street I lived on was closed by police, as they fired tear gas at retreating students. By the time the demonstrations abated, at least five students had died, hundreds were injured, and more than 1,300 protesters had been arrested, with at least 380 named suspects by police (CNN Indonesia 2019b). Many were detained without access to legal counsel or their families.

Speaking to crowds outside the national legislative complex on 24 September was legal aid lawyer Asfinawati (Asfin),[4] then director of the Indonesian Foundation of Legal Aid Institutes (*Yayasan Lembaga Bantuan Hukum Indonesia,* YLBHI).[5] In a searing speech to students from the back of a truck, Asfin said:

> "They refuse to pass laws that are needed by the people. They refuse to pass the bill on the eradication of sexual violence. They refuse to pass the bill on Indigenous Peoples. But in a short time, they will pass laws that benefit them, that benefit their cronies, that benefit the oligarchs. Are we going to stay silent, or are we going to fight, my friends? Fight!"
>
> (Trismana 2021).

It might seem unusual for legal aid lawyers to be engaging in acts of open political dissent or aligning themselves with popular protests like this, but YLBHI and its affiliated regional Legal Aid Institutes (*Lembaga Bantuan Hukum*, LBH)[6] have long had a deeply political understanding of their role. LBH pioneered legal aid and legal activism in Indonesia under Soeharto's brutal New Order (1966–98),[7] defending political detainees, bravely speaking out against abuses of state power, and advocating for democracy, human rights and the rule of law in profoundly challenging circumstances. Yet its more overtly political activities do not mean that LBH lawyers eschew legal action. When students were arrested, there was never any question about which organisation would be there to defend them. Throughout its history, LBH has worked tirelessly within Indonesia's dysfunctional legal system to try to secure better outcomes for Indonesia's poor and marginalised. It performs a vital and desperately needed service in an environment where the poor really "only have access to the legal system as criminals" (Meili 1998,

496).[8] This sentiment is so widely held that it is common for Indonesians to say the law is like a knife, "blunt upwards and sharp downwards" (*tumpul ke atas, tajam ke bawah*).

LBH was established by lawyer Adnan Buyung Nasution in 1970 to provide legal aid to poor and marginalised Indonesians. Its early leaders were also initially motivated by a struggle to limit state power through a largely lawyerly concern for instatement (or reinstatement) of the rule of law (Aspinall 2005, 100). But LBH quickly realised that simply providing pro bono legal assistance under the authoritarian system of the New Order was "hopelessly beside the point" (Lev 1987, 20). LBH devised its own ideology of legal aid and activism, which it termed structural legal aid' (Nasution 1985). This concept held that legal aid should be directed at addressing the structural causes of inequality that resulted in the legal problems faced by the poor. Litigation was combined with a variety of nonlitigation activities, including community legal education, community organising, research and media campaigns (Nasution 1985, 37).

LBH became involved in many of the monumental, path-breaking cases of the New Order period. To name just a handful, it defended families evicted for the creation of Beautiful Indonesia in Miniature Park (*Taman Mini Indonesia Indah*), a pet project of the wife of President Soeharto, Siti Hartinah;[9] in 1973, it supported transgender woman Vivian Rubiyanti in changing her gender on official documents following sex reassignment surgery;[10] it defended student council administrators in the wake of the Bandung Institute of Technology (*Institut Teknologi Bandung*, ITB) 1978 'White Book' controversy (Nasution 2011, 10);[11] it advocated for female students' right to wear headscarfs at school, at a time when the New Order regime sought to suppress public expressions of Islam (LBH Jakarta 2015, 27); it played a key role in advocacy following the death of labour activist Marsinah;[12] it represented communities evicted during the construction of the Kedung Ombo reservoir; and defended activists charged with subversion after they established the leftist Democratic People's Party (*Partai Rakyat Demokratik*, PRD).[13] In the most sensitive cases, hopes of victory in the courtroom were virtually nil. Yet these highly contentious cases often functioned as a form of "political theatre" (Lindsey and Crouch 2013, 624). They exposed abuses of power to broader public scrutiny and embarrassed the government on the domestic and international stages.

In the late New Order years, as calls for *reformasi* grew louder, LBH became even more political and confrontational. It developed into a key centre of civil society opposition to the Soeharto regime, and a meeting point for activists, students and other pro-democracy actors (Aspinall 2005, 108). At one point, it even labelled itself a "locomotive of democracy" (Nasution 2011, 12). While it would be overstating LBH's influence to suggest civil society was responsible for the fall of Soeharto, a crucial contribution of civil society – and LBH as one of its leading actors – was that it "eroded the ideological foundations of authoritarian rule" (Aspinall 2004, 82).

After Soeharto finally fell in May 1998, Indonesia transformed itself from an authoritarian regime to a lively constitutional democracy, and one of the freest countries in Southeast Asia (Aspinall 2015). One might assume that the demise of the New Order and the establishment of new democratic institutions would provide the conditions for legal and human rights organisations like LBH to thrive. To an extent, such assumptions are correct. There were indeed more opportunities for Indonesian lawyers to mobilise the law to promote social change after Soeharto left office. However, the experience of LBH tells a more complicated – and more interesting – story.

This book explores how democratic change affects lawyers and the strategies they use to promote social change, through a detailed case study of LBH, Indonesia's most prominent and influential legal and human rights organisation. How have lawyers adapted to the new legal opportunities and more open political structure available since the fall of the New Order in 1998? How have they adjusted their views about the state, the legal system, their causes and their roles as lawyers with democratic transition and, more recently, with democratic regression? Has LBH been more or less effective in its efforts to advance social justice reforms since 1998? What do LBH lawyers see as the most appropriate strategies to respond to recent declines in democratic quality in Indonesia? These are the core concerns of this book.

This book contributes to the body of scholarship on cause lawyering, a term that refers broadly to the use of law to advance or resist societal transformation (Marshall and Hale 2014, 303). In probing the questions above, the book aims to complicate the relationship between cause lawyering and democratic change. By tracing the trajectory of LBH from the New Order period through to the current day, I find that there is a clear relationship between the quality of democracy and the form of cause lawyering practiced. However, this relationship is not what one might expect. LBH struggled following the democratic transition, and it is thriving again now that Indonesian democracy is unravelling.

Cause lawyering

Cause lawyering is a concept that was introduced by Austin Sarat and Stuart A. Scheingold in 1998, and explored further through a book and series of edited volumes (Sarat and Scheingold 1998b; 2001a; 2005; 2006a; 2008; Scheingold and Sarat 2004). Most definitions of cause lawyering seek to capture the activist nature of the concept. Cause lawyers are said to differ from "conventional" lawyers in that they consciously use their legal skills in pursuit of moral or political causes in which they believe (Hajjar 2001, 68). In fact, moral or political commitment is said to be "an essential and distinguishing feature" of cause lawyering (Scheingold and Sarat 2004, 4).

Cause lawyering is often directed toward the politically and socially marginalised (Munger, Cummings, and Trubek 2013, 365), and this is especially the case in Indonesia. Yet it entails more than the simple provision of legal assistance to underrepresented individuals, such as with pro bono activities or conventional legal aid. Cause lawyers see their role as extending beyond client service to the advancement of certain causes (Scheingold and Sarat 2004, 3). This doesnot mean that cause lawyers do not take their duty toward clients seriously, but rather that they see client representation "as a means to their moral and political ends" (Scheingold and Sarat 2004, 7). The clients of cause lawyers are diverse, and may include poor and marginalised individuals facing direct legal threats, politically disempowered communities that have yet to develop a shared awareness about their rights, and even large, well organised social movement groups (Marshall and Hale 2014, 306–07). Sometimes, the cause being defended or advanced may be broad, and affect many groups in society, such as cause lawyering to defend the principle of freedom of expression. Cause lawyering does not preclude the representation of individual clients, but in taking on these cases, cause lawyers tend not to see their clients as atomised individuals. When legal efforts are directed toward correcting underrepresentation of individuals, such cases are viewed as having deeper political implications beyond the individual case. This also means that, inevitably, tensions occasionally arise when cause lawyers prioritise the affirmation of certain legal or political principles over their clients' immediate, sometimes more pragmatic, concerns (Marshall and Hale 2014, 311; Scheingold and Sarat 2004, 17; Luban 1988, 319).

In addition to the essential element of moral or political commitment, cause lawyering is further contrasted against conventional legal practice in that it is "frequently directed at altering some aspect of the social, economic, and political status quo" (Sarat and Scheingold 1998a, 4). Or, to use Boukalas's pithy description, cause lawyers are "essentially political actors – albeit ones whose work involves doing law" (Boukalas 2013, 396). For the purposes of this book, "cause lawyering" refers to "the set of social, professional, political and cultural practices engaged in by lawyers and other social actors to mobilise the law to promote or resist social change" (Marshall and Hale 2014, 303). This broad definition recognises that there are activities conducted by non-lawyers (such as paralegals or researchers) that also fall under the rubric of cause lawyering. Litigation is just one of many strategies associated with law that may be deployed in service of the cause.

Lawyers are involved in the administration of the law and are therefore sometimes considered part of the state justice system while also comprising an element of civil society (Liu and Halliday 2016, 4). The cause lawyers that are the focus of this book consider themselves to be firmly placed within civil society,[14] independent of – and often in opposition to – the state. Non-profit civil society organisations (CSOs) have historically been the main site for cause lawyering on behalf of poor and marginalised groups (Weisbrod, Handler, and Komesar 1978; Nielsen and Albiston 2006, 1596), and they are the dominant site for cause lawyering in Indonesia. However, it is worth noting that cause lawyering can be conducted from a variety of practice sites, including pro bono work in private firms, and within government structures

(Chen and Cummings 2013, 125–200; Klug 2001; NeJaime 2012), even if private firms and the state bureaucracy have yet to emerge as major practice sites for cause lawyering in Indonesia. An important consideration with cause lawyering from CSOs is that non-profit organisations are dependent on grants or donations from government, international donors, foundations, philanthropists or members of the public, which can shape the form that cause lawyering ultimately takes, as this book will show.

Cause lawyering, public interest law or something else?

Before moving on, I must justify my use of the term cause lawyering over other similar terms, such as the related concept of 'public interest law'. Indeed, the terms 'cause lawyering' and 'public interest law' are often used interchangeably with no attempt to make a distinction between the two or explain why one term is favoured over another.[15] Public interest law emerged in the United States in the 1960s and 1970s and involved using the law to advance the interests of politically vulnerable or marginalised groups, such as racial and ethnic minorities (Cummings 2008, 893).[16] It drew on a tradition of poverty and civil rights legal advocacy practiced by groups such as the American Civil Liberties Union (ACLU) and the National Association for the Advancement of Coloured People Legal Defence and Education Fund (NAACP LDF). The Ford Foundation was particularly influential in the 1970s in establishing and funding new public interest law initiatives in the United States (Trubek 2011, 418) and, eventually, abroad (Cummings and Trubek 2008, 28). In its early years, public interest law was strongly associated with 'legal liberalism', or the belief in the capacity of law to change society (Cummings and Trubek 2008, 7). It focused on test case or strategic impact litigation in US federal courts, aiming to bring about progressive reforms on behalf of marginalised groups (Cummings 2008, 900). The scope of public interest law expanded quickly to consumer rights and environmental protection cases, and it was not long before conservative groups also adopted the concept to further right-wing causes like restrictions on abortion, while claiming to be acting in the public interest (Southworth 2013, 498).

Dissatisfaction with the term 'public interest law' led scholars to propose alternative concepts (Cummings 2012, 519–20). Scheingold and Sarat justified their preference for the term 'cause lawyering' by stating that the term 'public interest lawyering' inevitably prompts questions "about what is, or is not, in the public interest" (Scheingold and Sarat 2004, 5). Cause lawyering avoids these questions by not "making distinctions between worthy and unworthy causes" (Scheingold and Sarat 2004, 5). But scholars also question whether the term 'cause lawyering' is too broad or contains some of the same moral assumptions as 'public interest law' (Cummings 2012, 522). A range of other terms have since emerged, such as political lawyering, activist lawyering, social justice lawyering, alternative lawyering and more (Menkel-Meadow 1998, 33). It is illustrative that a 2013

Wisconsin International Law Journal special issue on "lawyers for social justice" included articles not only on social justice lawyering but also "cause lawyering", "public interest lawyering", "public interest litigation", and "cause advocacy" (University of Wisconsin–Madison 2013).

In Indonesia, a variety of terms are used to describe the types of activities and practices conducted by Indonesians who use their legal skills to advocate for social and political change. LBH commonly refers to its legal aid lawyers as *pengabdi bantuan hukum*. The English translation is a little awkward but the word *pengabdi* emphasises service or devotion to a cause, while *bantuan hukum* means legal aid. A direct translation is 'legal aid servant'. Historically, some LBH offices have also used the term *pengacara publik*, or 'lawyer of the public', a term that suggests a reference to the public interest. However, there is no one term used consistently across Indonesia to describe people who deploy law to promote social reform and address underserved legal needs. Other commonly used terms include *aktivis hukum* (legal activist), *pengacara hak asasi manusia* (HAM) (human rights lawyer), *aktivis hukum dan HAM* (legal and human rights activist) or sometimes simply *pengacara dan aktivis HAM* (lawyer and human rights activist) or *advokat dan aktivis HAM* (advocate and human rights activist).

Given the array of terms and concepts with similar meanings, does it ultimately matter which is used to describe the practices engaged in by lawyers at LBH? 'Public interest law' and 'cause lawyering' are both imprecise terms and both come with their own baggage. 'Public interest law' is too restrictive for the Indonesian case, because of its association with the US experience and the historical emphasis on top-down impact litigation. There is some convergence between the activities of LBH and the scholarly literature on 'political lawyering' (Halliday, Karpik, and Feeley 2007; 2012), which focuses on the role of the legal complex – consisting of lawyers, judges and other actors with legal training – in struggles for political liberalism. Halliday, Karpik and Feeley define political liberalism as comprising basic legal rights, a moderate state and civil society (Halliday, Karpik, and Feeley 2007, 10–11). Although these are concerns of Indonesian lawyers at LBH, and LBH lawyers certainly understand their work to be political in nature, this focus also fails to capture the wide range of causes advanced by LBH lawyers. While there are valid criticisms associated with the term 'cause lawyering', it is broad enough to accommodate the practices engaged in by LBH lawyers and is therefore the term I have adopted for this book.

Democratisation and opportunities for cause lawyering

Much cause lawyering scholarship focuses on the experiences of cause lawyers in established liberal democracies, in particular the United States. Yet cause lawyering also exists in transitional and emerging democracies, and even under authoritarian regimes. There is growing scholarly attention being paid to legal mobilisation under authoritarianism (Moustafa 2007a; 2007b;

Ginsburg and Moustafa 2008; Tam 2013; Cheesman and San 2013; McEvoy and Bryson 2022; Mustafina 2022), and a substantial body of work on politically motivated lawyers in China (Fu and Cullen 2008; 2011; Fu 2011; 2018; Liu and Halliday 2011; 2016; Feng 2009; Benney 2013; Givens 2013; Pils 2014; Stern 2013; 2017; Nesossi 2015). This China scholarship has demonstrated that even in an authoritarian country lacking judicial independence and opportunities for political dissent, lawyers have mobilised to defend human rights and demand transformative change (Fu 2018; Liu and Halliday 2016; Pils 2014; Fu and Cullen 2011). There is also significant academic work looking at activist lawyers in other Southeast Asian countries, including in contexts with tighter controls over political opposition than in Indonesia, such as Malaysia and Singapore (Harding and Whiting 2012; Kanesalingam 2013; Rajah and Thiruvengadam 2013; Chua 2017; Soon and Wong 2023).

Even so, cause lawyering literature generally presents democratic regimes as offering more opportunities for mobilising the law to promote social change (Munger, Cummings, and Trubek 2013; Cummings and Trubek 2008, 33; Sarat and Scheingold 1998a, 5; Yap and Lau 2010, 2). Sarat and Scheingold in particular make the link between cause lawyering and democratisation quite explicit, declaring that there is a "natural affinity between cause lawyering and democratisation" (Sarat and Scheingold 2001b, 14) and that there is "no doubt" that cause lawyering is "more likely to flourish in liberal democratic settings" (Scheingold and Sarat 2004, 131).[17] Nevertheless, they do acknowledge that cause lawyering has an important role to play in other political regimes, stating that "the fewer the democratic alternatives, the more vital cause lawyering becomes" (Scheingold and Sarat 2004, 133).

Empirical accounts from countries including Argentina, Brazil, South Africa, South Korea and Taiwan describe changes associated with democratic transition as supporting the development of cause lawyering (Chang 2010; Baik 2010; Goedde 2009; Vieira 2008; Ginsburg 2007; Klug 2001; Meili 2009). While there are common elements in these narratives, the developments that have influenced legal mobilisation in these countries differ somewhat according to the domestic context. Here I highlight several structural and institutional factors that are often described as critical in providing opportunities for cause lawyering and shaping the form that it takes, thereby offering a framework for understanding the Indonesian case.

Constitutional and other legal guarantees of rights. The introduction of constitutional and legal rights frameworks is highlighted as a key factor in the development of cause lawyering in a broad range of contexts (Tam 2010; Cummings and Trubek 2008, 34; Baik 2010; Chang 2010; Goedde 2009; Ginsburg 2007; Meili 2009). In countries where rights have been formally recognised, cause lawyering can provide a means to "mobilise the newly minted rights regime" (Cummings and Trubek 2008, 34). Inconsistencies between written legal rights and state actions that encroach on these rights create openings for cause lawyering to develop (Fu and Cullen 2008, 123; Cummings and Eagly 2001, 468). This is particularly relevant in Indonesia, where one of the landmark reforms of the post-Soeharto era was the insertion of a new chapter on human rights into the amended constitution, establishing stronger on-paper rights protections than in many established democracies (Lindsey 2008, 29).

Judicial independence. Alongside rights protections, judicial independence is commonly described as an essential element in the development of cause lawyering (Munger, Cummings, and Trubek 2013, 376–78; Cummings and Trubek 2008, 33; Baik 2010, 119; Chang 2010, 137; Moustafa 2007a, 201; Sarat and Scheingold 1998a, 8). This seems relatively self-evident. An autonomous judiciary provides the means to mobilise the law to challenge political power. When courts are free from executive or legislative intervention and willing to rule against the interests of the state to protect or extend rights, cause lawyering should be on firmer ground. In many authoritarian countries around the world, however, legal systems end up acting more to protect the interests of the government rather than as an independent check on state power (Munger 2015, 304). Authoritarian governments may exert pressure over the courts through administrative and financial control, restrictions on judicial review, or even direct political pressure (Munger, Cummings, and Trubek 2013, 377; Pils 2014, 105–21). These pathologies were all present in Indonesia under Soeharto. Yet even in contexts with weak judicial autonomy, cause lawyering may be useful as a means of shining a spotlight on rights abuses and injustice, and attracting support for a cause (Meili 1998, 492; Ellmann 1998, 366; Pils 2014, 239). Under authoritarian regimes, courts can provide a forum for free speech, a space for activists to bring politically sensitive issues to public attention (Fu 2011, 355), or “speak truth” to the power of the state (Pils 2014, 121, 128). Indeed, LBH made savvy use of the courts in the highly repressive environment of the New Order, taking on some of the most politically sensitive cases with virtually no hope of success and using them to promote democratic values and aspirations for the rule of law.

Constitutional courts or other institutional channels where rights can be claimed. A connected institutional factor recognised as important for the development of cause lawyering is the presence of a court with the power to review the constitutionality of laws (Baik 2010, 119; Ginsburg 2007, 51; Chang 2010, 136). Indeed, constitutional litigation has been described as “the bedrock of cause lawyering” (Scheingold and Sarat 2004, 130), providing opportunities for cause lawyers to protect and extend human rights and keep the government in check. Even if constitutional litigation does not result in the affirmation of rights by the courts, it can be used strategically to raise awareness among the public, stimulate debate, or push the legislature to engage in reform (Chang 2010, 136). In Indonesia, the establishment of the Constitutional Court in 2003 created a new opportunity to challenge, and in some cases strike down, statutory provisions enacted under the previous authoritarian regimes. In some countries, administrative courts have also emerged as important forums for citizens to challenge government actions (Moustafa 2007b, 19–56; Chang 2010, 137; Baik 2010, 116; Ginsburg 2007, 51; Jigmiddash and Rasmussen 2013, 583). It is important to recognise, however, that even where progressive legal rights and new institutional structures like constitutional courts or administrative courts exist, cause lawyering may face challenges (Meili 2009). Courts must be at least somewhat receptive to rights claims and be willing to grant standing to civil society organisations and individual citizens if cause lawyering is to meet with success (Tam 2010, 679).

Political openness. Another factor influencing cause lawyering is political openness. Extending a point made by Munger, Cummings and Trubek (Munger, Cummings, and Trubek 2013, 378–79), political openness here refers to the space available for confronting or opposing the state outside the courts, as well as opportunities to engage directly with elements of the state. When governments are more tolerant of political dissent and where there is greater freedom for the public discussion of rights and injustice, cause lawyering becomes a more viable prospect (Cummings and Trubek 2008, 34). For example, stronger guarantees of freedom of expression can provide opportunities for public communication that may complement cause lawyering efforts (Munger, Cummings, and Trubek 2013, 400). In regimes that are more politically open, cause lawyers, activists and social movements may have greater freedom to engage in political advocacy or direct lobbying of the legislature, so they may find they need to use litigation less frequently, or supplement it with other strategies. Cause lawyers in South Africa, South Korea and Taiwan, for example, increasingly turned to legislative and bureaucratic strategies following democratic transition (Goedde 2009, 86; Chang 2010, 140; Klug 2001; Ginsburg 2007, 52). Following democratisation in Taiwan, legislative lobbying became "equally, if not more, important to public interest groups" (Chang 2010, 140). Conversely, in oppressive environments where opportunities for political participation are limited, civil society is weak and the media is tightly controlled, channelling grievances through the courts may be one of the safest options available for those seeking social or political change (Fu and Cullen 2011, 40; Stern 2017, 234). Indeed, there is now significant literature on the "judicialisation of politics" in more closed and authoritarian settings (Ginsburg and Moustafa 2008; Dressel 2012; Tam 2013).

Civil society. The development of civil society following democratic transition may play a significant role in shaping cause lawyering, both by providing new practice sites for cause lawyering and allowing expanded roles for advocacy organisations that can supplement legal mobilisation efforts. As noted, CSOs have historically been the main practice site for cause lawyering in the United States, and they are also the primary vehicle for cause lawyering in many countries of the Global South, including Indonesia. Democratic transition may reduce restrictions on civil society, fostering freer development of CSOs and subsequently practice sites for cause lawyering (Munger, Cummings, and Trubek 2013, 400, 413).

Democratisation, and the increased opportunities to mobilise law that often accompany it, may also lead to CSOs not outwardly focused on law reform adopting litigation strategies (Cummings and Trubek 2008, 35–36; Farid 2013, 425; Meili 2009, 46; Rekosh 2008, 70). Civil society can also perform important supportive functions. An active and free civil society sector allows the formation of more productive alliances that can strengthen the impact of legal mobilisation when the courts are weak (Munger, Cummings, and Trubek 2013, 400). Civil society organisations can, and often do, make rights claims on the state, collaborating with cause lawyers in activities to sustain the momentum of legal advocacy, such as advocating for legislative reform, conducting public media campaigns and demonstrations, and engaging in other policy advocacy activities (Goedde 2009, 78–79; Meili 1998; 2009; Jigmiddash and Rasmussen 2013; Chang 2010, 137; Ginsburg 2007; Baik 2010, 130; Hsu 2019). By the same token, cause lawyers can provide legal representation for and support civil society actors when they face legal threats, for example from participating in direct action activities that provoke the ire of authorities (Sarat and Scheingold 2006b, 9).

The legal profession. If cause lawyers are to challenge government policies, or bring rights violations to public attention without fearing for their safety, the legal profession must be able to operate with a degree of independence (Chang 2010, 137; Munger, Cummings, and Trubek 2013, 399). Bar associations may act to defend the autonomy of the profession from state interference, and sometimes get involved in the direct provision of legal aid (Abel 1998, 98; Ellmann 1998, 367). Yet support from the organised profession for cause lawyering efforts is by no means assured. In many countries around the world, professional associations may not encourage cause lawyering, or even pro bono activities for underserved communities (Vieira 2008, 257; Cummings and Trubek 2008, 40–41). Indonesia has never had a unified bar association (Crouch 2011). During the late Soekarno and early New Order years, the Indonesian Advocates Union (*Persatuan Advokat Indonesia*, Peradin) emerged as an independent and influential professional organisation, and was central to LBH's establishment in 1969, as Chapter 2 discusses. But the New Order regime's subsequent efforts to repress the profession eventually resulted in a weak and divided legal profession, and in the post-Soeharto period the organised profession has provided little leadership or even coordinated support to law reform efforts (Lev 2007, 410).

Funding. Availability of sufficient financial resources is critical in providing opportunities for cause lawyering. Indeed, funding has long been identified as a key element of the "support structure" essential for legal mobilisation, along with rights advocacy lawyers and rights advocacy organisations (Epp 1998). As stated, in many countries, including Indonesia, cause lawyering is most commonly conducted from CSOs (Yap and Lau 2010, 7), which are largely dependent on donor funding. This funding is more likely to be drawn from foreign sources than domestic governments. When domestic governments do provide funding, it is often for less politically sensitive activities, such as conventional legal aid (Ellmann 1998, 353).[18] Democratic transition may influence financial resources available to cause lawyering organisations, for example by easing restrictions on the foreign funding of CSOs, or bringing about shifts in the priorities of international donors. The proliferation of cause lawyering across the globe has been connected to trends in the international development sector (Cummings and Trubek 2008, 18; Ellmann 1998, 351–54). Cause lawyers may find that donors only wish to offer grants for initiatives to strengthen civil and political rights, rather than economic and social rights (Halliday 1999, 1048), or for activities carried out in line with neoliberal values (Scheingold 2001, 396–98). Indeed, foreign funders' rule of law orientation has sometimes raised concerns about western legal imperialism (Cummings 2008, 964), similarly to the now discredited law and development movement (Lindsey 2004, 16). Furthermore, a dependence on foreign money can compromise CSOs' domestic standing, particularly among government figures who are suspicious of foreign influence (Abel 2011, 314). This has often been the case for LBH, which regularly faces questions about the sources of its funding when it raises sensitive issues (CNN Indonesia 2019a).

Legal education. Formal legal education has been of only limited importance in the development of cause lawyering in the Indonesian context, and of the factors described it is the one least connected to democratic change. Nevertheless, it is worth mentioning briefly because of both international and local factors shaping cause lawyering in Indonesia. The training of foreign lawyers in US (and other Global North) law schools has been described as a critical factor in the spread of cause lawyering in countries of the Global South (Cummings and Trubek 2008, 37–39; Meili 2009, 47). I do not want to overstate the influence of foreign legal education in Indonesia or suggest that Indonesian cause lawyering has been a 'transplant' from abroad, but it is important to acknowledge that LBH was founded by Adnan Buyung Nasution after he returned to Indonesia from studies in Australia, as Chapter 2 discusses. In many countries, university legal education may be formalistic and actively resistant to efforts to promote social change (Abel 2011, 314). This is also true of Indonesia, where law schools prioritise formal legal education and place little emphasis on critical thinking skills, socio-legal studies, human rights, or anti-corruption (Kelly and Susanti 2014, 39). In some countries, foreign donor support for clinical legal education programmes at local law schools has been influential in the development of cause lawyering (Rekosh 2008, 83–92). In Indonesia, too, there is a growing clinical legal education movement, developed with the support of the US Agency for International Development (USAID), but only in a fraction of Indonesia's 300-plus law schools (Kelly and Susanti 2014, 39). While law schools have had a limited impact on the development of cause lawyering in Indonesia, LBH has developed its own intensive training programme to encourage cause lawyering practice, which it calls Kalabahu (*Karya Latihan Bantuan Hukum*, or Legal Aid Training). Discussed in Chapter 3, it has arguably been of greater influence in encouraging the development of a certain style of cause lawyering than the nascent clinical legal education programme.

Cause lawyering is of course also dependent on the lawyers themselves. Institutional structures like those outlined above are essential but there must be lawyers who are willing to use these structures to advocate for rights. This is sometimes referred to as a matter of "agency", or the "subjective" conditions necessary for legal mobilisation in the literature on cause lawyering and legal mobilisation, (Tam 2013; Chua 2022, 24). The ways in which cause lawyers respond to changes in the political and institutional factors described above can have significant influence over the tactics and strategies they deploy. A major focus of this book is the exploration of the tactical decisions cause lawyers make in response to changing structural conditions, even if I do not always refer to these choices in terms of agency.

Strategies of cause lawyering

Cause lawyering might be commonly associated with litigation, but cause lawyers typically adopt multifaceted strategies, encompassing a wide range of legal and non-legal practices, from litigation to public policy advocacy and legislative lobbying, community organising, street protests, media campaigning and more. The discussion above touched on some of these tactics, but to understand how Indonesian cause lawyers and cause lawyering has been affected by democratic change it is worth briefly reviewing the types of strategies cause lawyers have at their disposal.

Cause lawyering has historically been most closely associated with litigation (Epp 1998; Handler 1978). It is not surprising that lawyers may be predisposed to litigation strategies, given their education and experiences (McCann and Silverstein 1998, 262). Litigation may take a variety of forms, including the conventional public interest law strategy of test case or strategic impact litigation designed to reform the law or change policy, collective or class action cases, and even individual case defence (Scheingold and Sarat 2004, 19). A classic dilemma faced by cause lawyers is whether they should defend individual cases or pursue legal changes intended to have impacts for a broader range of people. This is a problem LBH has grappled with throughout its history, and was a key consideration in the development of its ideology of legal aid – 'structural legal aid' – discussed in Chapter 2. Yet even when cause lawyers represent individual cases, such cases are often used to make broader points about the cause. In the difficult Chinese context, for example, some cause lawyers intentionally select sensitive criminal cases to promote liberal values like basic legal rights and constraints on state power (Liu and Halliday 2016; Fu and Cullen 2011, 557; 2010, 22; Feng 2009, 160–61).

Litigation strategies might dominate the popular imagination but scholars have long recognised that formal legal action may have important "indirect" effects (McCann 1994). McCann has argued that even if litigation does not result in legal victory, it can be important in accelerating the growth of social movements, rallying public support, or in complementing other political strategies (McCann 2006, 8). Litigation, or the threat of litigation, may act as a source of leverage, to force "uncooperative foes into making concessions or compromises" (McCann 2006, 19). McCann's work added to an influential critique of public interest law made by Scheingold, who argued that lawyers were too often captured by "the myth of rights", that is, the view that securing legal rights in the courts could lead to meaningful social change (Scheingold 2004, 5). While Scheingold was critical of an excessive focus on litigation, he also considered that the myth of rights could be used as a "resource" to stimulate political action. "It is possible", he said, "to capitalise on the perceptions of entitlement associated with rights to initiate and to nurture political mobilisation" (Scheingold 2004, 131). The degree to which litigation is prioritised, and its deployment alongside other non-litigation strategies, are key considerations of this book.

Since the early days of public interest law, advocating for policy change through legislative lobbying and drafting new laws and regulations have also been core tactics of cause lawyers (Marshall and Hale 2014, 304). As discussed, democratisation may provide new avenues for cause lawyers to contribute to policy change by working with the legislature and executive (Goedde 2009; Klug 2001; Ginsburg 2007; Hsu 2019; Chang 2010). However, a consequence of working in such proximity to the state is that it will inevitably involve a degree of compromise (Scheingold 2001, 390). It is not hard to imagine how cause lawyers may be tempted to attenuate or discard more radical demands in order to maintain a productive working relationship with state agencies. In China, for example, public interest lawyers with more moderate goals have tended to have more success with legislative lobbying and law reform-focused strategies (Fu and Cullen 2010, 25). Although working within or in close proximity to the state may lead to cause lawyers ratcheting down their more ambitious goals, it can also bring considerable advantages. These include improving access to state institutions for cause lawyers and other social movement actors, and shaping government policy to benefit the cause being advocated or the broader operating environment for civil society (NeJaime 2012).

While litigation may be the most widely recognised form of cause lawyering, many cause lawyers spend little time doing "traditional" lawyering (Levitsky 2006, 147). Cause lawyers are often acutely aware of the "myth of rights" (McCann and Silverstein 1998, 267–72), including cause lawyers at LBH, and may instead deploy litigation in concert with movement building approaches, including activities such as community organising, community legal education, coalition building and political strategising (Cummings 2020; Cummings and Eagly 2001; Sarat and Scheingold 2006a; McCann 2004; McCann and Silverstein 1998). Grassroots community organising and community legal education activities are broadly designed to stimulate "rights consciousness" among community members, supporting them in developing a shared understanding of their problems as rights issues and taking action to advocate for the needs of their communities (McCann 2004, 510–11). Many cause lawyers involved in movement building activities may even stand alongside the social movements and communities they represent, and become actively involved in more disruptive direct action tactics like mass demonstrations and street protests (Levitsky 2006, 147). These forms of extra-legal action have been an important part of LBH's approach to cause lawyering throughout its history, as this book will show.

'Types' of cause lawyers

To bring these considerations of democratic change, institutional structures and strategies together, this book draws on literature identifying certain 'types' of cause lawyering. Scholars have proposed various taxonomies for thinking about and analysing cause lawyers and their practices (Kilwein 1998; Fu and Cullen 2008; Luban 2012, 706–07; Liu and Halliday 2016; Fu 2018), which often share broad elements. Rather than adding another set of subtly different terms to the list, this book adapts the framework proposed by Hilbink (Hilbink 2004). Hilbink identifies three versions of cause lawyering: 'proceduralist', 'elite/vanguard' and 'grassroots', differentiating lawyers based on their visions of the system, the cause, and their role as lawyers (Hilbink 2004, 663). While the typology was not expressly developed to examine cause lawyers' views and strategies in relation to democratic change, it is a highly useful lens through which to examine how LBH has responded to shifts in the quality of Indonesian democracy.

Lawyers in the **proceduralist** category believe in an ideal vision of the law as being above politics – that it is impartial and objective. They view the legal system as "essentially fair and just" (Hilbink 2004, 665). Proceduralist lawyers trust the legal system, and therefore believe that once procedural flaws are addressed, justice can be achieved. This type of lawyer tends to believe the legal system can act as a stabilising force, helping to dampen tense political or social disputes. For proceduralist lawyers, the cause is the legal system and the rule of law. They aim to strengthen the legal system so that it functions in a dependable, equitable manner, providing adequate representation to all clients who require it. This form of lawyering would seem to be at odds with definitions of cause lawyering that seek to alter some element of the status quo. Rather, proceduralist lawyers would appear to be working to legitimise the existing system and structures of power. When the state routinely neglects its obligations, however, a procedural emphasis can be focused on altering elements of the status quo, by pushing the state towards meeting its promises to uphold rights. Proceduralist lawyers' vision of their role as lawyers is strongly influenced by traditional understandings of professionalism. Proceduralist lawyers tend to view clients as individual cases and avoid personal identification with the causes of their clients. A proceduralist approach to cause lawyering is therefore commonly associated with conventional legal aid, with its focus on expanding representation (Hilbink 2004, 665–72). As this book will show, LBH lawyers have often been uncomfortable with this kind of approach to cause lawyering, particularly during periods when Indonesian democracy has been weak or absent.

The **elite/vanguard** form of cause lawyering is consistent with liberal legal views about the capacity of strategic impact or test case litigation to lead to social change. Elite/vanguard lawyers do not consider law and politics to be separate, and in fact acknowledge their legal activism as a form of politics – albeit one they consider to be superior to demonstrations, protests or other forms of direct political activism. Given elite/vanguard lawyers believe that the legal system is capable of engendering change, they typically hold the courts and judges in high esteem. In contrast to proceduralist lawyers' concern with strengthening the legal system, elite/vanguard lawyers are focused on substantive justice. As such, their causes may be manifold, and include environmental rights, labour rights, the rights of ethnic and religious minorities, and more. Their primary strategy is litigation, although they may often engage in public policy advocacy work directed towards the legislature or the executive, usually pursuing their cause through the use of legal arguments or strategies when they do so. Elite/vanguard cause lawyering is therefore directed 'up' towards elites, and lawyers in this category are often only loosely connected to their grassroots constituents. Elite/vanguard lawyers tend to have respect for formal procedures, such as the use of legislative challenges. However, they also transgress traditional understandings of professionalism to personally identify with their clients' causes. In doing so, elite/vanguard lawyers may proactively select clients and cases in line with the cause they are advocating, meaning that the interests of individual clients can sometimes be subordinated to the broader political or ideological goals they seek to achieve (Hilbink 2004, 673–80). As this book will show, the elite/vanguard form of cause lawyering only emerged as a serious cause lawyering strategy in Indonesia after the reforms implemented following the collapse of the authoritarian New Order in 1998.

The final category is **grassroots** cause lawyering. Grassroots lawyers reject the notion that the law is above politics, and often view the legal system as "corrupt, unjust or unfair – an oppressive force" (Hilbink 2004, 681). Judges and the courts are viewed as acting in the interests of the powerful. Grassroots lawyers may pursue a diverse range of social and political causes, but in contrast to elite/vanguard lawyers, they view change as being driven from the grassroots. They do not always pursue action through the courts, and in fact often question the value of litigation. They regularly combine legal approaches with other non-legal strategies, such as community organising, public education and media engagement. Grassroots lawyers emphasise client empowerment. They work in an intimate and collaborative manner with clients and social movements, complicating traditional understandings of the relationship between lawyers and their clients. Grassroots lawyers actively and consciously reject traditional notions of legal professionalism and moral neutrality to identify with their clients' causes. Their personal moral and political beliefs also influence the cases or clients they choose to take on. While proceduralist lawyers might accept any case in line with their organisational priorities, and elite/vanguard lawyers might choose to represent an 'uncivil' civil society organisation in order to defend the higher principle of freedom of association, grassroots lawyers generally only take on clients they agree with or causes that reflect their values (Hilbink 2004, 681–90).

To summarise, cause lawyering scholarship emphasises that liberal democratic systems involving factors such as strong constitutional and legal protections of human rights, judicial independence, constitutional courts or other legal forums where these rights can be claimed, legislative or bureaucratic opportunities to collaborate with the state on reform, an open civil society sector and adequate funding should provide more legal and political opportunities for cause lawyering. At the same time, however, there are many examples of cause lawyering in different (and often more limited) forms in authoritarian contexts. This book draws on this work on cause lawyering and legal mobilisation, strengthening its analytical framework by adapting Hilbink's three typologies of cause lawyers – 'proceduralist', 'elite/vanguard' and 'grassroots' – for the exploration of cause lawyering and democratic change. Although these typologies are not sharply demarcated, and cause lawyers may adopt strategies from multiple categories simultaneously, they are helpful when thinking about how democratic change has affected the choices made and strategies deployed by cause lawyers in Indonesia. I draw on these categories throughout this book as I examine the extent to which cause lawyers have adjusted their views and changed their approaches as different opportunities became available with democratic transition, and closed again with democratic regression.

LBH in an age of democratic reform and regression

There is arguably no more appropriate organisation for exploring how cause lawyering responds to democratic change than LBH. The work of LBH under the New Order has been documented by scholars including Daniel Lev (1987; 1996; 1998; 2007), Edward Aspinall (2005), Takeshi Kohno (2010) and LBH's two most important figures during the New Order years, Adnan Buyung Nasution (1985) and Todung Mulya Lubis (1985b; 1985a). Tim Lindsey and Melissa Crouch (2013) have also published a paper on cause lawyering in Indonesia that looked at the work of LBH. Despite this, there have been no detailed academic works in English examining how this trailblazing organisation that once labelled itself a "locomotive of democracy" has conducted its cause lawyering practice since the transition to democracy occurred in 1998. On the rare occasions that foreign legal scholars discuss cause lawyering in Indonesia, they continue to cite Daniel Lev, whose excellent, unparalleled work focused primarily on cause lawyers and LBH under Soeharto. However, this work is now out of date, leading to sometimes inaccurate characterisations of the work of contemporary Indonesian cause lawyers (Munger,

Cummings, and Trubek 2013, 384; Hilbink 2004, 672). For such a remarkable and influential organisation, the limited academic attention given to LBH in the post-authoritarian era is a significant oversight.

This book, therefore, serves as an institutional history of LBH after 1998, though it is more than that. The emphasis on democratic change is crucial. The democratic transition in Indonesia involved the development of many of the political and institutional factors described in the literature as supportive of cause lawyering. The constitution was amended to guarantee an extensive array of rights; a constitutional court with powers of judicial review was established; judicial independence was improved, with the Ministry of Justice handing control of the judiciary to the Supreme Court; and restrictions on civil society and the press were lifted. The government was more open to contributions from civil society, and there were opportunities for lawyers to collaborate with and even enter the state. While these factors provided new opportunities for Indonesian cause lawyers, the past five to ten years have seen changes that undermine the country's democratic gains. There is now a broad academic consensus that a serious decline in Indonesia's democratic quality has occurred under President Joko "Jokowi" Widodo, who came to power in 2014 (Power 2018; Aspinall et al. 2020; Aspinall and Mietzner 2019; Diprose, McRae, and Hadiz 2019; Schäfer 2019; Warburton and Aspinall 2019; Power and Warburton 2020; Mietzner 2021; Mujani and Liddle 2021). The 2019 *Reformasi Dikorupsi* protests discussed above were a clear manifestation of student and civil society anger over this democratic backsliding.

Indonesia is not alone in experiencing threats to its democracy. Democratic regression in Indonesia is consistent with a trend toward democratic backsliding that has been observed around the world (Haggard and Kaufman 2021; Repucci 2020). The Indonesian example is particularly interesting for exploring how cause lawyers have been affected by democratic change because it involves a dramatic shift to democracy and then a slide toward regression over the past decade. This book examines how adaptable Indonesian cause lawyers are to these shifts in contextual dynamics, and how their views of the legal and political system have changed. The Indonesian case demonstrates that simple assumptions about democratisation providing more opportunities for cause lawyering do not hold in the case of LBH. At the same time, the book contributes to broader understandings about the role of cause lawyers in defending liberal democratic gains in an increasingly illiberal democracy.

At its core, this book is a story about LBH finding its feet again. I argue that the trajectory of YLBHI and its regional LBH branches from the authoritarian period through to the current day demonstrates that there is a strong association between the quality of democracy and the form of cause lawyering practiced. In contrast to expectations, LBH struggled following the democratic transition, and it is only thriving again now that Indonesian democracy is fraying. The book will show that as Indonesia's democracy has come under pressure, LBH has rediscovered its raison d'être. When the state was more open, LBH made some forays into elite-focused forms of cause lawyering, but it struggled to reinvent itself in the more open political environment in which collaboration with the state became possible. It experimented but was never entirely comfortable with elite-focused versions of cause lawyering. More recent democratic regression has led to a hardening of the oppositional identity LBH forged under the New Order. LBH stands firmly on the side of victims, prioritising a confrontational form of cause lawyering that keeps the state at a distance and involves close collaboration with grassroots social movements. It has started to once again play a convening, coalition-building role among pro-democratic groups in civil society, similar to the role it played under the New Order, even if it no longer has the same extent of influence over civil society. Rather than simply demonstrating the challenges of cause lawyering in states where liberal democracy is on unsteady ground, this book underscores the important roles it can play.

Methodology

This book is a detailed case study of LBH, cause lawyering and democratic change. Yet it is crucial to acknowledge that there are many other cause lawyering organisations in Indonesia other than those in the LBH network. For example, 619 civil society legal aid organisations were accredited as legal aid providers by the Ministry of Law and Human Rights for the 2022–24 period.[19] There are also many law reform-focused CSOs that do not engage in direct service provision. I focus only on LBH because of its rich history, influence, and size. LBH remains the most prominent and active cause lawyering organisation in Indonesia and is the only one with a long history of operation under both the authoritarian New Order as well as in democratic Indonesia. Any examination of cause lawyering in Indonesia would be incomplete if it did not examine LBH.

Two points on terminology are worth making here. First, YLBHI is the Indonesian Foundation of Legal Aid Institutes.[20] It is the Jakarta-based umbrella body for the network of 18 LBH offices based in the capital cities of many Indonesian provinces. LBH offices are run independently, but they are guided by the YLBHI organisational by-laws and code of conduct. Given these similarities, many academic discussions of YLBHI/LBH blur the lines between these organisations. I am also guilty of this to a degree. I use the generic term LBH to refer to the entire YLBHI-LBH network, simply to avoid using the clumsy longer acronym. If I refer to YLBHI or name a specific LBH office, such as LBH Bandung, or LBH Surabaya, I am referring to that organisation specifically.

Second, LBH has been so influential as the pioneer of legal aid in Indonesia that the acronym 'LBH' is now used as a common term to refer to a legal aid organisation. There are many other Indonesian legal aid providers that use the acronym LBH in their organisational names but are not part of the

YLBHI network. Prominent examples include LBH Apik (Legal Aid Institute of the Indonesian Women's Association for Justice, or *Lembaga Bantuan Hukum Asosiasi Perempuan Indonesia untuk Keadilan*), LBH Masyarakat (Community Legal Aid Institute, or *Lembaga Bantuan Hukum Masyarakat,* often styled as LBHM), and LBH Pers (Legal Aid Institute for the Press, or *Lembaga Bantuan Hukum Pers*). Some of the organisations that use the LBH term are progressive cause lawyering organisations in the same mould as YLBHI-LBH, while others may simply offer conventional legal aid services.

The majority of data for this book was collected through more than 70 in-depth, semi-structured interviews with 51 current and former LBH staff, and eight representatives from domestic and international nongovernmental organisations (NGOs) and private development contractors that fund LBH's work. Most interviews were conducted during an extended period of fieldwork in Indonesia, from June to December 2019. Several senior YLBHI and LBH staff were interviewed more than once, with follow-up interviews completed over voice and video call from 2020–23. YLBHI's annual national work meeting in Jakarta in late 2019 provided an opportunity to meet with directors of all the regional LBH offices, including from areas that are difficult if not impossible to travel to, such as Papua.

Interviews generally lasted for 30–90 minutes, and aside from the handful of non-Indonesian respondents, all were conducted in Indonesian. All interviews were recorded, transcribed in English from the original Indonesian, then coded based on the main themes identified. Given the sometimes sensitive nature of the issues being discussed, all interviewees were offered the opportunity to remain anonymous. Most interview participants are activists who are accustomed to providing interviews to the media, and the majority declined this offer. In select circumstances where attributing statements might have led to organisational tension, I also elected to keep the identities of some interviewees confidential.

Empirical data was also collected through participant observation. During 2019, I spent several months with the leadership team of YLBHI, as well as shorter periods in the offices of LBH Bandung, LBH Semarang, LBH Surabaya and LBH Makassar. I attended press conferences, public seminars, paralegal training sessions and community legal education activities, including with some regional LBH offices. Public events were held almost every week at the YLBHI/LBH Jakarta office. These events were often quite revealing in terms of how cause lawyers at LBH viewed the state and conceptualised their role, and their personal and institutional values.

In addition to interviews and observation, I examined a broad range of written documents like organisational bylaws and internal organisational policies, training manuals and guidebooks, public annual reports (extending back to the early 2000s), public policy and research papers, press releases, and workshop and conference proceedings – all avenues through which LBH articulates its contribution to democratisation, human rights and the rule of law. I also collected and analysed formal laws and regulations, published court decisions, and English- and Indonesian-language media reporting on LBH and its cases.

Overview of the book

This book is divided into three parts. Part One consists of two chapters, including this introductory chapter. Chapter 2 examines the rich Indonesian tradition of cause lawyering practiced by LBH, focusing on Soeharto's New Order period. The chapter establishes how cause lawyering was conducted by LBH under an authoritarian regime, examines LBH's structural legal aid approach, and places it in the context of the analytical framework outlined above. This history provides an important basis for examination of how cause lawyers' practices and views about the state changed with democratic transition and, more recently, regression.

Part Two of the book focuses on LBH's approach to cause lawyering following the democratic transition, with an emphasis on the current period of democratic decline. It consists of four chapters and begins with a brief introductory section. The Introduction to Part Two describes the democratic transition and key political and institutional changes that occurred after Soeharto fell in 1998, as well as the threats to democracy that have emerged over the past decade, providing the necessary context for the discussions in Part Two. The democratic changes outlined in this introductory section had significant implications for LBH and the form of cause lawyering it has practiced, as the chapters that form Part Two demonstrate.

Chapter 3 focuses on the often overlooked organisational issues that affect the shape of cause lawyering. Changes to these internal organisational factors – all of which were precipitated by the democratic transition in some way – help to explain why LBH did not immediately thrive following the democratic transition. Chapter 4 then examines the litigation strategies deployed by LBH in the post-authoritarian era. Political and institutional changes after 1998 delivered new opportunities for Indonesian cause lawyers to mobilise the law, for example, with strategies like constitutional litigation, litigation in the administrative courts and citizen lawsuits. Chapter 4 focuses on these more elite-focused strategies, exploring how LBH's approach to these tactics have shifted with democratic erosion. The next chapter, Chapter 5, examines LBH's approach to working with social movements, describing how LBH has placed greater emphasis on these grassroots strategies over recent years, as it has begun to lose faith in litigation-focused approaches to social change. Chapter 6 focuses on a different kind of non-litigation strategy – direct engagement with the state through public policy advocacy and legislative lobbying. The chapter examines LBH's experimentations with this accommodative, elite-focused form of cause lawyering, and describes how disappointment with these efforts has led to LBH prioritising a confrontational approach that avoids close entanglement with the state.

The final section of the book, Part Three, consists of two chapters. Chapter 7 synthesises discussions in Chapters 4, 5, and 6, by revisiting the concept of structural legal aid and looking at how it has been applied in the post-Soeharto period. The chapter also documents LBH's return to playing the kind of convening and coalition-building role in civil society that it played in the late New Order period as Indonesian democracy has begun to deteriorate. Finally, the concluding chapter (Chapter 8) reviews the main findings of the book and discusses the implications of the Indonesian experience of cause lawyering practiced by LBH for broader understandings of cause lawyering and democratic change.

Notes

1 *Rancangan Undang-Undang Kitab Undang-Undang Hukum Pidana.*
2 *Rancangan Undang-Undang Tentang Tindak Pidana Kekerasan Seksual*, RUU TPKS.
3 The term *reformasi* can be used to refer to both the movement that had a role in Soeharto's downfall and the period of democratic reform that followed.
4 Asfinawati uses only one name.
5 The most common English translation for *Yayasan Lembaga Bantuan Hukum Indonesia* is 'Indonesian Legal Aid Foundation', and as such it is the one used in the title of this book. I prefer the slightly awkward but direct translation 'Indonesian Foundation of Legal Aid Institutes' to emphasise that YLBHI is the central body that oversees 18 Legal Aid Institutes (*Lembaga Bantuan Hukum*, LBH) in the provinces. This is discussed further below.
6 There are important differences between YLBHI and LBH, which are explained further below. In this book I use the term LBH to refer to the organisation in general terms, although I also specify YLBHI or individual regional LBH offices, such as LBH Surabaya, when necessary.
7 Although Soeharto was officially installed as president in 1968, the New Order is conventionally considered to have begun on 11 March 1966, when Indonesia's first president, Soekarno, was forced into signing the *Supersemar* (*Surat Perintah Sebelas Maret,* 11 March Instruction), which effectively transferred presidential authority to Soeharto (Lindsey and Santosa 2008, 3).
8 I am borrowing an observation made about Brazil, but one that could be equally applied to Indonesia.
9 See Chapter 2.
10 For this support of Vivian, LBH faced harsh criticism from conservative Islamic groups. In 1973, the court ruled in favour of Vivian – the first time an Indonesian court allowed a change of gender to occur on official documents (LBH Jakarta 2015, 23–24).
11 The 'White Book of the Students' Struggle 1978' (*Buku Putih Perjuangan Mahasiswa 1978*) was issued by ITB students, and involved a searing critique of New Order rule. The book was quickly banned and prompted a clampdown on student political activity (Bourchier and Hadiz 2013, 119–26).
12 Marsinah was a worker and labour organiser at a watch factory in East Java who was found murdered days after leading a demonstration over her employer's failure to provide workers with the minimum wage. LBH was one of the most prominent voices in the Solidarity Committee for Marsinah (*Komite Solidaritas Untuk Marsinah*, KSUM), a civil society coalition that formed to demand an investigation into her death, (Avonius 2008, 104–07).

13 Both the Kedung Ombo and PRD cases are discussed in Chapter 2.
14 This book uses Larry Diamond's commonly cited definition of civil society as "the realm of organised social life that is open, voluntary, self-generating, at least partially self-supporting, autonomous from the state and bound by a legal order or set of shared rules" (Diamond 1999, 221). Although this book is mainly concerned with pro-democratic civil society, I recognise that there are many actors and organisations in Indonesian civil society that are 'uncivil' or do not necessarily promote democratic values and ideals (Hefner 2000, 35–36; Beittinger-Lee 2013).
15 See Trubek (2011, 420) for an example of this.
16 Influential works that described the emergent public interest law industry included a 'Comment' in the *Yale Law Journal* (1970) and the books *Public Interest Law* (Weisbrod, Handler, and Komesar 1978) and *Lawyers and the Pursuit of Legal Rights* (Handler, Hollingsworth, and Erlanger 1978).
17 When Sarat and Scheingold refer to liberal democracy, they state that it entails "representative government, the rule of law, individual rights (including private property), and an autonomous, open and pluralistic civil society" (Scheingold and Sarat 2004, 102). This book follows a similarly conventional understanding of liberal democracy, involving free and fair elections, the rule of law and guarantees of civil and political rights, including for minorities (Huq and Ginsburg 2018, 87; Warburton and Aspinall 2019, 263; Peerenboom 2004, 4).
18 This was the case with LBH, which was funded by the Jakarta government in its early years.
19 Law and Human Rights Ministerial Decision No. M.HH-02.HN.03.03 of 2021.
20 As stated, I prefer the slightly awkward but direct translation 'Indonesian Foundation of Legal Aid Institutes', to emphasise that YLBHI is the central body that oversees the 18 LBH offices in the provinces.

References

Abel, Richard L. 1998. "Speaking Law to Power: Occasions for Cause Lawyering." In *Cause Lawyering: Political Commitments and Professional Responsibilities*, edited by Austin Sarat and Stuart A. Scheingold, 69–117. New York; Oxford: Oxford University Press. https://doi.org/10.1093/oso/9780195113198.003.0003.

Abel, Richard L. 2011. "Epilogue: Just Law?" In *The Paradox of Professionalism: Lawyers and the Possibility of Justice*, edited by Scott L. Cummings, 296–318. Cambridge: Cambridge University Press. https://doi.org/10.1017/CBO9780511921506.

Akbari, Anugerah Rizki. 2019. "Indonesians Better Get Ready for Jail, as Flawed New Criminal Code Looks Set to Pass." *Indonesia at Melbourne*, 3 September 2019. https://indonesiaatmelbourne.unimelb.edu.au/indonesians-better-get-ready-for-jail-as-flawed-new-criminal-code-looks-set-to-pass/.

Aspinall, Edward. 2004. "Indonesia: Transformation of Civil Society and Democratic Breakthrough." In *Civil Society and Political Change in Asia: Expanding and Contracting Democratic Space*, edited by Muthiah Alagappa, 61–96. Stanford, California: Stanford University Press. https://doi.org/10.1515/9780804767545-008.

Aspinall, Edward. 2005. *Opposing Suharto: Compromise, Resistance, and Regime Change in Indonesia*. East-West Center Series on Contemporary Issues in Asia and the Pacific. Stanford, California: Stanford University Press. https://doi.org/10.1515/9780804767316.

Aspinall, Edward. 2015. "The Surprising Democratic Behemoth: Indonesia in Comparative Asian Perspective." *The Journal of Asian Studies* 74 (4): 889–902. https://doi.org/10.1017/s0021911815001138.

Aspinall, Edward, Diego Fossati, Burhanuddin Muhtadi, and Eve Warburton. 2020. "Elites, Masses, and Democratic Decline in Indonesia." *Democratization* 27 (4): 505–526. https://doi.org/10.1080/13510347.2019.1680971.

Aspinall, Edward, and Marcus Mietzner. 2019. "Indonesia's Democratic Paradox: Competitive Elections Amidst Rising Illiberalism." *Bulletin of Indonesian Economic Studies* 55 (3): 295–317. https://doi.org/10.1080/00074918.2019.1690412.

Baik, Tae-Ung. 2010. "Public Interest Litigation in South Korea." In *Public Interest Litigation in Asia*, edited by Po Jen Yap and Holning Lau, 115–135. London: Routledge. https://doi.org/10.4324/9780203842645.

Benney, Jonathan. 2013. *Defending Rights in Contemporary China: Reserving the Right*. London and New York: Routledge. https://doi.org/10.4324/9780203108307.

Boukalas, Christos. 2013. "Politics as Legal Action/Lawyers as Political Actors: Towards a Reconceptualisation of Cause Lawyering." *Social & Legal Studies* 22 (3): 395–420. https://doi.org/10.1177/0964663912471552.

Butt, Simon. 2019. "Amendments Spell Disaster for the KPK." *Indonesia at Melbourne*, 18 September 2019. https://indonesiaatmelbourne.unimelb.edu.au/amendments-spell-disaster-for-the-kpk/.

Butt, Simon, and Tim Lindsey. 2018. *Indonesian Law*. Oxford; New York: Oxford University Press. https://doi.org/10.1093/oso/9780199677740.001.0001.

Cahya, Gemma Holliani. 2019. "Knocking on the House's Door: Victims, Activists Send Flowers for the Sexual Violence Bill." *The Jakarta Post*, 7 September 2019. https://www.thejakartapost.com/news/2019/09/06/knocking-on-the-houses-door-victims-activists-send-flowers-for-the-sexual-violence-bill.html.

Chang, Wen-Chen. 2010. "Public Interest Litigation in Taiwan: Strategy for Law and Policy Changes in the Course of Democratisation." In *Public Interest Litigation in Asia*, edited by Po Jen Yap and Holning Lau, 136–160. London: Routledge. https://doi.org/10.4324/9780203842645.

Cheesman, Nick, and Kyaw Min San. 2013. "Not Just Defending; Advocating for Law in Myanmar." *Wisconsin International Law Journal* 31 (3): 702–733.

Chen, Alan K., and Scott L. Cummings. 2013. *Public Interest Lawyering: A Contemporary Perspective*. New York: Wolters Kluwer Law & Business.

Chua, Lynette J. 2017. "Collective Litigation and the Constitutional Challenges to Decriminalizing Homosexuality in Singapore." *Journal of Law and Society* 44 (3): 433–455. https://doi.org/10.1111/jols.12037.

Chua, Lynette J. 2022. *The Politics of Rights and Southeast Asia*. Elements in Politics and Society in Southeast Asia. Cambridge: Cambridge University Press. https://doi.org/10.1017/9781108750783.

CNN Indonesia. 2019a. "Dituduh Perkeruh Papua, Kantor LBH Surabaya Digeruduk Massa." *CNN Indonesia*, 29 August 2019. https://www.cnnindonesia.com/nasional/20190829164448-20-425832/dituduh-perkeruh-papua-kantor-lbh-surabaya-digeruduk-massa.

CNN Indonesia. 2019b. "Demonstran Pembawa Bendera Merah Putih Segera Disidang." *CNN Indonesia*, 27 November 2019. https://www.cnnindonesia.com/nasional/20191127064459-12-451814/demonstran-pembawa-bendera-merah-putih-segera-disidang.

Crouch, Melissa. 2011. "Cause Lawyers, the Legal Profession and the Courts in Indonesia: The Bar Association Controversy." *Lawasia Journal*2011: 63–86.

Cummings, Scott L. 2008. "The Internationalization of Public Interest Law." *Duke Law Journal* 57 (4): 891–1036.

Cummings, Scott L. 2012. "The Pursuit of Legal Rights - And Beyond." *UCLA Law Review* 59 (3): 506–549.

Cummings, Scott L. 2020. "Movement Lawyering." *Indiana Journal of Global Legal Studies* 27 (1): 87–130. https://doi.org/10.2979/indjglolegstu.27.1.0087.

Cummings, Scott L., and Ingrid V. Eagly. 2001. "A Critical Reflection on Law and Organizing." *UCLA Law Review* 48 (3): 443–517.

Cummings, Scott L., and Louise Trubek. 2008. "Globalizing Public Interest Law." *UCLA Journal of International Law and Foreign Affairs* 13 (1): 1–53. https://doi.org/10.2139/ssrn.1338304.

Diprose, Rachael, Dave McRae, and Vedi R. Hadiz. 2019. "Two Decades of Reformasi in Indonesia: Its Illiberal Turn." *Journal of Contemporary Asia* 49 (5): 691–712. https://doi.org/10.1080/00472336.2019.1637922.

Dongoran, Hussein Abri. 2019. "Mahasiswa Bergerak." *Tempo Magazine*, 28 September 2019. https://majalah.tempo.co/read/laporan-utama/158486/mahasiswa-ber gerak.

Dressel, Björn, ed. 2012. *The Judicialization of Politics in Asia*. London: Routledge. https://doi.org/10.4324/9780203115596.

Ellmann, Stephen. 1998. "Cause Lawyering in the Third World." In *Cause Lawyering: Political Commitments and Professional Responsibilities*, edited by Austin Sarat and Stuart A.Scheingold, 349–430. New York; Oxford: Oxford University Press. https://doi.org/10.1093/oso/9780195113198.003.0012.

Epp, Charles R. 1998. *The Rights Revolution: Lawyers, Activists, and Supreme Courts in Comparative Perspective*. Chicago: University of Chicago Press. https://doi.org/10.7208/chicago/9780226772424.001.0001.

Farid, Cynthia. 2013. "New Paths to Justice: A Tale of Social Justice Lawyering in Bangladesh." *Wisconsin International Law Journal* 31 (3): 421–461.

Feng, Chongyi. 2009. "The Rights Defence Movement, Rights Defence Lawyers and Prospects for Constitutional Democracy in China." *Cosmopolitan Civil Societies: An Interdisciplinary Journal* 1 (3): 150–169. https://doi.org/10.5130/ccs.v1i3.1076.

Fu, Hualing. 2011. "Challenging Authoritarianism Through Law: Potentials and Limit." *National Taiwan University Law Review* 6 (1): 339–365.

Fu, Hualing. 2018. "The July 9th (709) Crackdown on Human Rights Lawyers: Legal Advocacy in an Authoritarian State." *Journal of Contemporary China* 27 (112): 554–568. https://doi.org/10.1080/10670564.2018.1433491.

Fu, Hualing, and Richard Cullen. 2008. "Weiquan (Rights Protection) Lawyering in an Authoritarian State: Building a Culture of Public-Interest Lawyering." *The China Journal*, no. 59: 111–127. https://doi.org/10.1086/tcj.59.20066382.

Fu, Hualing, and Richard Cullen. 2010. "The Development of Public Interest Litigation in China." In *Public Interest Litigation in Asia*, edited by Po Jen Yap and Holning Lau, 9–34. London: Routledge. https://doi.org/10.4324/9780203842645.

Fu, Hualing, and Richard Cullen. 2011. "Climbing the Weiquan Ladder: A Radicalizing Process for Rights-Protection Lawyers." *The China Quarterly* 205: 40–59. http s://doi.org/10.2139/ssrn.1487367.

Ginsburg, Tom. 2007. "Law and the Liberal Transformation of the Northeast Asian Legal Complex in Korea and Taiwan." In *Fighting for Political Freedom: Comparative Studies of the Legal Complex and Political Liberalism*, edited by Terence C. Halliday, LucienKarpik, and Malcolm M. Feeley, 43–63. Oxford: Hart Publishing. https://doi.org/10.5040/9781472560179.ch-002.

Ginsburg, Tom, and Tamir Moustafa, eds. 2008. *Rule by Law: The Politics of Courts in Authoritarian Regimes.* Cambridge: Cambridge University Press. https://doi.org/10.1017/cbo9780511814822.
Givens, John Wagner. 2013. "Sleeping with Dragons; Politically Embedded Lawyers Suing the Chinese State." *Wisconsin International Law Journal* 31 (3): 734–775.
Goedde, Patricia. 2009. "From Dissidents to Institution-Builders: The Transformation of Public Interest Lawyers in South Korea." *East Asia Law Review* 4 (1): 63–89. https://doi.org/10.1017/9781108893947.009.
Haggard, Stephan, and Robert Kaufman. 2021. *Backsliding: Democratic Regress in the Contemporary World.* Elements in Political Economy. Cambridge: Cambridge University Press. https://doi.org/10.1017/9781108957809.
Hajjar, Lisa. 2001. "From the Fight for Legal Rights to the Promotion of Human Rights: Israeli and Palestinian Cause Lawyers in the Trenches of Globalisation." In *Cause Lawyering and the State in a Global Era*, edited by Austin Sarat and Stuart A.Scheingold, 68–95. New York: Oxford University Press. https://doi.org/10.1093/0195141172.003.0003.
Halliday, Terence C. 1999. "Politics and Civic Professionalism: Legal Elites and Cause Lawyers." *Law & Social Inquiry* 24 (4): 1013–1060. https://doi.org/10.1086/492707.
Halliday, Terence C., Lucien Karpik, and Malcolm M. Feeley. 2007. "The Legal Complex in Struggles for Political Liberalism." In *Fighting for Political Freedom: Comparative Studies of the Legal Complex and Political Liberalism*, edited by Terence C. Halliday, Lucien Karpik, and Malcolm M. Feeley, 1–40. Oxford: Hart Publishing. https://doi.org/10.5040/9781472560179.ch-001.
Halliday, Terence C., Lucien Karpik, and Malcolm M. Feeley. eds. 2012. *Fates of Political Liberalism in the British Post-Colony: The Politics of the Legal Complex.* Cambridge: Cambridge University Press. https://doi.org/10.1017/cbo9781139002981.
Handler, Joel F. 1978. *Social Movements and the Legal System: A Theory of Law Reform and Social Change.* New York: Academic Press.
Harding, Andrew, and Amanda Whiting. 2012. "'Custodian of Civil Liberties and Justice in Malaysia': The Malaysian Bar and the Moderate State." In *Fates of Political Liberalism in the British Post-Colony: The Politics of the Legal Complex*, edited by Lucien Karpik, Malcolm M. Feeley, and Terence C. Halliday, 247–304. Cambridge: Cambridge University Press. https://doi.org/10.1017/cbo9781139002981.012.
Hilbink, Thomas M. 2004. "You Know the Type: Categories of Cause Lawyering." *Law & Social Inquiry* 29 (3): 657–698. https://doi.org/10.1086/430155.
Hsu, Ching-Fang. 2019. "The Currency Exchanger: Taiwanese Public Interest Lawyers in the 21st Century." *UCLA Pacific Basin Law Journal* 36 (1): 33–63. https://doi.org/10.5070/p8361042635.
Jigmiddash, Bayartsetseg, and Jennifer Rasmussen. 2013. "Protecting Community Rights: Prospects for Public Interest Lawyering in Mongolia." *Wisconsin International Law Journal* 31 (3): 566–585.
Kanesalingam, Shanmuga. 2013. "Monkey in a Wig: Loyarburok, Undimsia, Public Interest Litigation and Beyond." *Wisconsin International Law Journal* 31 (3): 586–619.
Kelly, Linda, and Bivitri Susanti. 2014. "*Independent Progress Review of DFAT Law and Justice Assistance in Indonesia.*" Prepared for the Australia Indonesia Partnership for Justice (AIPJ).
Kilwein, John. 1998. "Still Trying: Cause Lawyering for the Poor and Disadvantaged in Pittsburgh, Pennsylvania." In *Cause Lawyering: Political Commitments and*

Professional Responsibilities, edited by Austin Sarat and Stuart A. Scheingold, 181–200. New York; Oxford: Oxford University Press. https://doi.org/10.1093/oso/9780195113198.003.0006.

Klug, Heinz. 2001. "Local Advocacy, Global Engagement: The Impact of Land Claims Advocacy on the Recognition of Property Rights in the South African Constitution." In *Cause Lawyering and the State in a Global Era*, edited by Austin Sarat and Stuart A. Scheingold, 264–286. New York: Oxford University Press. https://doi.org/10.1093/0195141172.003.0010.

Kohno, Takeshi. 2010. *The Emergence of the Legal Aid Institute in Authoritarian Indonesia: How a Human Rights Organization Survived the Suharto Regime and Became a Cornerstone for Civil Society in Indonesia*. Saarbrücken: VDM Verlag Dr. Müller.

LBH Jakarta, ed. 2015. *Rentang Jejak LBH Jakarta: Kisah-Kisah Penanganan Kasus*. Jakarta: LBH Jakarta.

Lev, Daniel. 1987. *Legal Aid in Indonesia*. Working Papers (Monash University. Centre of Southeast Asian Studies) 44. Clayton, Victoria: Monash University.

Lev, Daniel. 1996. "Between State and Society: Professional Lawyers and Reform in Indonesia." In *Making Indonesia*, edited by Daniel Lev and Ruth McVey, 144–163. Ithaca, New York: Cornell University Press. https://doi.org/10.7591/9781501719370-009.

Lev, Daniel. 1998. "Lawyers' Causes in Indonesia and Malaysia." In *Cause Lawyering: Political Commitments and Professional Responsibilities*, edited by Austin Sarat and Stuart A. Scheingold, 431–452. New York; Oxford: Oxford University Press. https://doi.org/10.1093/oso/9780195113198.003.0013.

Lev, Daniel. 2007. "A Tale of Two Legal Professions: Lawyers and State in Malaysia and Indonesia." In *Raising the Bar: The Emerging Legal Profession in East Asia*, edited by William P. Alford, 383–414. Cambridge, Massachusetts: Harvard University Press.

Levitsky, Sandra R. 2006. "To Lead with Law: Reassessing the Influence of Legal Advocacy Organisations in Social Movements." In *Cause Lawyers and Social Movements*, edited by Austin Sarat and Stuart A. Scheingold, 145–163. Stanford, California: Stanford University Press. https://doi.org/10.1515/9780804767965-009.

Lindsey, Tim. 2004. "Legal Infrastructure and Governance Reform in Post-Crisis Asia: The Case of Indonesia." *Asian-Pacific Economic Literature* 18 (1): 12–40. https://doi.org/10.1111/j.1467-8411.2004.00142.x.

Lindsey, Tim. 2008. "Constitutional Reform in Indonesia: Muddling Towards Democracy." In *Indonesia: Law and Society*, edited by Tim Lindsey, 2nd Edition, 23–47. Sydney, Australia: The Federation Press.

Lindsey, Tim, and Melissa Crouch. 2013. "Cause Lawyers in Indonesia: A House Divided." *Wisconsin International Law Journal* 31 (3): 620–645.

Liu, Sida, and Terence C. Halliday. 2011. "Political Liberalism and Political Embeddedness: Understanding Politics in the Work of Chinese Criminal Defense Lawyers." *Law & Society Review* 45 (4): 831–865. https://doi.org/10.1111/j.1540-5893.2011.00458.x.

Liu, Sida, and Terence C. Halliday. 2016. *Criminal Defense in China: The Politics of Lawyers at Work*. Cambridge Studies in Law and Society. Cambridge: Cambridge University Press. https://doi.org/10.1017/9781316677230.

Luban, David. 1988. *Lawyers and Justice: An Ethical Study*. Princeton: Princeton University Press. https://doi.org/10.2307/j.ctv346rrr.

Luban, David. 2012. "The Moral Complexity of Cause Lawyers Within the State." *Fordham Law Review* 81 (2): 705–714.
Lubis, Todung Mulya. 1985a. "Legal Aid in the Future (A Development Strategy for Indonesia)." *Third World Legal Studies* 4 (1): 133–148.
Lubis, Todung Mulya. 1985b. "Legal Aid: Some Reflections." In *Access to Justice: Human Rights Struggles in South East Asia*, edited by Laurie S. Wiseberg and Harry M. Scoble, 40–45. London: Zed Books.
Maharani, Tsarina. 2019. "Masa Transisi 3 Tahun, RUU KUHP Baru Efektif Berlaku 2022." *Detik*, 3 September 2019. https://news.detik.com/berita/d-4691192/masa-transisi-3-tahun-ruu-kuhp-baru-efektif-berlaku-2022.
Marshall, Anna-Maria, and Daniel Crocker Hale. 2014. "Cause Lawyering." *Annual Review of Law and Social Science* 10 (1): 301–320. https://doi.org/10.1146/annurev-lawsocsci-102612-133932.
McCann, Michael. 1994. *Rights at Work: Pay Equity Reform and the Politics of Legal Mobilization*. Chicago Series in Law and Society. Chicago: University of Chicago Press.
McCann, Michael. 2004. "Law and Social Movements." In *The Blackwell Companion to Law and Society*, edited by Austin Sarat, 506–522. Malden, Massachusetts, USA; Oxford, UK; Carlton, Australia: Blackwell Publishing. https://doi.org/10.1002/9780470693650.ch27.
McCann, Michael. 2006. "Legal Mobilization and Social Reform Movements: Notes on Theory and Its Application." In *Law and Social Movements*, edited by Michael McCann, 3–32. The International Library of Essays in Law and Society. London and New York: Routledge. https://doi.org/10.4324/9781315091983-1.
McCann, Michael, and Helena Silverstein. 1998. "Rethinking Law's 'Allurements': A Relational Analysis of Social Movement Lawyers in the United States." In *Cause Lawyering: Political Commitments and Professional Responsibilities*, edited by Austin Sarat and Stuart A. Scheingold, 261–292. New York; Oxford: Oxford University Press. https://doi.org/10.1093/oso/9780195113198.003.0009.
McEvoy, Kieran, and Anna Bryson. 2022. "Boycott, Resistance and the Law: Cause Lawyering in Conflict and Authoritarianism." *The Modern Law Review* 85 (1): 69–104. https://doi.org/10.1111/1468-2230.12671.
Meili, Stephen. 1998. "Cause Lawyers and Social Movements: A Comparative Perspective on Democratic Change in Argentina and Brazil." In *Cause Lawyering: Political Commitments and Professional Responsibilities*, edited by Austin Sarat and Stuart A. Scheingold, 487–522. New York; Oxford: Oxford University Press. https://doi.org/10.1093/oso/9780195113198.003.0015.
Meili, Stephen. 2009. "Staying Alive: Public Interest Law in Contemporary Latin America." *International Review of Constitutionalism* 9 (1): 43–71.
Menkel-Meadow, Carrie. 1998. "The Causes of Cause Lawyering: Toward an Understanding of the Motivation and Commitment of Social Justice Lawyers." In *Cause Lawyering: Political Commitments and Professional Responsibilities*, edited by Austin Sarat and Stuart A. Scheingold, 31–68. New York; Oxford: Oxford University Press. https://doi.org/10.1093/oso/9780195113198.003.0002.
Mietzner, Marcus. 2021. "Sources of Resistance to Democratic Decline: Indonesian Civil Society and Its Trials." *Democratization* 28 (1): 161–178. https://doi.org/10.4324/9781003346395-9.
Moustafa, Tamir. 2007a. "Mobilising the Law in an Authoritarian State: The Legal Complex in Contemporary Egypt." In *Fighting for Political Freedom: Comparative*

Studies of the Legal Complex and Political Liberalism, edited by Terence C. Halliday, Lucien Karpik, and Malcolm M. Feeley, 193–218. Oxford: Hart Publishing. https://doi.org/10.5040/9781472560179.ch-006.

Moustafa, Tamir. 2007b. *The Struggle for Constitutional Power: Law, Politics, and Economic Development in Egypt*. Cambridge: Cambridge University Press. https://doi.org/10.1017/cbo9780511511202.

Mujani, Saiful, and R. William Liddle. 2021. "Indonesia: Jokowi Sidelines Democracy." *Journal of Democracy* 32 (4): 72–86. https://doi.org/10.1353/jod.2021.0053.

Munger, Frank. 2015. "Thailand's Cause Lawyers and Twenty-First-Century Military Coups: Nation, Identity, and Conflicting Visions of the Rule of Law." *Asian Journal of Law & Society* 2 (2): 301. https://doi.org/10.1017/als.2015.18.

Munger, Frank, Scott L. Cummings, and Louise Trubek. 2013. "Mobilising Law for Justice in Asia: A Comparative Approach." *Wisconsin International Law Journal* 31 (3): 353–420.

Mustafina, Renata. 2022. "Turning on the Lights? Publicity and Defensive Legal Mobilization in Protest-Related Trials in Russia." *Law & Society Review* 56 (4): 601–622. https://doi.org/10.1111/lasr.12631.

Nasution, Adnan Buyung. 1985. "The Legal Aid Movement in Indonesia: Towards the Implementation of the Structural Legal Aid Concept." In *Access to Justice: Human Rights Struggles in South East Asia*, edited by Laurie S. Wiseberg and Harry M. Scoble, 31–39. London: Zed Books.

Nasution, Adnan Buyung. 2011. "*Towards Constitutional Democracy in Indonesia*." 1. Papers on Southeast Asian Constitutionalism. Melbourne: Asian Law Centre, Melbourne Law School.

NeJaime, Douglas. 2012. "Cause Lawyers Inside the State." *Fordham Law Review* 81 (2): 649–704.

Nesossi, Elisa. 2015. "Political Opportunities in Non-Democracies: The Case of Chinese Weiquan Lawyers." *International Journal of Human Rights* 19 (7): 961–978. https://doi.org/10.1080/13642987.2015.1075305.

Nielsen, Laura Beth, and Catherine R. Albiston. 2006. "The Organization of Public Interest Practice: 1975–2004." *North Carolina Law Review* 84 (5): 1591–1622.

Pils, Eva. 2014. *China's Human Rights Lawyers: Advocacy and Resistance*. Routledge Research in Human Rights Law. London and New York: Routledge. https://doi.org/10.4324/9780203769061.

Power, Thomas. 2018. "Jokowi's Authoritarian Turn and Indonesia's Democratic Decline." *Bulletin of Indonesian Economic Studies* 54 (3): 307–338. https://doi.org/10.1080/00074918.2018.1549918.

Power, Thomas, and Eve Warburton, eds. 2020. *Democracy in Indonesia: From Stagnation to Regression?* Singapore: ISEAS Publishing. https://doi.org/10.1355/9789814881524.

Putsanra, Dipna Videlia. 2019. "Isi RUU Bermasalah Didemo Mahasiswa Hari Ini di Jakarta & Kota Lain." *Tirto.id*, 24 September 2019. https://tirto.id/isi-ruu-bermasalah-didemo-mahasiswa-hari-ini-di-jakarta-kota-lain-eiCs.

Rajah, Jothie, and Arun K. Thiruvengadam. 2013. "Of Absences, Masks and Exceptions: Cause Lawyering in Singapore." *Wisconsin International Law Journal* 31 (3): 646–671.

Rekosh, Edwin. 2008. "Constructing Public Interest Law: Transnational Collaboration and Exchange in Central and Eastern Europe." *UCLA Journal of International Law and Foreign Affairs* 13 (1): 55.

Repucci, Sarah. 2020. "The Freedom House Survey for 2019: The Leaderless Struggle for Democracy." *Journal of Democracy* 31 (2): 137–152.

Ristianto, Christoforus. 2019. "RKUHP Soal Penghinaan Presiden, Kumpul Kebo, Hingga Unggas, Ini Penjelasan Menkumham." *Kompas.com*, 21 September 2019. https://nasional.kompas.com/read/2019/09/21/08562241/rkuhp-soal-penghinaan-presiden-kumpul-kebo-hingga-unggas-ini-penjelasan.

Sarat, Austin, and Stuart A. Scheingold. 1998a. "Cause Lawyering and the Reproduction of Professional Authority: An Introduction." In *Cause Lawyering: Political Commitments and Professional Responsibilities*, edited by Austin Sarat and Stuart A. Scheingold, 3–28. New York; Oxford: Oxford University Press. https://doi.org/10.1093/oso/9780195113198.003.0001.

Sarat, Austin, and Stuart A. Scheingold. eds. 1998b. *Cause Lawyering: Political Commitments and Professional Responsibilities.* New York; Oxford: Oxford University Press. https://doi.org/10.1093/oso/9780195113198.001.0001.

Sarat, Austin, and Stuart A. Scheingold. eds. 2001a. *Cause Lawyering and the State in a Global Era.* New York: Oxford University Press.

Sarat, Austin, and Stuart A. Scheingold. 2001b. "State Transformation, Globalization, and the Possibilities of Cause Lawyering: An Introduction." In *Cause Lawyering and the State in a Global Era*, edited by Austin Sarat and Stuart A. Scheingold, 3–31. New York: Oxford University Press. https://doi.org/10.1093/0195141172.003.0001.

Sarat, Austin, and Stuart A. Scheingold. eds. 2005. *The Worlds Cause Lawyers Make: Structure and Agency in Legal Practice.* Stanford, California: Stanford University Press. https://doi.org/10.1515/9781503625440.

Sarat, Austin, and Stuart A. Scheingold. eds. 2006a. *Cause Lawyers and Social Movements.* Stanford, California: Stanford University Press. https://doi.org/10.1515/9780804767965.

Sarat, Austin, and Stuart A. Scheingold. 2006b. "What Cause Lawyers Do for, and to, Social Movements: An Introduction." In *Cause Lawyers and Social Movements*, edited by Austin Sarat and Stuart A. Scheingold, 1–34. Stanford, California: Stanford University Press. https://doi.org/10.1515/9780804767965-003.

Sarat, Austin, and Stuart A. Scheingold. eds. 2008. *The Cultural Lives of Cause Lawyers.* Cambridge; New York: Cambridge University Press. https://doi.org/10.1017/cbo9780511619786.

Schäfer, Saskia. 2019. "Democratic Decline in Indonesia: The Role of Religious Authorities." *Pacific Affairs* 92 (2): 235–255.

Scheingold, Stuart A. 2001. "Cause Lawyering and Democracy in Transnational Perspective: A Postscript." In *Cause Lawyering and the State in a Global Era*, edited by Austin Sarat and Stuart A. Scheingold, 287–304. New York: Oxford University Press. https://doi.org/10.1093/0195141172.003.0015.

Scheingold, Stuart A. 2004. *The Politics of Rights: Lawyers, Public Policy and Political Change.* Ann Arbor, United States: University of Michigan Press. https://doi.org/10.3998/mpub.6766.

Scheingold, Stuart A., and Austin Sarat. 2004. *Something to Believe In: Politics, Professionalism, and Cause Lawyering*. Stanford, California: Stanford University Press.

Soon, Edmund Bon Tai, and Pui Yi Wong. 2023. "Vernacularising Human Rights in Southeast Asia." In *Routledge Handbook of Civil and Uncivil Society in Southeast Asia*, edited by Eva Hansson and Meredith Weiss, 312–328. London and New York: Routledge. https://doi.org/10.4324/9780367422080-23.

Southworth, Ann. 2013. "What Is Public Interest Law? Empirical Perspectives on an Old Question." *DePaul Law Review* 62 (2): 493–518.

Stern, Rachel E. 2013. *Environmental Litigation in China: A Study in Political Ambivalence*. New York: Cambridge University Press. https://doi.org/10.1017/cbo9781139096614.
Stern, Rachel E. 2017. "Activist Lawyers in Post-Tiananmen China." *Law & Social Inquiry* 42 (1): 234–251. https://doi.org/10.1111/lsi.12225.
Tam, Waikeung. 2010. "Political Transition and the Rise of Cause Lawyering: The Case of Hong Kong." *Law & Social Inquiry* 35 (3): 663–687. https://doi.org/10.1111/j.1747-4469.2010.01199.x.
Tam, Waikeung. 2013. *Legal Mobilization Under Authoritarianism: The Case of Post-Colonial Hong Kong*. Cambridge: Cambridge University Press. https://doi.org/10.1017/cbo9781139424394.
Trismana, Ari, dir. 2021. *LBH - Meniti Jalan Terjal Demokrasi*. Documentary. Watchdoc.
Trubek, Louise. 2011. "Public Interest Law: Facing the Problems of Maturity." *University of Arkansas at Little Rock Law Review* 33 (4): 417. https://doi.org/10.2139/ssrn.1881889.
University of Wisconsin-Madison. 2013. "Symposium on a Comparative Perspective on Social Justice Lawyering in Asia: Conditions, Practices, and Possibilities." *Wisconsin International Law Journal* 31 (3).
Vieira, Oscar Vilhena. 2008. "Public Interest Law: A Brazilian Perspective." *UCLA Journal of International Law and Foreign Affairs* 13: 219.
Warburton, Eve, and Edward Aspinall. 2019. "Explaining Indonesia's Democratic Regression: Structure, Agency and Popular Opinion." *Contemporary Southeast Asia* 41 (2): 255–285. https://doi.org/10.1355/cs41-2k.
Weisbrod, Burton Allen, Joel F. Handler, and Neil K. Komesar. 1978. *Public Interest Law: An Economic and Institutional Analysis*. Berkeley, Los Angeles and London, England: University of California Press. https://doi.org/10.2307/jj.5233077.
Yap, Po Jen, and Holning Lau. 2010. "Public Interest Litigation in Asia: An Overview." In *Public Interest Litigation in Asia*, edited by Po Jen Yap and Holning Lau, 1–8. London: Routledge. https://doi.org/10.4324/9780203842645.

2 The making of a ‘locomotive of democracy’

Cause lawyers under Soeharto’s New Order

The Indonesian legal aid movement was born amid and in response to authoritarianism. To understand the complex ways in which Indonesian cause lawyers have responded to democratic transition and regression, it is important to first appreciate the rich tradition of legal aid and activism pioneered by LBH in the difficult New Order setting. This chapter investigates that history.

Established in 1970 as the Jakarta Legal Aid Institute, LBH is one of the oldest and most prominent CSOs in Indonesia. LBH was founded to provide ‘conventional’ case-based legal aid for poor and marginalised Indonesians, but its role quickly expanded as the realities of life under Soeharto’s New Order became clear. Emphasising ‘modernisation’ and political stability, the New Order provided an oversized role for the military in social and political life, articulated through the doctrine of *dwifungsi* (dual function) (Aspinall and Fealy 2010, 5).[1] Political participation was restricted, and civil society was constrained. Elections were held, but victory for the government’s political vehicle, Golkar (*Golongan Karya*, Functional Groups), was guaranteed. The regime permitted only two other political parties from the early 1970s, and they were prevented from opening offices or campaigning at the local level, except in the period just before elections (Hadiwinata 2003, 55–56). Despite early promises to observe the rule of law, the executive dominated the political system, as it maintained control over both the legislative and judicial branches (Lindsey and Santosa 2008, 10). Civil society was also under considerable pressure. Public gatherings and discussions were restricted (Hadiwinata 2003, 56–57). The media was tightly regulated, with media producers required to hold permits that could be easily revoked if they angered the government (Kakiailatu 2007, 63). One of the defining features of the New Order regime was significant political repression and violence, and this was a major factor in its longevity (Aspinall and Fealy 2010, 6). It was in this hostile environment that LBH came into existence.

This chapter examines LBH’s first three decades, focusing on how cause lawyering was practiced by LBH under the authoritarian New Order, in the absence of many of the structural factors considered supportive of cause lawyering. The chapter also explores the concept of ‘structural legal aid’, which LBH developed to explain its approach to legal aid and political

DOI: 10.4324/9781003486978-5

action. The chapter describes LBH's evolution from a legal aid organisation to "the most prominent source of social-legal criticism in the country" (Lev 1987, 23). LBH's legacy as a site of resistance to the New Order provides important context for understanding the organisation's approach to cause lawyering following the fall of Soeharto. The chapter will also briefly discuss internal conflicts that emerged in the organisation during the 1990s, which resulted from tensions between lawyers who favoured a more conservative, strictly litigation-focused role, and those committed to a more overtly political approach to cause lawyering. The shadow of these conflicts can be seen in the leadership tensions that afflicted the organisation in the post-Soeharto period.

Birth of the Indonesian legal aid movement

LBH was founded by Adnan Buyung Nasution (1934–2015), one of Indonesia's most distinguished lawyers and statespeople. A skilled advocate, Buyung was known for being "voluble, articulate, flamboyant" and liked "to call a spade a spade" (Jenkins 1994). He was also well-connected politically and maintained ties with senior military and government officials throughout his career (Aspinall 2005, 111), even if he often drew their ire.

LBH's origins extend back to the late 1950s, when Buyung, then a young assistant public prosecutor, studied at Melbourne Law School, University of Melbourne. During his time in Australia, Buyung undertook an internship with the Supreme Court of Victoria and spent time with the office of the Public Solicitor in Sydney,[2] where he learned about the provision of free legal aid there (Nasution 2011, 4). Inspired by his experiences in Australia, Buyung returned to Indonesia determined to introduce free legal aid and a public defender's office in Indonesia (Nasution 1994, 114).

When Buyung departed for Australia in 1958, Indonesia's first president, Soekarno, was still in power and Indonesia was at the end of its brief period of parliamentary democracy (1950–57). During the parliamentary democracy period, constitutionalism and the rule of law (known in Indonesia as *negara hukum,* although the German term *Rechtsstaat* is also used) seemed assured. While they were only few in number, lawyers enjoyed a relatively comfortable status (Lev 1987, 5). On 5 July 1959, however, while Buyung was still in Australia, Soekarno issued a decree reinstating the brief, original 1945 Constitution,[3] which dismantled the separation of powers and provided the president with considerable control. This put an end to the parliamentary democracy period and marked the beginning of his so-called 'Guided Democracy',[4] a period when "Soekarno was the guide, and there was no democracy" (Palmier 1973, 296).

Soekarno dismantled many of the structures that lawyers held dear (Lev 2000, 8). Law enforcement institutions, including the police, Prosecutor's Office and the Supreme Court, were brought under Soekarno's authority (Nasution 2011, 6–8). Soekarno even passed a decree allowing him to intervene in the legal process in matters of national interest (Lev 1987, 6). The legal profession became quickly disillusioned with the status of the *negara hukum*. When he arrived back at the Public Prosecutor's Office in 1959, Buyung was deeply disappointed by the erosion in political freedom and judicial independence that had occurred in his absence. He increasingly came into conflict with prosecutors he felt were abusing their power and was eventually dismissed from the Prosecutor's Office in 1969 (Nasution 2011, 8).

Soekarno's 'Guided Democracy' lasted less than a decade, with the horrific events of 1965 putting an end to the first president's rule by March the following year.[5] In the early hours of 1 October 1965, a group of military officers known as the '30 September Movement' killed six military generals in an apparent coup attempt (Melvin 2018, 2). General Soeharto, then the head of the Army's Strategic Reserve Command (*Komando Strategis Angkatan Darat*, Kostrad), blamed the attempted coup on the Indonesian Communist Party (*Partai Komunis Indonesia*, PKI) and used it as a pretext to gradually assume the presidency (Roosa 2006). What followed was a campaign of mass violence, which the military had been preparing for some time (Melvin 2018), with 500,000–1,000,000 alleged members of the PKI and its sympathisers killed and at least a million imprisoned. Despite the violence that marked Soeharto's ascent to the presidency, the New Order initially made promises to undo some of the excesses of the Guided Democracy period. There was renewed enthusiasm for bringing Indonesia closer to the *negara hukum* ideal that had been abandoned under Soekarno, with *negara hukum* ideas discussed regularly in the media, and among professionals and academics (Lev 2000, 309).

But it quickly became clear that the New Order regime and private lawyers had very different understandings of *negara hukum* (Lev 1987, 10–11). While lawyers interpreted the concept to involve limitations on state power and the implementation of legal process, New Order leaders viewed law as "a state resource" (Lev 2000, 288). The New Order's notion of rule of law was more like rule *by* law, with the law expected to serve the interests of the state, rather than acting as a check on its power (Carothers 1998, 97). Disenchantment with ongoing state repression – and the fact that New Order efforts to move Indonesia closer toward the *negara hukum* were more rhetoric than substance – were driving forces that inspired the birth of the legal aid movement.

Soon after being dismissed from the Prosecutor's Office, Buyung set about establishing a legal aid service in Indonesia. In 1969, he took his proposal to the third annual congress of the Indonesian Advocates Union (*Persatuan Advokat Indonesia*, Peradin), the leading professional association for lawyers in Indonesia at the time. Established by about a dozen prominent lawyers in 1963 (Lev 2008, 55), Peradin had become known for its commitment to judicial independence, due process and reform (Lev 2000, 286–87). Some members of the association were reportedly reluctant to provide support for Buyung's proposal, fearing pro bono legal aid could encroach on their activities (Lev 1987, 13). In the end, however, Buyung received the backing of Peradin head Lukman Wiriadinata and secretary Suardi Tasrif (Kusumah 1995) and LBH was established in October 1970 as a pilot project of Peradin, with Buyung as its first director. It quickly became clear there was a great underserved need for legal services. In its first year of operation, in 1971, LBH handled 595 cases (LBH Jakarta 2015, 13). According to one estimate, by 1985, more than 100,000 individuals had come to LBH seeking its assistance, and 30,000 of these cases had been represented in court (Lubis 1985a, 135).[6]

The primary motivation for the establishment of LBH might have been to provide legal assistance to poor and marginalised Indonesians but even from its early days, the organisation viewed its role more expansively. Another powerful motivational factor for lawyers at LBH was limiting abuses of state power through the (re)instatement of the *negara hukum* (Aspinall 2005, 100).

In setting up LBH, Buyung also sought and secured the support of progressive Jakarta Governor Ali Sadikin, who agreed to provide initial funding of about Rp 300,000 per month (Winarta 2013, 49) (about US$2,800 today), and issued a decision (*surat keputusan*) formalising the establishment of the institute (Nasution 2011, 8). Buyung also sought the endorsement of President Soeharto through his powerful personal assistant, Major General Ali Moertopo. At the time, Moertopo was deputy head of Bakin, the State Intelligence Coordination Agency (*Badan Koordinasi Intelijen Negara*), and head of the intelligence mechanism Opsus (Special Operations, *Operasi Khusus*), which both played leading roles in stifling political opposition to the New Order regime. Not only did Moertopo offer his backing, but he also donated ten motorcycles to support the organisation (Nasution 2011, 9). Buyung later noted this support was likely window dressing, designed to placate international critics concerned about the democratic and human rights situation in Indonesia. In fact, just a few years later, Moertopo had Buyung arrested during the *Malari* riots of 1974,[7] along with several other activists and government critics (Nasution 2011, 9). Buyung was wrongly accused of playing a role in instigating the rioting, and spent nearly two years in prison before being released without charge (Tempo 2014).

Although LBH was established in part as a response to growing repression under the New Order, it was initially somewhat sympathetic to the government's modernisation project and Aspinall notes that early organisational documents included references to New Order-style terms like *modernisasi hukum* (legal modernisation), and *pembangunan hukum* (legal development) (Aspinall 2005, 101). This was also reflective of the dominance of modernisation theories of development, which were prominent from the post-World War II period until at least the 1970s. Indeed, when established under Peradin in 1970, LBH even included a reference to modernisation in its organisational mission. LBH was founded to:

- Provide pro bono legal aid to all citizens who are unable to afford legal fees
- Develop and increase holistic legal awareness in the community and increase citizens' awareness of their rights
- Develop the law and law enforcement in accordance with the modernisation programme (Winarta 2013, 50).

In accordance with these modernisation tendencies, Aspinall also noted that during these early years, LBH leaders often stated that deficiencies in the rule of law were linked to "feudal ideas" in the community, caused by "economic backwardness" (Aspinall 2005, 102). Documents produced around this time contain somewhat paternalistic language consistent with modernisation theories of development. While the term is in less popular use now, many earlier LBH documents refer to the target of legal aid as being poor Indonesian citizens who are *buta hukum* (ignorant of the law) (Lubis 1985b, 41; Nasution 1985, 32).

Despite these references to modernisation, LBH was not directly connected to the short-lived and now discredited law and development movement that was also popular around the same time. The law and development model was anchored in modernisation theory and was based on a belief that a stronger legal profession would promote development (Trubek and Galanter 1974, 1075). One of the major backers of the movement was the Ford Foundation, which established the International Legal Centre (ILC) in New York to fund law and development initiatives in the Global South (Frühling 2000, 56). In Indonesia, the Ford Foundation supported the placement of teaching fellows in the law faculties of several state universities, in an attempt to reform teaching and research practices (Linnan 1999, 10). Several years after LBH was founded, Buyung was invited to attend ILC annual meetings, where he and then Indonesian Minister of Justice Mochtar Kusumaatmadja discussed the development of the rule of law and human rights in Indonesia (Nasution 2004, 31).

But even if Buyung and other influential figures within LBH were aware of, and conversant with, debates around law and development and public interest law occurring in the United States, the legal aid model developed in Indonesia by LBH was not a 'project' of the Ford Foundation or other foreign donors.[8] As noted, LBH emerged from Peradin, the main professional organisation for lawyers at the time, and was largely funded by domestic sources for the first decade of its operation. Even though nods to modernisation and legal development may have been present in early LBH documents, this type of language soon disappeared.[9] In fact, several years later, senior YLBHI lawyer Todung Mulya Lubis rejected the law and development model, stating that such programmes may really be "intended to preserve order, security, and/or capitalism" (Lubis 1985a, 136). This rejection also reflected the growing influence of the dependency theory of development in postcolonial societies like Indonesia. Despite this rejection of law and development as an approach, LBH continued to retain a point on legal reform in its mission statement, in addition to its strong focus on the struggle for social justice, the *negara hukum*, and democracy.

It is true that LBH founders like Buyung may have been motivated in part by self-interest in their aspirations for the *negara hukum* – its restoration would enable lawyers to regain some of the security and status they had enjoyed in the parliamentary democracy period (Lev 1987, 11). As Lev noted, "strong autonomous courts, binding procedural codes that allow space and leverage to attorneys – litigators above all – the *Rechtsstaat* itself and the public values that surround it, make for a lawyer's paradise" (Lev 1998, 447). But even if self-interest played a role, lawyers also had an ideological commitment to the *negara hukum* as a force for the realisation of justice for poor and marginalised Indonesians (Aspinall 2005, 101). Placing these ideas within the context of cause lawyering, LBH's primary 'cause' in these early years was the rule of law and the legal system – a strongly proceduralist orientation. Proceduralist lawyers believe "in the fundamental soundness of the legal system itself (if not at present, then in the system's future promise)" (Hilbink 2004, 666). As discussed in Chapter 1, they believe that if the legal process functions as it should, it is capable of delivering justice. Notably, given their proceduralist focus, lawyers in this category define their cause in procedural terms too. Indeed, as Lev once observed of LBH lawyers, "it is largely procedural justice, fairness of institutional treatment that dominates their imaginations" (Lev 1998, 447). Proceduralist lawyers tend to view clients as individuals, and consider that the cause is advanced through the act of individual client representation, consistent with most mainstream understandings of legal professional norms and ethics (Hilbink 2004, 672). It is perhaps unsurprising LBH had such a strong proceduralist orientation in its early days, given its inception as a legal aid project of the professional association for lawyers, Peradin.

It is overly simplistic, however, to represent LBH as ever having been solely proceduralist in its approach. Although many lawyers within LBH remained sympathetic to and were motivated by proceduralist ideas (Lev 1987, 20), LBH also took on highly controversial, politically imbued cases from the beginning. LBH was founded at a time when Soeharto's New Order was beginning to undertake large-scale investments in development projects, many of which had significant and damaging consequences for local people. Two of LBH's most important early cases challenged the forced eviction of families for the Simprug luxury housing estate in South Jakarta and the Beautiful Indonesia in Miniature Park (*Taman Mini Indonesia Indah*, TMII) in Lubang Buaya, East Jakarta, both of which occurred within just a few years of LBH's founding and put it in direct conflict with the Jakarta government.

In the Simprug case, LBH defended 108 families (or about 700 people) who were to be displaced by the Simprug housing estate (Saptono 2012, 2). This was the first time LBH lawyers "got their boots dirty" (literally as well as figuratively) and put their professionalism on the line (Saptono 2012, 1). As such, it occupies a special place in LBH folklore. LBH lawyers transgressed conventional understandings of lawyerly professionalism and openly expressed solidarity with their clients – they even stood alongside the community, while bulldozers, accompanied by the military, rolled in (Nasution 2011, 9). It was a clear example of the moral and political commitment and identification with the causes of clients that are key attributes of cause lawyering. It was also one of the first cases where LBH went beyond representing individuals to use an approach more typically associated with class actions, advocating on behalf of many individual clients (Lev 1987, 18). In addition to standing alongside the community, LBH also made efforts to educate locals about their rights, and there were early attempts at community organising – LBH supported the establishment of a People's Defence Committee (*Komite Pembelaan Rakyat*) to facilitate discussion of advocacy approaches (Nasution 2011, 9). LBH's work, while not successful in preventing the Simprug development from going ahead, was able to force the government to engage in more meaningful dialogue with the community, and secured relocation land and compensation for the affected families (Saptono 2012, 2).

Taman Mini, meanwhile, was a pet project of Soeharto's wife, Siti Hartinah (known as Tien Soeharto), designed to be a theme park showcasing Indonesian culture. About 500 families were evicted from their land in the Lubang Buaya area when construction began in 1972, and were offered minimal compensation. The case put LBH in confrontation not only with the president's wife (and, by extension, the regime) but also with Jakarta Governor Sadikin, who Tien had asked to be project director (Lev 2000, 292). Again, LBH's efforts were not able to prevent the project from going ahead, but they were able to force the government to increase the amount of compensation offered, and provide residents with priority access to job opportunities (LBH Jakarta 2015, 22).

Despite a high demand for its services, LBH had only one office, in Jakarta, for nearly a decade. The New Order resisted Peradin's attempts to establish regional LBH offices, partly because the government had been caught off guard by LBH's community organising efforts during the Simprug case, and because of its success in securing increased compensation for the affected families (Saptono 2012, 18). In 1972, the Soeharto-era security and intelligence body Kopkamtib[10] reportedly published an instruction ordering its regional officers to put an end to LBH efforts to establish offices in cities including Yogyakarta, Solo, Tegal, Bandung and Palembang (Saptono 2012, 19). It was not until 1978, after the government relaxed this initial resistance, that a second office was established in Medan, North Sumatra. LBH then expanded quickly, and by 1990, had 13 regional offices located in major cities across the country, as well as legal aid posts in 'hotspots', such as Aceh and Lampung (Nasution 1994, 117; Kusumah 1990, 134).[11]

LBH Jakarta was well supported by the Jakarta administration over its first decade. In the 1977/1978 Jakarta administration budget, it was reportedly allocated Rp 2.5 million per month (equivalent to about US$7,200 today) (Sadikin 2012, 257), a significant sum at a time when the starting salary for civil servants was just Rp 12,000 per month.[12] Other regional LBH offices, for example in Medan (North Sumatra), Padang (West Sumatra), Palembang (South Sumatra), Bandung (West Java), Semarang (Central Java) and Surabaya (East Java), were also able to secure some local government funding (Kohno 2010, 82). While these funds were crucial for covering LBH's operations, they were also a sign of the political protection the organisation enjoyed at this early stage.

Receiving government funds did not mean LBH was co-opted or that their more political demands were blunted (Lev 1987, 13). Perhaps to prove as much, LBH vigorously pursued the Jakarta administration – and not only in the landmark Simprug and Taman Mini cases. In his biography, Sadikin commented that he was taken to court "about 200 times" by members of the community over the state's development policies, with these cases facilitated by LBH lawyers (Sadikin 2012, 258). Although he "was often annoyed with Buyung Nasution", Sadikin said that, as governor, he thought it was important there was an organisation like LBH to "control" the government (Sadikin 2012, 259).

One might ask why the New Order government not only tolerated LBH but actively funded it when it was so confrontational. But LBH served a useful role for the government – it allowed the New Order to claim it was tolerant of (a degree of) political dissent, and it directed potentially bothersome conflicts into the courts. LBH recognised this, with senior YLBHI lawyer Lubis commenting that the government had allowed LBH to survive because it was a means of reducing conflict (Lubis 1986, 88). This recognition was also reflected in Lubis's eventual push for a more activist interpretation of the role of the legal aid movement, discussed further below. Whatever part LBH might have played in absorbing political dissent, however, local governments eventually showed they were not able to submit to such sustained and spirited criticism, and government support for most LBH offices had dried up by the late 1980s (Kohno 2010, 82).

In 1980, LBH separated from Peradin and the Indonesian Foundation of Legal Aid Institutes (YLBHI) was established to serve as the central 'umbrella' organisation responsible for coordinating and distributing funds among the various regional LBH offices. While this aimed to improve institutional strength and efficiency, it was also a political strategy. In its efforts to domesticate the legal profession, the New Order moved to force Peradin and other professional legal associations to join a government-sponsored bar association, Ikadin (*Ikatan Advokat Indonesia*, Indonesian Advocates League) (Lev 2000, 315–16). Establishing a *yayasan* (foundation) structure separate from Peradin allowed YLBHI to act with relative autonomy, free of government intervention (Nasution 1994, 115).[13]

Under the *yayasan* structure, which remains in place today, an unelected voluntary Board of Trustees, comprised of prominent intellectuals and statespeople, oversees a Board of Directors (*Dewan Pengurus*), which is responsible for the daily management of YLBHI (Nasution 1994, 118). The Board of Directors is led by an executive director (a position now held by Muhamad Isnur), who sets the broad programmatic priorities for the organisation, and can decide on budget allocations to the regional offices. Importantly, the Board of Directors also has the power to approve (or reject) staffing changes in the regional offices. The Board of Trustees is responsible for issuing policy guidance and has the final say over decision making. It was also historically (and at times controversially) responsible for selecting the executive director – a policy that has recently changed to allow the directors of regional LBH offices to vote for this position alongside the board members (Riana 2016).

Structural legal aid

In the late 1970s, a time when the New Order's authoritarian nature had become increasingly clear, LBH lawyers began to reflect more broadly about the meaning and function of legal aid, and the type of organisation LBH wanted to be. There was growing frustration with a legal system they viewed as only responsive to the needs of the wealthy, and largely immune to the problems of the poor (Nasution 1985, 36). These frustrations were well expressed by senior YLBHI lawyer Lubis:

> Legal aid… can itself be trapped into doing the work of preserving the status quo, of keeping social, economic, political, legal and cultural resources out of reach of the people. Legal aid will be a rubber stamp or an instrument of the Establishment if it fails to evaluate the varieties of its work, and fails to identify the laws that are responsive to the needs of the majority. Legal aid of this kind no longer has the right to call itself legal aid.
>
> (Lubis 1985a, 136).

It was these ideas that underpinned the development of 'structural legal aid', a concept that gained popularity among LBH lawyers in the late 1970s and early 1980s and has since become LBH's trademark approach to legal aid. Senior LBH lawyers, including Buyung, Lubis, Mulyana W. Kusumah, Abdul Hakim Garuda Nusantara, and Fauzi Abdullah, produced a great deal of writing during the 1980s that was conversant with emerging international theories around legal aid, public interest law and development, and sought to articulate LBH's distinctive approach to legal aid (Kusumah and Nusantara 1981; Nasution 1981; Lubis 1986). Buyung and Lubis, in particular, did considerable work in building the intellectual and theoretical heft behind the structural legal aid approach, through their influential works *Bantuan Hukum di Indonesia* ('Legal Aid in Indonesia') (Nasution 1981) and *Bantuan Hukum dan Kemiskinan Struktural* ('Legal Aid and Structural Poverty') (Lubis 1986). LBH documents state that it was, in fact, Netherlands-based Indonesian academic Paul Moedikdo Moeliono who first introduced the term 'structural legal aid' to Buyung in the 1970s. Buyung travelled to the Netherlands on a fundraising trip in 1976, where he met with Moedikdo and discussed LBH's approach to the Simprug case. Moedikdo reportedly said that LBH was putting structural theory into practice, even if LBH did not describe it in these terms (Saptono 2012, 14).

The structural legal aid concept has never been neatly defined but Buyung comes close to a succinct definition when he describes structural legal aid as "a series of programmes aimed at bringing about changes, both through legal means and in other lawful ways, in the relationships which form the bases of social life, towards more parallel and balanced patterns" (Nasution 1985, 36). Lubis's explanation expands on the concept:

> The concept of structural legal aid is tied to the destruction of injustices in the social system. It is directed not merely towards aiding individuals in specific cases, but towards an emphasis on cases which have a structural impact. Legal aid becomes the power to move in the direction of restructuring social order so that a more just social order can be established. Law, in and of itself, cannot address the question of injustice. Indeed, it often strengthens the status quo because it is obliged to be neutral. Once one accepts that society is neither just nor egalitarian, then one must reject neutrality in favour of supporting the weak who form the majority of the population. The legal struggle must be oriented towards creating a law that is supportive of the weak.
>
> (Lubis 1985b, 42).

Lubis posited structural legal aid in contrast to a traditional or mainstream proceduralist understanding of legal aid. In his view, 'individualistic' legal aid that viewed the legal system as neutral or egalitarian would not be able to deliver justice in Indonesia. In advocating for a structural approach to legal aid, Lubis, Buyung and their contemporaries aimed to "transform the legal aid movement from being legalistic-professional, as if it were simply charity from the professional class, to a more political movement, an oppositional movement to change the system" (Saptono 2012, 49). Structural legal aid, therefore, was a distinctly political understanding of legal aid. Clearly, elements of the approach emphasised distributive justice, but the approach also had more overt oppositional connotations. As the Soeharto regime was considered the source of many of the problems faced by the poor and marginalised, it had a strong focus on changing the structure of the authoritarian New Order state to a more democratic structure.

While LBH lawyers did not necessarily situate structural legal aid within law and society studies, or connect it to dependency theory, there was significant correlation with the theoretical trends of the time. In articulating the concept of structural legal aid, Buyung described the law as an "ever-changing superstructure" that is influenced by, and results from, patterns of relations in society. As long as patterns of relationships in society remain unequal, he wrote, it will be difficult to produce just law. According to this view of the law, injustice and oppression was not solely the result of individuals who consciously abused human rights but was structurally determined (Nasution 1985, 36). There was a clear rejection of the notion of law as being autonomous and apolitical, and increasing attention to the ways in which law and the legal system could perpetuate inequality. Whether LBH leaders were aware of it or not, these ideas closely mirrored the arguments of the critical legal studies (CLS) scholars, which were emerging during the same period (Galanter 1974; Hunt 1986; Kelman 1987). The ideas behind structural legal aid also paralleled Galanter's influential work observing that, despite being formally neutral, the legal system often perpetuated the advantages of privileged groups (Galanter 1974, 104). Notably, although LBH lawyers made few references to these theoretical texts at that point,[14] LBH staff now connect structural legal aid to CLS.[15] The links to dependency theory were clearer. Lubis was heavily influenced by Norwegian sociologist Johan Galtung's "structural theory of imperialism" (Galtung 1971), which was part of the broader field of dependency theory popular at the time. Galtung was concerned about inequalities in the distribution of power within peripheral and core countries, and this was particularly influential for Lubis in articulating the goal of structural legal aid in redistributing power resources to the periphery (Lubis 1986, 58).

Although considerable work was done in developing the conceptual foundations of structural legal aid, its meaning in an operational sense was never precisely articulated and that has allowed it to continue to evolve over the course of LBH's existence. It can sometimes appear as quite a vague or loose concept. For example, former senior LBH Jakarta staff member Henny Supolo Sitepu (1980–84) described it as being about a humanitarian approach to legal aid, with an emphasis on client education (Hukum Online 2015). Meanwhile, Bambang Widjojanto (head of YLBHI from 1996–2001) emphasised how structural legal aid demanded LBH lawyers develop a comprehensive understanding of the social and political roots of the problems faced by their clients (Hukum Online 2015). One of the clearer descriptions was provided by Lubis, who described seven characteristics of structural legal aid:

1 Directed toward groups in society, rather than individuals. This does not mean that there is no room for the representation of individuals, but individual cases must contain a structural element.
2 Increased focus on the problems of the rural poor, rather than urban residents.
3 Active rather than passive approach to case identification.
4 A greater focus on extra-legal approaches.
5 Legal aid organisations must be willing to collaborate with non-legal social organisations, including organisations for labourers, farmers and fishermen, and the press.
6 There must be an effort to advocate for legal reform to allow for class actions and representation of structural cases. Procedural regulations must ensure the community can obtain cheap, timely, simple and transparent resolutions to their disputes.
7 Legal aid must become a "social movement" and empower people at the local level (Lubis 1986, 55–57).

While structural legal aid no longer has an explicit focus on the rural poor, and there is less emphasis on procedural concerns, these characteristics continue to be influential for how the concept is understood. Under the New Order, a structural legal aid approach to case management typically involved educating community members about their rights and responsibilities, then often described as 'conscientisation' (Nasution 1985, 37; Lubis 1985b, 44), as well as community organising, and efforts to encourage the participation of community members in the cases affecting them (Kusumah 1990, 140).[16] These community education and organising efforts were combined with other nonlitigation activities, such as research and publication, often with savvy use of the media, aiming to generate widespread publicity about the injustices faced by the community (Nasution 1985, 37). As Buyung explained:

> LBH is less concerned about drafting winning briefs and racking up legal precedents than about choosing cases that have a major 'structural' element… 'winning' this or that case on the crazily tilted playing field of the New Order's legal system is likely to be impossible or meaningless anyway. Therefore the most prudent thing to do is to choose cases that offer an entrée into a pressing issue and provide occasions for community organising and education. The goal is not to achieve a technical victory in court (which would in all likelihood be unenforceable anyway, given the impunity with which the regime and its officials operate), but to focus domestic and international attention on abuses and to mobilise people to press for the reform of laws or procedures that discriminate against the poor and vulnerable.
>
> (Nasution 1994, 119).

While one could certainly argue about the extent to which LBH was ever solely proceduralist in its approach, structural legal aid marked a clear shift towards a grassroots approach. To recap, lawyers of the grassroots type are motivated by a desire to achieve social justice but place less emphasis on litigation as a means to reach this goal. In fact, they are sceptical about the ability of the legal system to bring about meaningful change (Hilbink 2004, 683). Grassroots cause lawyers see themselves as active participants in social movements and openly express solidarity with clients and their causes (Hilbink 2004, 689). These characteristics correlate closely with structural legal aid.

Proceduralist cause lawyers generally speak favourably of the law's ability to be a stabilising force. Proponents of the structural legal aid approach argued against this conservative function of law and legal assistance. Mulyana W. Kusumah, for example, wrote that traditional legal aid organisations do not threaten the status quo, and "may even localise structural conflict so that it does not become a national level conflict, accommodate the needs of privileged groups, even stem or suppress social conflicts and divert them into the legal system" (Kusumah 1990, 137). Lubis also argued for "continually creating conflict" to break down patterns of social relations that were becoming more established (Lubis 1986, 60). There was therefore a shift to a more activist, even radical, understanding of the legal aid movement's purpose and goals. This view essentially held that, in the face of an oppressive state, dissipating conflict by channelling it into the legal system would only serve to perpetuate inequality. There was instead a deliberate effort to bring local level conflicts and injustice to national attention, and to use these cases to make broader points about government hypocrisy and injustice, a strategy discussed further below. The New Order would have been happy with a proceduralist LBH, but it got a grassroots version, and that eventually led to increasing conflict between them.

The HR Dharsono case

One of the most famous cases handled by LBH during the New Order period was that of former military general turned government critic, Hartono Rekso Dharsono, commonly known as HR Dharsono. The pretence for the case against Dharsono was the so-called 'Tanjung Priok tragedy' but the former general had become a persistent source of irritation for the government long before this event.

On 12 September 1984, in Tanjung Priok, North Jakarta, the military opened fire on a large group of Islamic protesters who had gathered to hear a sermon rejecting the Soeharto government's plan to require all Indonesian social and political organisations to adopt the national ideology, *Pancasila*, as their sole ideological basis (Bourchier and Hadiz 2013, 140),[17] and demanding the release of four local men arrested a few days earlier. The men had been arrested following a dispute with a military official, who they accused of entering a local mosque without removing his shoes (Bourchier and Hadiz 2013, 140). The National Commission on Human Rights (*Komisi Nasional Hak Asasi Manusia*, Komnas HAM) has reported 24 people were killed and 54 others injured in the Tanjung Priok tragedy, although other reports suggest as many as 400 individuals died (Woodward 2007, 90; Butt and Lindsey 2018, 276). The commander of the Indonesian Armed Forces at the time, General LB (Benny) Moerdani, claimed only nine had died (Pratama 2019).

Soon after the Tanjung Priok incident, on 4 October, bombs targeting ethnic Chinese-owned banks and a small grocery store were detonated across Jakarta (Hail 1985). The bombings were widely seen as retaliation for the Tanjung Priok massacre. Ethnic Chinese Indonesians have often been made scapegoats throughout Indonesian history, particularly during times of political unrest and economic stress. General Benny Moerdani was not ethnically Chinese but he was a Christian, like many Chinese Indonesians. The Tanjung Priok massacre fuelled conspiracy theories popular among some Indonesian Muslims about a Catholic-Chinese conspiracy to weaken the Muslim community (Human Rights Watch 1998b). Dharsono, a Muslim, was accused of having a hand in these bombings.

Although he played a key role in Soeharto's rise to power, Dharsono had quickly expressed doubts about New Order rule (Vatikiotis 1990). He was closely associated with the 'Petition of 50', a group of 50 prominent figures including other former military leaders, politicians and academics, who signed an 'Expression of Concern' in 1980 about Soeharto's use of *Pancasila* to suppress dissent. In September 1984, Dharsono and several other members of the Petition of 50 signed a 'white paper' (*lembaran putih*) challenging the military's version of events at Tanjung Priok. Soon after the Tanjung Priok incident, Dharsono also attended a meeting at which he reportedly discouraged Islamist radicals from seeking revenge for the military's actions (Cribb 1986, 4). However, officials claimed the meeting was used to plan the bombings (The Times 1985).

On 8 November 1984, Dharsono was arrested and charged with subversion over his role in drafting the white paper and attending the meeting. When Dharsono was arrested, LBH founder Buyung was undertaking doctoral study in the Netherlands. Following some convincing from other LBH staff, Buyung returned to take on the case, with the head of YLBHI at the time, Todung Mulya Lubis, as a junior member of the legal team (Nasution 2004, 48). The Dharsono case was the kind of individual case that fit LBH's 'structural' criteria. Although the case involved a single defendant, it responded directly to New Order repression.

The trial began on 19 August 1985, and immediately attracted huge public and media attention. The defence was given "unusual freedom" in presenting its case (Cribb 1986, 4). It called on several signatories of the white paper to testify in Dharsono's defence, including former Jakarta Governor Sadikin (Nasution 2004, 51). Rather than simply addressing the subversion charges in a narrow legal sense, the defence used the trial to expose the real events of the Tanjung Priok incident and publicly excoriate the Soeharto regime (Cribb 1986). The testimony by the much-respected Sadikin, on 31 October 1985 in the 13th hearing, attracted massive public attention (Nasution 2004, 58). More than a thousand spectators gathered outside the courtroom, cheering as he criticised President Soeharto for failing to meet his promise to govern constitutionally (Routledge 1985).

Despite the sensational nature of the trial, as Cribb wrote at the time, Dharsono was always going to be convicted under the loose provisions of the 1963 Anti-Subversion Law.[18] Dharsono had long been a thorn in the government's side over his association with the Petition of 50, and his criticism was seen as particularly potent because of his key role in bringing Soeharto to power (Cribb 1986, 5).[19] On 8 January 1986, Dharsono was duly sentenced to ten years in prison for helping to draft and disseminate the white paper, and for addressing Muslim youths at a meeting where he "might have" encouraged them to take further action against the state (The Times 1986).

Dharsono's trial was to have long-lasting consequences for Buyung. Reading out the decision, one of the members of the panel of judges,

Achmad Intan, said that by painting the government in a negative light, the defence team had behaved inappropriately and unethically. Buyung was incandescent. He grabbed the microphone in front of him and interrupted the judge, asking why the allegations of unethical behaviour had not been raised earlier in the trial. When a policeman entered the courtroom, carrying a firearm and walkie-talkie, and told Buyung to stay quiet, he exploded, shouting "Hey you… get out! Out! This room is under the authority of judges, not police!" (Nasution 2004, 59). For this outburst, Buyung was accused of having committed contempt of court, despite no such charge existing under Indonesian law.[20]

Furious over the way Buyung had used the trial to expose government hypocrisy, Minister of Justice Ismail Saleh and judicial officials pursued him with great vigour (Lev 2000, 317). On 11 May 1987, by which time Buyung had returned to the Netherlands, Saleh revoked his licence to practice for a year. A couple of months later, Saleh and Supreme Court Chief Justice Ali Said signed a joint ministerial decision on the supervision of the legal profession,[21] clearly designed to address the issue of contempt of court and tighten restrictions on the profession. Article 3(c) of the regulation stated that measures could be taken against legal counsel who "act, behave, bear themselves, speak, or issue statements that indicate lack of respect for the law, statutes, public authority, the courts or their officials" (Lev 2000, 318). Although his licence was only suspended for one year, in light of the threats he faced while in Indonesia, Buyung remained in the Netherlands until 1993, where he completed a doctorate (discussed below) (Nasution 2011, 10).

Dharsono, meanwhile, had his sentence reduced to seven years on appeal to the Jakarta High Court. He was eventually released in 1990, following remissions (Sitompul 2018). He courageously continued to promote democratic values following his release, establishing the People's Sovereignty Purification Forum (*Forum Pemurnian Kedaulatan Rakyat*, FPKR) in 1991, but died before the end of authoritarian rule in 1996 at age 70, reportedly of a brain tumour (UPI 1996).

While the Dharsono case was one of the highest profile cases taken on by LBH in the New Order period, it was in many ways typical of its approach to political cases. These show trials, in which chances of securing an acquittal were virtually non-existent, were used as a forum for free speech, as a platform to expose government hypocrisy and advance notions of the rule of law and human rights. In his autobiography, Buyung recalled that either the Straits Times or Asian Wall Street Journal wrote that in his defence of Dharsono, Buyung had broadened the scope of the trial. By exposing the real events of Tanjung Priok, the accused was no longer Dharsono – it was Soeharto (Nasution 2004, 63).

The Dharsono trial was consistent with patterns of cause lawyering in other authoritarian contexts, with the legal arena converted into a site of political resistance (Ginsburg and Moustafa 2008, 2; McEvoy and Bryson 2022). With civil society constrained, demonstrations banned or restricted, and the media tightly controlled, the courts may be one of the only avenues available for publicly advancing principles such as the rule of law and human rights. Under the New Order, there were so few opportunities for political contestation that the courts provided an almost unique platform for direct confrontation with the regime. But in circumstances like these, where the courts are politically compliant, the case itself becomes of almost secondary importance to the broader cause. Buyung was aware that the likelihood of Dharsono's acquittal was negligible. Some have estimated that the New Order prosecuted more than 1,000 individuals using the Anti-Subversion Law after 1966, and no defendant was ever acquitted in the first instance (Heryanto 2005, 110).[22] Buyung and LBH instead used the Dharsono case as a springboard to launch a highly public critique of New Order rule. Adapting Fu's point on activist lawyers in China (Fu 2011, 355), if the courts are an instrument of state power in authoritarian regimes, Buyung was exercising dissent in the heart of the state.

In using the trial to make a passionate critique of Soeharto and the New Order, Buyung was in fact deploying a tactic of performance used by Indonesia's founding fathers. In 1928, Mohammad Hatta (who went on to serve as both vice president and prime minister) and other prominent students in the Indonesia Association (*Perhimpoenan Indonesia*, PI), a nationalist organisation for Indonesian students in the Netherlands, faced trial in The Hague for sedition (Ricklefs 2001, 220). During the trial, Hatta gave a rousing defence speech, later published as *Indonesia Vrij* (Indonesia Free), in which he described Dutch repression in detail and said that Indonesian independence could only ever be achieved through violence (Elson 2008, 60; Legge 2010, 47). Hatta and the students were eventually released without charge. Similarly, when Indonesia Nationalist Party (PNI) leader (and soon to be founding president) Soekarno was arrested by the Dutch colonial government in late 1929, and eventually put on trial in Bandung in August 1930, his defence speech did much more than address the charges against him. In fact, he dealt with the charges in a rather cursory manner. Instead, Soekarno used his very lengthy defence oration, *Indonesia Menggugat* (Indonesia Accuses) (Soekarno 1951) – which in one published version ran to more than 140 pages – as a platform to launch a powerful and articulate public condemnation of imperialism and colonialism (Frederick 1977, 387). The speech became a landmark document in the Indonesian independence struggle (Rush 2014, 181–83), and activists continue to draw inspiration from it decades later. Indeed, the word *menggugat* has become shorthand for standing up to the abuse of power by state authorities.[23] Both Hatta's and Soekarno's speeches were typical of the cause lawyering approach of "speaking law to power" (Abel 1998). Their show trials served as a platform to "indict the government before the court of public opinion" (Abel 1998, 95). *Indonesia Menggugat* also set a template for heroic yet futile performances of resistance,[24] which LBH lawyers continue to draw on today.

The primary audience for these displays of political theatre is the public. In authoritarian environments where legal autonomy is weak, the political environment is constrained, and there is almost no hope of victory, the true significance of cases like the Dharsono case lies in the opportunities they provide for making ideological points, exposing rights abuses to public scrutiny, and undermining the government's pretensions to legality (McEvoy and Bryson 2022, 91; Moustafa 2007, 40). LBH lawyers understand they are unlikely to win in the courtroom, but as McCann has pointed out, victory may not necessarily be needed for litigation tactics to yield results (McCann 2006, 31). This form of judicial activism can only go so far in authoritarian environments. As discussed, the New Order may have tolerated LBH's cause lawyering because it was a way of modulating politically sensitive issues. However, the Dharsono case demonstrated that the New Order's tolerance had its limits. When Buyung transgressed the bounds of what was politically acceptable (Munger, Cummings, and Trubek 2013, 378), the New Order retaliated with vigour. In any case, as the New Order period progressed, LBH became convinced it would not be able to achieve change through the judiciary and became increasingly political in its outlook, turning to extra-legal approaches.

Increasing activism

From the 1980s, lawyers including Todung Mulya Lubis (YLBHI chair from 1983–87), Abdul Hakim Garuda Nusantara (chair from 1987–93), and Mulya W. Kusumah (executive director from 1993–96) pushed LBH in a more activist direction (Collins 2007, 29). LBH's political activities were not restricted to the community organising and community legal empowerment work associated with 'structural legal aid'. As the 1980s and 1990s progressed, LBH held regular public seminars and discussions about human rights and democratisation, arguing for rights such as freedom of expression, assembly and association, and often airing vocal criticism of the New Order regime. LBH evolved from an organisation founded to provide case-based legal aid to a major and influential human rights organisation (Nasution 2011, 10–11; Aspinall 2005, 103).

From its establishment through to the early 1990s, LBH was divided into "litigation" and "nonlitigation" divisions (Khor and Lim 2002, 227–28).[25] Lubis, who served as the head of the nonlitigation division before becoming YLBHI chair, played a critical role in LBH's transformation into an expressly political, human rights advocacy organisation. In 1979, LBH set up a human rights division, with Lubis as its head. The division began producing an influential Human Rights Report, which documented the rights violations of the regime (Aspinall 2005, 103). Lubis soon became one of the most prominent voices helping to shape the national conversation on human rights. His activities (including his involvement in the defence of Dharsono) increasingly provoked the ire of the New Order regime. He was eventually prevented from teaching, speaking at public events, writing in the media, and raising funds for LBH, at a time when he was also YLBHI director. As a form of political exile, in 1987, then-US Ambassador to Indonesia Paul Wolfowitz supported Mulya in securing a position at Harvard Law School, where he undertook a second Master of Laws (he had completed his first masters at University of California, Berkeley in 1978).[26] He then went on to complete a PhD at the University of California, Berkeley, in 1990, drawing largely on experiences and data collected at LBH.[27]

By the 1980s, LBH was attracting some of Indonesia's best and brightest activists. While it was once staffed almost exclusively by lawyers, it began to attract a broader spectrum of politically engaged young people. According to former LBH lawyer Luhut Pangaribuan, non-lawyers were recruited because the organisation recognised that it would not achieve its goals through a strictly legal approach (Saptono 2012, 83). Many were recruited from the student movement, the political activities of which had been suppressed on campuses since the late 1970s.[28] Prominent non-lawyers included Fauzi Abdullah, Mulyana W. Kusumah and Hendardi. Abdullah, in particular, had a reputation as a skilled labour organiser, and made a major contribution to the practice of structural legal aid on the ground. He devoted considerable efforts to promoting legal awareness among workers, and supporting them to form and lead their own organisations (Saptono 2012, 22–25). Part of the reason that LBH thrived was precisely because of the limited alternative opportunities for expressing political dissent. As Lev observed, "as political parties were hemmed in to the point of meaninglessness, legal aid blossomed as an attractive outlet for young (and not so young) men and women committed to change and at all inclined to political and social activism" (Lev 1987, 27).

One case considered influential in LBH's transition to a more activist organisation was the Kedung Ombo reservoir case. Kedung Ombo was a World Bank-funded project that involved the flooding of a valley in Central Java for a massive dam. Reportedly, 5,268 families were displaced for its construction and offered only minimal compensation of Rp 800 per square metre for their land (Pompe 2005, 149). While the Simprug case included early efforts at community organising, the Kedung Ombo case involved deeper connections with social movement groups. The LBH office in nearby Semarang took up the case, building alliances with farmers and organising protests at the local level, including in collaboration with other prominent CSOs, like the Indonesian Forum for the Environment (*Wahana Lingkungan Hidup Indonesia,* Walhi). This emphasis on community organising has since become a trademark of LBH Semarang's approach to cause lawyering (see also Chapter 5). To amplify advocacy, YLBHI drew on its connections in the recently formed International NGO Forum on Indonesia (INGI), which linked Indonesian organisations with NGOs from Indonesia's foreign donors, as well as international NGOs with a focus on Indonesia (Uhlin 1993, 526). International NGOs directed global attention toward the role of the World Bank in displacing local farmers, while LBH Semarang supported affected families in launching a class action. Unsurprisingly, their efforts were unsuccessful at the district court and Semarang High Court on appeal. Miraculously, though, in July 1993, the Supreme Court decided in the farmers' favour, ruling they should be offered compensation of Rp 80,000 per square metre, well above the Rp 10,000 the families had requested (Pompe 2005, 150).[29] The decision was a rare display of judicial independence at a time when courts were subservient to the ruling regime.

The decision did not stand for long. Soon after it was made public, President Soeharto called the head of the Supreme Court, Purwoto Gandasubrata, to the Presidential Palace, and said he hoped the court would make "the most just decision" when the case went up for reconsideration (*peninjauan kembali*) to a different bench at the Supreme Court (Pompe 2005, 151). Just two days before his retirement, in October 1994, Purwoto heard and decided the case. The Supreme Court nullified the earlier decision on the suspect grounds that the court should not have awarded compensation greater than the amount requested (Pompe 2005, 152).[30] The message to LBH was clear. The organisation became increasingly convinced it was near futile to try to achieve meaningful change through the New Order's rigged legal system (Collins 2007, 30).

The shift to greater human rights activism and a bolder stance in support of democratisation (in addition to its ongoing litigation work) was assisted by the fact that, by the late 1970s, LBH had also started to attract significant funding from foreign donors, which funded CSOs in Indonesia as one of the few outlets for supporting human rights under the repressive New Order (Aspinall 2010, 2; Winarta 2013, 76). LBH Jakarta started receiving funds from Novib (the Netherlands Organisation for International Assistance, a Dutch government-funded international NGO, now part of Oxfam) in the late 1970s (Lev 1987, 13–14). By 1992, Novib grants accounted for about 88% of YLBHI's US$650,000 operating budget (Nasution 1994, 120). This reliance on Novib funding eventually proved problematic for LBH. In April 1992, not long after the Netherlands temporarily suspended development assistance to Indonesia over human rights concerns,[31] Indonesian Home Affairs Minister Rudini responded by announcing that CSOs like LBH could no longer accept funds from the Netherlands (Schwarz 1992), and LBH faced a funding crisis (Bunnell 1996, 184). However, foreign donors were eventually able to get around such restrictions by channelling funds through international human rights organisations (Ford Foundation 2003, 80).

In fact, foreign funding increased throughout the 1990s, along with growing international support for human rights-focused organisations in the Global South. According to YLBHI data, between 1994 and 1998, LBH received Rp 8.18 billion (about US$4.2 million today) in grants, from donors including Novib, the National Endowment for Democracy (NDI), USAID, the Canadian Government, the Swedish Development Coordination Agency, the 'Triple 11' collective from Belgium, and the Quaker Service of Australia. From the late 1970s, it also received funding from the Ford Foundation (Ford Foundation 2003, 80). The expansion in funding by international donors from the late 1970s provided LBH with greater freedom to become more adversarial (Aspinall 2005, 93; Aspinall and Mietzner 2008, 19). Donors, too, increasingly looked to LBH because of its integrity, strong focus on human rights, and growing reputation as a site of resistance to the Soeharto regime. Notably, much of the funding it (and many other CSOs) received during the New Order period was in the form of 'block grants', providing the organisation with considerable autonomy in determining its programmatic priorities (Zen 2004).[32] This funding was delivered to YLBHI as the central organisation, which then distributed it to the regional offices.

Notably, LBH's growing political and human rights advocacy occurred during a brief window of relaxation in the Soeharto regime's social control, known as the *keterbukaan* (openness) period. Beginning in the late 1980s, conflicts among members of the ruling elite provided more room for open expressions of opposition and demands for democratisation (Uhlin 1993, 519). Soeharto allowed a degree of economic liberalisation and political freedom (Beittinger-Lee 2013, 65). The New Order loosened its control over the press and was willing to tolerate a degree of political debate and criticism, allowing protests from students, labour and farmers' groups, including a few at which there were open calls for Soeharto's removal (Beittinger-Lee 2013, 66; Aspinall 2005, 137). In 1993, Soeharto permitted the establishment of the National Commission on Human Rights (Komnas HAM). This was mainly done to deflect growing international criticism of Indonesia's human rights record, and was timed just one week before the World Conference on Human Rights in Vienna (Butt and Lindsey 2012, 191–92). Media reports at the time even suggested Komnas HAM was established in part to create an alternative to LBH (Aspinall 2005, 115), a sign of LBH's growing influence. However, the period of openness did not last long. *Keterbukaan* ended in 1994, with the government's revocation of the publishing licences of critical news magazines *Tempo, Detik*, and *Editor* (Beittinger-Lee 2013, 66).[33]

Although LBH staff like Lubis, Abdul Hakim and Hendardi ramped up criticism of the New Order and helped to push the organisation in a more activist direction, it was ultimately Buyung who explicitly positioned LBH as a force for democratic change (Aspinall 2005, 106). As mentioned, Buyung had completed a doctorate during his time in the Netherlands, a detailed analysis of the work of the Indonesian Constituent Assembly from 1956–59. Buyung's dissertation resulted in a highly influential book, *The Aspiration for Constitutional Government in Indonesia* (1992), which argued for urgent constitutional reform in Indonesia. Buyung returned to Indonesia in 1993, and was thrilled to see LBH had taken a leading role in the growing opposition to the New Order regime (Nasution 2011, 10). LBH labelled itself a "locomotive of democracy" and began determined efforts to further elaborate its ideas for democratic change. In 1994, YLBHI prepared a four-year plan in which it laid out its plans for a more democratic system (Aspinall and Mietzner 2008, 20) and Buyung set about promoting the ideas elaborated in his book about the need for constitutional reform, again putting him into conflict with the regime. During this period in the early to mid-1990s, working alongside LBH executive director Mulyana W. Kusumah, Buyung toured the country, delivering discussions at universities and CSOs on democratic change, and building networks with students and activists. The New Order banned Buyung's book almost immediately and often frustrated his attempts to hold public discussions (Nasution 2011, 11).

In the 1990s, LBH became Indonesia's most prominent human rights group, a symbol of the civil society movement, and an icon of resistance against the Soeharto regime (Saptono 2012, 90). It was a hub of reform activity (Lev 1998, 437), a place where activists, students, and other pro-democracy actors came to meet, coordinate and discuss ideas (Nasution 2011: 12). In the wake of Kedung Ombo, YLBHI sought to strengthen networks between students who had been mobilised by the highly public Kedung Ombo campaign, and connect them to labour activists and farmers' groups (Collins 2003, 158). Former senior YLBHI figure Luhut Pangaribuan described how in the constrained New Order environment, LBH attempted to provide an outlet for political action.

> In the past, LBH was highly political in its use of structural legal aid. We did not want an authoritarian government that practiced rule by law, rather than the rule of law… We organised masses to add pressure. So, in other words, we were taking on the role of political parties.[34]

Along with several other CSOs, YLBHI promoted collective action to challenge the Soeharto regime, including through protests and street demonstrations (Hadiwinata 2003, 100–01). With the support of foreign donors,[35] YLBHI also held regular discussions, workshops and seminars, during which it fleshed out and refined its approach to democratic transformation, covering issues like elections, political party reform, human rights, the rule of law and more (Nasution 2011, 12).

YLBHI also formed alliances with (and recruited from) elements of the leftist and more radical youth movements (Aspinall 2005, 108). Notably, the leftist Democratic People's Party (*Partai Rakyat Demokratik,* PRD) was established at the LBH office on 22 July 1996. Dissident poet Wiji Thukul even read his famous poem *Peringatan* (Warning) at the event – reportedly his last public performance before he went into hiding and was 'disappeared' in early 1998 (Tempo 2013, 177). The fact that such an open display of resistance was tolerated by the regime was a sign of how important LBH had become. But this tolerance was ephemeral. PRD was blamed for violence that occurred during the 27 July 1996 Incident (*Peristiwa 27 Juli*), which involved demonstrations in support of opposition figure Megawati Soekarnoputri that degenerated into rioting.[36] The New Order government labelled PRD a "subversive, neo-communist organisation" (US Department of State 1997), and arrested its leader, Budiman Sudjatmiko, several other senior PRD figures, and Muchtar Pakpahan, head of the independent Indonesian Workers

Welfare Union (*Serikat Buruh Sejahtera Indonesia,* SBSI), who had also long been an irritant to the regime (Ismartono 1996).

As LBH assumed a more prominent role in civil society resistance to the New Order, it was assisted in its efforts by a receptive media. Invigorated by *keterbukaan*, the media provided greater room for alternative and critical voices that deviated from the official government position (Romano 1996, 159). LBH cultivated productive relationships with major national media organisations *Kompas, Sinar Harapan* and *Tempo* (Winarta 2013, 63) and *Tempo*'s influential dissident journalists Goenawan Mohamad and Bambang Harymurti. Such was LBH's importance as a gathering point for activists and anti-regime resistance activities, for certain periods journalists were even assigned to cover LBH activities on a daily basis. There was, in effect, an 'LBH beat'. As former YLBHI lawyer Robertus Robet explains:

> Even though the media was tightly controlled, there were still major organisations that wanted to cover LBH. Anything that happened at LBH was reported on by the big papers. Even the reporting of Kompas [which has a reputation for being cautious] at that time had a political impact.[37]

With its frequent coverage in the media, LBH was also able to rely on considerable public support, in particular from the growing middle class (Lev 1987, 29). As Lev noted, the middle class were mainly supportive of LBH's efforts to advance the rule of law, even if they were not always the most enthusiastic backers of its human rights campaigns. Nevertheless, without this public support, there would be little stopping the government from simply squashing LBH "in a moment of impatience" (Lev 1987, 29).

Not all LBH lawyers were happy with the new activist direction of the organisation. On two occasions, in 1993 and 1996, tensions between activists and those who continued to favour a proceduralist approach (including, significantly, the conservative Board of Trustees), erupted into open conflict. Both conflicts coincided with leadership transition.[38] These conflicts have been covered in considerable detail by Aspinall (2005) and Kohno (2010). [39] In 1993, a group of former LBH lawyers, members of LBH regional offices, and non-LBH civil society activists protested against the undemocratic nature of the selection of the new chairperson by the Board of Trustees, with some CSOs even demanding that organisations external to LBH be allowed to elect the new chair (Kohno 2010, 106). The fact that such a demand emerged was a sign of the status LBH enjoyed in civil society at the time. The conflict was influenced in part by the nature of the *yayasan* structure YLBHI had adopted in 1980 (Aspinall 2005, 109). The *yayasan* structure means frequently "self-perpetuating" boards of trustees (unelected, and typically senior) have significant control over organisations (Eldridge 1996, 28). Recognising this, regional LBH offices demanded to have a greater say in how YLBHI was run. The Board eventually selected Buyung, who was still considered to represent a relatively more proceduralist approach, to return as chair. To ease tensions, Buyung created a new executive director position underneath the chair, and installed his more activist-inclined rival for the chair, Mulyana W. Kusumah (Kohno 2010, 107). The two ended up working well together, as discussed above.

During the conflict in 1996, Buyung aligned himself with the more conventional, proceduralist group. By this point he was reportedly already beginning to express some doubts about the locomotive of democracy idea (Aspinall 2005, 110). While this could be interpreted as a reversal of his previous position, it likely also reflected the degree to which the organisation had shifted in a more radical direction (Kohno 2010, 107–08). The conflict in 1996 was particularly acrimonious and extended over several months (Aspinall 2005, 109). In early March 1996, the Board of Trustees selected Bambang Widjojanto as the next leader, Buyung's preferred candidate (Aspinall 2005, 112). Several LBH offices boycotted the results of the election, arguing the method of election was undemocratic. The conflict was finally resolved in September, with significant losses for the organisation. By this time, some of LBH's most respected and vigorous campaigners, including Mulyana W. Kusumah, Hendardi, Benny K. Harman, Abdul Hakim Garuda Nusantara and Nursyahbani Katjasungkana, had left the organisation (Kohno 2010, 112). They went on to found and lead some of the most influential and important CSOs in the late New Order and early democratic era. Mulyana W. Kusuma established the Independent Election Monitoring Committee (*Komite Independen Pemantau Pemilu*, KIPP) in 1996, Hendardi founded the Indonesian Legal Aid Association (*Perhimpunan Bantuan Hukum Indonesia*, PBHI) in 1996 and is now the director of the influential Setara Institute for Democracy and Peace, Benny K. Harman joined Hendardi at PBHI and is now a Democratic Party politician, Abdul Hakim Garuda Nusantara established the prominent human rights organisation Elsam (*Lembaga Studi dan Advokasi Masyarakat*, Institute for Policy Research and Advocacy) in 1993, and Nursyahbani founded the women's rights-focused legal aid organisation LBH Apik in 1995.

Cause lawyering organisation or incubator of civil society?

In this period of the mid to late 1990s, LBH began to consider itself – in the words of the one-time labour division head Surya Tjandra – "the centre of Indonesia's civil society movement" (Botz 2001, 136). LBH relished the convening role it was able to play and viewed itself as an incubator of civil society. As the student movement, prominent pro-democracy activists, and CSOs began making louder demands for *reformasi*, LBH felt that it occupied an important middle ground (Isnur, n.d.). Although LBH offices continued to provide legal aid throughout this period, LBH documents demonstrate it no longer saw its primary business as providing legal aid, see Figure 2.1 on p.58. As the figure demonstrates, from its establishment in 1970 through to the leadership crisis of 1996, LBH was operating much as it always had, focusing on implementing its strategy of structural legal aid. But from 1996–98, at a time when opposition to the Soeharto regime was growing, LBH believed its primary role was to cultivate civil society resistance to the New Order. In doing so, it still aimed to ensure resistance to the regime was expressed in line with principles of democracy and human rights (Isnur, n.d.). During the crisis phase of the mid-1990s, LBH encouraged and facilitated the establishment of five specialised CSOs, which it felt would be more agile and effective in responding to the particular challenges of the time.[40] According to then-YLBHI director Bambang Widjojanto, it was important that LBH escape from the "trap" of its initial format (Saptono 2012, 109). There was a need to diminish the dominant role of YLBHI as the central umbrella organisation, and provide greater autonomy to the regional offices, and other specialised organisations.

Recognising the huge potential for corruption in government and the lack of effective oversight mechanisms, LBH felt there was a need for an institution that could expose dirty officials and increase public pressure on the executive. This led to the birth of Indonesia Corruption Watch (ICW), now Indonesia's highest profile anti-corruption CSO. YLBHI labour division leader Teten Masduki became its first leader, while other respected figures were recruited to serve on its founding board. During 1997–98, several pro-democracy activists were kidnapped or disappeared. In response, YLBHI helped to establish the Commission for the Disappeared and Victims of Violence (*Komisi untuk Orang Hilang dan Korban Tindak Kekerasan*, KontraS).[41] KontraS, too, remains one of the most influential human rights organisations in Indonesia. Indonesian human rights icon Munir Thalib Said,[42] then head of information and documentation at YLBHI, was selected as its first leader (Saptono 2012, 109). The third major civil society organisation supported by YLBHI at this time was the National Consortium for Law Reform (*Konsorsium Reformasi Hukum Nasional*, KRHN). As demands for *reformasi* intensified through the 1990s, LBH recognised civil society needed to contribute to law reform proposals. But LBH felt its main capacity was in legal aid and activism and was not confident about its ability to formulate democratic legal and policy proposals. It gathered respected legal academics under the KRHN, and the result of their discussions were then delivered to the legislature, including proposals for the development of a Constitutional Court and a Judicial Commission (Saptono 2012, 110).[43]

Although LBH might have seen itself more as an incubator of civil society opposition to the New Order in the late 1990s, it also represented some of the highest profile political cases of the late Soeharto years, using them to put forward its case for the rule of law, democracy and strengthened protection of human rights. While the organisation continued to have a grassroots cause lawyer's scepticism of the courts, they were used as a site of resistance. Individual cases were used to make broader points about the injustices of the New Order regime, just as occurred in the Dharsono case. The courts were used

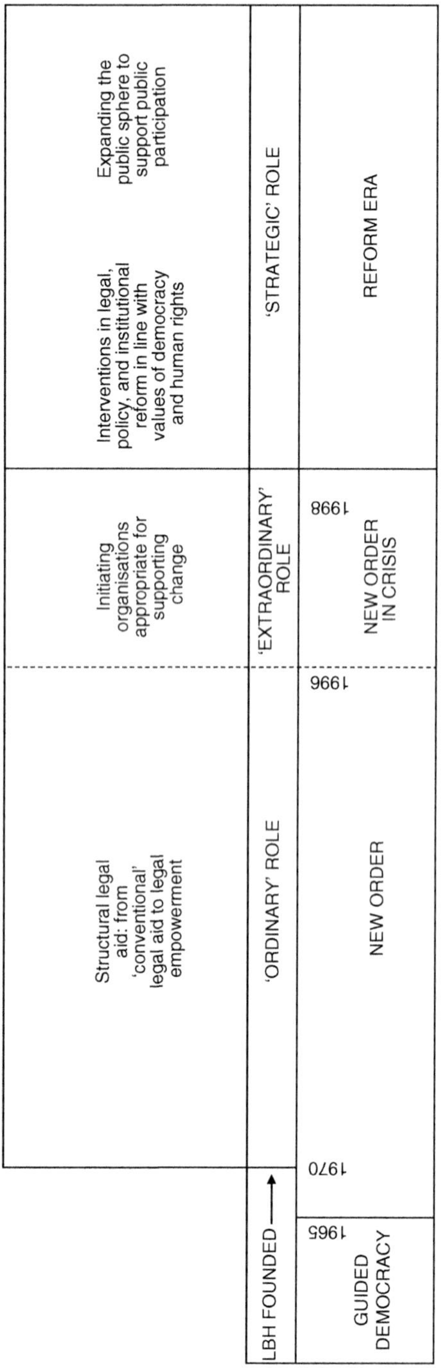

	Structural legal aid: from 'conventional' legal aid to legal empowerment	Initiating organisations appropriate for supporting change	Interventions in legal, policy, and institutional reform in line with values of democracy and human rights / Expanding the public sphere to support public participation
LBH FOUNDED ⟶	'ORDINARY' ROLE	'EXTRAORDINARY' ROLE	'STRATEGIC' ROLE
GUIDED DEMOCRACY (1965)	NEW ORDER (1970–1996)	NEW ORDER IN CRISIS (1996–1998)	REFORM ERA

Figure 2.1 The changing role of LBH (Isnur, n.d.)

"not as a legal forum, but as a political forum" (Hilbink 2004, 687). Even if it knew it had no chance of success, LBH used the courts as a stage to educate the public and expose the often-ridiculous nature of the charges made against political dissidents. These politically sensitive cases often resulted in LBH lawyers facing considerable threats and reprisals, as the Dharsono case also showed.

Among the many consequential political cases LBH took on during the late New Order years was the case against PRD leader Budiman Sudjatmiko and the 11 other PRD members charged with subversion following the 27 July Incident. The defence team was pointedly named "Team of Defenders of Law and Justice in Indonesia" (*Tim Pembela Hukum dan Keadilan Indonesia, TPHKI*). (These flowery names for defence teams were common at the time, reflecting the fact that lawyers knew that chances of legal victory were slim, and the cases were used to make broader points about democracy and the rule of law). Senior YLBHI lawyers Buyung and Bambang Widjojanto were part of the team that took on the high-profile case of Muchtar Pakpahan, the labour leader charged with subversion along with the PRD members after the 27 July Incident. Buyung and Widjojanto also defended outspoken politician Sri Bintang Pamungkas, a former member of the national legislature who in 1996 established his own political party, the Indonesian United Democracy Party (*Partai Uni Demokrasi Indonesia, PUDI*). Pamungkas was eventually sentenced to nearly three years in prison after being accused of calling Soeharto a dictator during a lecture in Germany (The Jakarta Post 1996).[44]

Senior YLBHI figures Buyung and Lubis also represented *Tempo* in one of the landmark cases of the Soeharto era, when the magazine challenged Information Minister Harmoko's decision to revoke its publishing licence in June 1994. YLBHI supported the establishment of an advocacy coalition called Indonesian Solidarity for Press Freedom (*Solidaritas Indonesia Untuk Pembebasan Pers,* SIUPP),[45] which included organisations closely affiliated with YLBHI. To the surprise of many, on 3 May 1995, the State Administrative Court (*Pengadilan Tata Usaha Negara,* PTUN) found the minister had acted unlawfully and ordered that *Tempo's* licence be reinstated (Tapol 1995). The decision was upheld by the State Administrative High Court in November 1995 but, unsurprisingly, reversed by a compliant Supreme Court in June 1996 (Tapol 1996).

In addition to providing a stage for LBH to exercise dissent and challenge New Order repression, these political trials also performed another important function. Scholars have described the important ways in which courts can function as sites of memory work (McEvoy and Bryson 2022, 94–99; Pils 2014, 122). Through its defence of *Tempo* and political detainees such as the PRD activists, Pakpahan, Pamungkas and others, YLBHI made a concerted effort to maintain the 'memory' of constitutional democracy (Butt and Lindsey 2012, 16–19), which Indonesia had briefly experienced from 1950 until 1957, when Soekarno declared martial law. In criticising executive pressure on the courts, as they did in these political trials, Buyung and others invoked arguments about democracy and a *negara hukum* involving true separation of powers. Lawyers like Buyung recalled fondly – and even celebrated – the liberal democratic period of the 1950s, attacking the New Order for "failing to fulfil the promises it made after Soekarno fell to deliver the *negara hukum*" (Butt and Lindsey 2018, 18).[46] The New Order might have made claims about implementing the *negara hukum* but its understanding of the concept was never anything more than rule by law. LBH lawyers used these highly controversial political trials to keep an alternative vision of the *negara hukum* alive. The massive media and public attention these cases attracted helped to disseminate ideas about constitutionalism and separation of powers during a time of significant state repression. This history also helps to explain why cause lawyering strategies involving theatrical displays of defiance continue to resonate for LBH cause lawyers. As subsequent chapters will also show, LBH continues to draw on these techniques as the quality of Indonesian democracy has deteriorated.

Conclusions

LBH was a product of authoritarianism. It was founded in 1970 in part as a response to state repression, and the realisation that the New Order was not going to fulfil its early promises to restore the rule of law. Under the New Order, the courts were subservient to executive pressure, there were few guaranteed legal rights, with even the few protections outlined in the Criminal Procedure Code (*Kitab Undang-Undang Hukum Acara Pidana*, KUHAP) routinely ignored by authorities, and there was little safe public space for discussion of political issues. Yet LBH showed cause lawyering could still exist, and arguably even thrive, in this repressive environment.

Although it engaged with international theoretical and intellectual debates around public interest law occurring at the time, LBH developed its own understanding and practice of cause lawyering. In LBH's very early years, it practiced a more conventional version of cause lawyering, one that could be described as proceduralist. It was focused on extending representation to poor and marginalised Indonesians and strengthening the legal system and rule of law. As the New Order period progressed, however, LBH developed a more grassroots style of cause lawyering, articulated through the concept of 'structural legal aid'. The notion of rights was used to educate and mobilise communities and encourage them to take action to make demands on the state. LBH cause lawyers believed that demands for the rule of law had to come from the people.

As the New Order clung to power and ratcheted up pressure on civil society, LBH in turn tended toward a highly political, oppositional version of cause lawyering. Throughout the 1990s, LBH became increasingly activist, as it built alliances with broader civil society, students, labour groups and farmers, and its role expanded into more straightforward human rights and democratic advocacy. To the extent that such activities were still interpreted through the prism of structural legal aid, structural legal aid became more focused on changing the structure of the authoritarian state to a more democratic one. There were already indications of a relationship between the openness of the state and the form of cause lawyering practiced. Rather than retreating into 'safer' versions of cause lawyering in response to state repression, or moderating its approach, LBH was invigorated by its opposition to the Soeharto regime, and became more openly adversarial.

This is particularly important to consider in reflecting on how LBH has been affected by democratic change over the past two decades. Sweeping changes occurred in the *reformasi* period that followed the fall of Soeharto in 1998, unimaginable to most while the New Order was in power, and there were new opportunities for LBH to contribute to the restructuring of Indonesian state and society. These changes forced LBH to begin to rethink its entire reason for being and the form of cause lawyering it wanted to pursue. Deterioration in democratic quality over the past five to ten years has prompted further reflection on the most appropriate form of cause lawyering. It is to these changes, and how they affected LBH, that I turn in parts Two and Three of this book.

Notes

1 The idea that the military had both a security and political role, used to justify the appointment of military officials to the legislature and senior positions in the civil service (Hadiwinata 2003, 52).
2 The office of the Public Solicitor was the precursor to what is now known as Legal Aid NSW (Legal Aid New South Wales 2020).
3 Presidential Decree No. 150 of 1959 on Return to the 1945 Constitution.
4 Although Soekarno had declared martial law in 1957, 'Guided Democracy' is generally considered to have begun with the passage of the 1959 decree (Bourchier and Hadiz 2013, 5).
5 When Soekarno was compelled into signing the *Supersemar* in March 1966.
6 Todung Mulya Lubis, chair of YLBHI from 1983 to 1987, acknowledged that this 30,000 figure was a rough calculation, based on LBH Jakarta's average annual caseload of 2,000 cases over a period of 13 years, plus 4,000 cases from LBH's provincial offices (Lubis 1985a, 135).
7 The 'Malari' incident (Malari is a contraction of the Indonesian phrase '*Malapetaka Lima Belas Januari*', 'The Fifteenth of January Disaster') began as a student protest against the influence of foreign capital in Indonesia coinciding with the visit to Indonesia of Japanese Prime Minister Kakuei Tanaka. The demonstration quickly got out of hand, however, and devolved into anti-Chinese and anti-Japanese rioting. The military eventually put an end to the rioting, shooting dead 11 people. The Soeharto administration depicted the protests as an attempt by student and anti-regime protesters (especially those associated with the Indonesian Socialist Party (PSI)) to overthrow it, although it later emerged that infighting among New Order elites played a role in inciting the violence (Leifer 2001, 173).

8 LBH did eventually receive some Ford Foundation funding (Ford Foundation 2003, 80).

9 See, for example, Decree of the Board of Trustees of the Indonesian Foundation of Legal Aid Institutes No.: TAP 01/V/1985/YLBHI on the Guide to Core Struggle Values of the Indonesian Foundation of Legal Aid Institutes.

10 Command for the Restoration of Security and Order (*Komando Pemulihan Keamanan dan Ketertiban*).

11 Aceh and Lampung were both sites of military violence against civilians during the New Order era.

12 Government Regulation No. 7 of 1977 on Regulation of Salaries for Civil Servants.

13 Perhaps influenced by YLBHI's separation from Peradin, but also because of the political conditions at the time, in the 1980s and 1990s the organisation began recruiting more non-lawyers, including many from the student movement. Many of these non-lawyers had a distinctly political, oppositional vision of the organisation's role (Aspinall 2005, 110).

14 Similarly, I could not find any references to Galanter's work in LBH documents produced during the New Order period, although later documents do refer to Galanter, for example, YLBHI Report No. 10, November 2005, 18.

15 Interview, Muhammad Rasyid Ridha, November 2019.

16 These nonlitigation aspects of structural legal aid are discussed in significant depth in Chapter 5.

17 Soeharto pushed for this requirement from the early 1980s, but it was eventually passed into law as Law No. 8 of 1985 on Mass Organisations (Bourchier and Hadiz 2013, 98).

18 Law No. 11/PNPS/1963 on the Eradication of Subversive Acts.

19 Dharsono provided troops that Soeharto used to overcome the leaders of the coup attempt, who were using the main Jakarta airport as a base.

20 Point 4 of the general elucidation to Law 14 of 1985 on the Supreme Court, passed in December 1985, stated that a contempt of court law should be made to regulate acts that could denigrate or undermine the authority or dignity of the court. At the time of the Dharsono trial, however, no such law had been formulated. Contempt of court was finally introduced into Indonesian law with the passage of the revised Criminal Code in December 2022.

21 KMA/005/SKB/VII/1987 and M.03-PR.08.05 of 1987, as cited in Lev 2000, 318.

22 Reportedly, only one defendant ever had a conviction for subversion overturned by the Supreme Court.

23 For example, one-time playwright and activist Ratna Sarumpaet created a political performance in the late New Order period in support of murdered labour activist Marsinah titled *Marsinah Menggugat* (Marsinah Accuses) (Hatley 1998). In a recent example, workers and students protesting against the so-called Omnibus Law on Job Creation in Semarang, Central Java, gathered under the banner of *Aliansi Gerakan Rakyat Menggugat (Geram)* (The Alliance of People's Movements Accuses) (Mulyono 2020).

24 While such performances may be futile in terms of immediate legal outcomes, they often have deeper significance beyond the trial, as *Indonesia Menggugat* demonstrates so clearly.

25 These divisions were occasionally sources of significant tension in the organisation (Winarta 2013, 60; Saptono 2012, 30–31).

26 Interview, Todung Mulya Lubis, March 2023.

27 Lubis' doctoral dissertation, *In Search of Human Rights: Legal-Political Dilemmas of the New Order 1966–1990*, published as a book in 1993, proved to be a major and influential text in Indonesia.

28 Following widespread student demonstrations in 1978, Minister of Education and Culture Daoed Jusuf introduced Decision No. 0156/U/1978 on the Normalisation

of Campus Life (*Normalisasi Kehidupan Kampus*, NKK). This decision banned student political activity and expression on campuses, and limited student activities to those associated with student welfare or strictly academic discussions. Later regulations put student activities under the direct supervision of university rectors (Director General of Higher Education Instruction/002/DK/Inst/1978) and banned campus-wide student councils (Minister of Education and Culture Instruction No. 1/U/1978 and Minister of Education and Culture Decision No. 037/U/1979) (Human Rights Watch 1998a, 148).

29 Supreme Court Decision No. 2263.K/Pdt/1991 dated 18 July 1993.

30 Supreme Court Decision No. 650.PK/Pdt/1994 dated 26 October 1994.

31 The Dutch temporarily suspended assistance following the execution of four political prisoners in 1990 and the Dili massacre in 1991 (Rich 2004, 325). The Soeharto regime accused the Dutch of interfering with Indonesian sovereignty by attempting to attach human rights conditions to its assistance.

32 This type of block grant funding was relatively common at the time, particularly for CSOs that resisted New Order repression.

33 See discussion on Tempo below.

34 Interview, Luhut Pangaribuan, November 2019.

35 Such as the International Commission of Jurists (ICJ), the National Democratic Institute (NDI), Human Rights Watch (HRW), Friederich Ebert Stiftung (FES), and Friederich Naumann Stiftung (FNS).

36 On 27 July 1996, not far down the road from the LBH office, military-backed thugs attacked protestors who had gathered at the headquarters of the opposition Indonesian Democratic Party (*Partai Demokrasi Indonesia*, PDI) to support Megawati Soekarnoputri, daughter of the first president, Soekarno. As a result of regime interventions, Megawati had been removed a month earlier from the leadership of the PDI, and replaced with a government-supported candidate (Hadiwinata 2003, 73–74). The headquarters were burned and riots and confrontations broke out in several areas of Jakarta. Five people were killed, 23 went missing and 149 were injured in the violence.

37 Interview, Robertus Robet, April 2020.

38 Leadership transition has consistently been a significant source of tension within YLBHI (see Chapter 3).

39 Kohno depicts the conflict as occurring between "legalism" and "activism" arms of YLBHI, while Aspinall describes the conflict as occurring between the "litigational" and "political" oriented poles of LBH. The terms may differ but the arguments are essentially the same.

40 The five organisations established were ICW, KontraS, KRHN, Lerai and Voice of Human Rights. Only ICW, KontraS and KHRN are discussed here. Lerai focused on conflict resolution and was established in response to growing tensions in Maluku. Voice of Human Rights, meanwhile, was a media organisation.

41 The capital 'S' is intentional. Reflecting the tensions of the time, the acronym had a double meaning – Kontra S could also be read as "against Soeharto".

42 Munir was killed by arsenic poisoning in September 2004 on a flight to the Netherlands, where he was to undertake a master's degree. See Chapter 3.

43 Notably, two of the academics who served on the KRHN, Abdul Mukthie Fadjar and Harjono, went on to serve as judges at the Constitutional Court.

44 See Zifcak 1999 for a detailed discussion of this case.

45 The acronym also mirrored the acronym for publishing licences, *Surat Izin Usaha Penerbitan Pers.*

46 In a similar manner, Pils has documented how Chinese human rights lawyers find strength in the memory of the more hopeful Republican period in China (1912–49) (Pils 2014, 37).

References

Abel, Richard L. 1998. "Speaking Law to Power: Occasions for Cause Lawyering." In *Cause Lawyering: Political Commitments and Professional Responsibilities*, edited by Austin Sarat and Stuart A. Scheingold, 69–117. New York; Oxford: Oxford University Press. https://doi.org/10.1093/oso/9780195113198.003.0003.

Aspinall, Edward. 2005. *Opposing Suharto: Compromise, Resistance, and Regime Change in Indonesia*. East-West Center Series on Contemporary Issues in Asia and the Pacific. Stanford, California: Stanford University Press. https://doi.org/10.1515/9780804767316.

Aspinall, Edward. 2010. "*Assessing Democracy Assistance: Indonesia*." FRIDE Project Report: Assessing Democracy Assistance. FRIDE.

Aspinall, Edward, and Greg Fealy. 2010. "Introduction: Soeharto's New Order and Its Legacy." In *Soeharto's New Order and Its Legacy: Essays in Honour of Harold Crouch*, edited by Edward Aspinall and Greg Fealy, 1–14. Asian Studies Series Monograph 2. Canberra: ANU Press. https://doi.org/10.22459/SNOL.08.2010.

Aspinall, Edward, and Marcus Mietzner. 2008. "*From Silkworms to Bungled Bailout: International Influences on the 1998 Regime Change in Indonesia*." Working Paper 85. CDDRL Working Papers. Stanford, California: Center on Democracy, Development, and The Rule of Law (CDDRL), Freeman Spogli Institute for International Studies.

Beittinger-Lee, Verena. 2013. *(Un)Civil Society and Political Change in Indonesia: A Contested Arena*. Routledge Studies on Civil Society in Asia. London and New York: Routledge. https://doi.org/10.4324/9780203868799.

Botz, Dan La. 2001. *Made in Indonesia: Indonesian Workers Since Suharto*. Cambridge, Massachusetts: South End Press.

Bourchier, David, and Vedi Hadiz. 2013. *Indonesian Politics and Society: A Reader*. London and New York: Routledge. https://doi.org/10.4324/9780203987728.

Bunnell, Frederick. 1996. "Community Participation, Indigenous Ideology, Activist Politics: Indonesian NGOs in the 1990s." In *Making Indonesia*, edited by Daniel Lev and Ruth McVey, 180–202. Studies on Southeast Asia. Ithaca, New York: Cornell University Press. https://doi.org/10.7591/9781501719370-011.

Butt, Simon, and Tim Lindsey. 2018. *Indonesian Law*. Oxford; New York: Oxford University Press. https://doi.org/10.1093/oso/9780199677740.001.0001.

Butt, Simon, and Tim Lindsey. 2012. *The Constitution of Indonesia: A Contextual Analysis*. Constitutional Systems of the World. Oxford; Portland: Hart Publishing. https://doi.org/10.5040/9781509955732.

Carothers, Thomas. 1998. "The Rule of Law Revival." *Foreign Affairs*, 1 March 1998. https://www.foreignaffairs.com/articles/1998-03-01/rule-law-revival.

Collins, Elizabeth Fuller. 2003. "The Struggle for Political Reform in South Sumatra." In *Autonomy and Disintegration in Indonesia*, edited by Damien Kingsbury and Harry Aveling, 157–176. London and New York: Routledge. https://doi.org/10.4324/9780203060292.

Collins, Elizabeth Fuller. 2007. *Indonesia Betrayed: How Development Fails*. Honolulu: University of Hawaii Press.

Cribb, Robert. 1986. "The Trials of HR Dharsono." *Inside Indonesia*, May 1986.

Eldridge, Philip. 1996. "Development, Democracy and Non-Government Organizations in Indonesia." *Asian Journal of Political Science* 4 (1): 17–35. https://doi.org/10.1080/02185379608434070.

Elson, Robert. 2008. *The Idea of Indonesia: A History.* Cambridge; New York: Cambridge University Press.

Ford Foundation. 2003. *Celebrating Indonesia: Fifty Years with the Ford Foundation, 1953–2003*. Jakarta: Ford Foundation.

Frederick, William H. 1977. "Book Review: Indonesia Accuses! Soekarno's Defence Oration in the Political Trial of 1930. Roger K. Paget." *The Journal of Asian Studies* 36 (2): 386–387. https://doi.org/10.2307/2053770.

Frühling, Hugo. 2000. "From Dictatorship to Democracy: Law and Social Change in the Andean Region and the Southern Cone of South America." In *Many Roads to Justice: The Law-Related Work of Ford Foundation Grantees Around the World*, edited by Mary McClymont and Stephen Golub, 55–87. New York: Ford Foundation.

Fu, Hualing. 2011. "Challenging Authoritarianism Through Law: Potentials and Limit." *National Taiwan University Law Review* 6 (1): 339–365.

Galanter, Marc. 1974. "Why the 'Haves' Come out Ahead: Speculations on the Limits of Legal Change." *Law & Society Review* 9 (1): 95–160. https://doi.org/10.2307/3053023.

Galtung, Johan. 1971. "A Structural Theory of Imperialism." *Journal of Peace Research* 8 (2): 81–117. https://doi.org/10.1177/002234337100800201.

Ginsburg, Tom, and Tamir Moustafa. 2008. "Introduction: The Functions of Courts in Authoritarian Politics." In *Rule by Law: The Politics of Courts in Authoritarian Regimes*, edited by Tamir Moustafa and Tom Ginsburg, 1–22. Cambridge: Cambridge University Press. https://doi.org/10.1017/cbo9780511814822.001.

Hadiwinata, Bob S. 2003. *The Politics of NGOs in Indonesia: Developing Democracy and Managing a Movement*. London and New York: RoutledgeCurzon. https://doi.org/10.4324/9780203200124.

Hail, John. 1985. "Moslem Sentenced in Bombing Plot." *UPI*, 8 August 1985. https://www.upi.com/Archives/1985/08/08/Moslem-sentenced-in-bombing-plot/1228492321600/.

Hatley, Barbara. 1998. "Ratna Accused, and Defiant." *Inside Indonesia*, September 1998. https://www.insideindonesia.org/ratna-accused-and-defiant.

Heryanto, Ariel. 2005. *State Terrorism and Political Identity in Indonesia: Fatally Belonging*. Florence, United States: Routledge. https://doi.org/10.4324/9780203099827.

Hilbink, Thomas M. 2004. "You Know the Type: Categories of Cause Lawyering." *Law & Social Inquiry* 29 (3): 657–698. https://doi.org/10.1086/430155.

Hukum Online. 2015. "Bantuan Hukum Struktural di Mata Tiga 'Pentolan' LBH Jakarta." *Hukum Online*, 2 December 2015. https://www.hukumonline.com/berita/baca/lt565ea035305l4/bantuan-hukum-struktural-di-mata-tiga-pentolan-lbh-jakarta.

Human Rights Watch. 1998a. *Academic Freedom in Indonesia: Dismantling Soeharto-Era Barriers.* Human Rights Watch.

Human Rights Watch. 1998b. "*Indonesia Alert: Economic Crisis Leads to Scapegoating of Ethnic Chinese*," 18 February 1998. https://www.hrw.org/report/1998/02/18/indonesia-alert/economic-crisis-leads-scapegoating-ethnic-chinese-february-1998.

Hunt, Alan. 1986. "The Theory of Critical Legal Studies." *Oxford Journal of Legal Studies* 6 (1): 1–45. https://doi.org/10.1093/ojls/6.1.1.

Ismartono, Yuli. 1996. "Indonesia: Accused Riot Ringleaders May Face Subversion Charges." *Inter Press Service*, 14 August 1996. http://www.ipsnews.net/1996/08/indonesia-accused-riot-ringleaders-may-face-subversion-charges/.

Isnur, Muhamad. n.d. "*Memaknai Bantuan Hukum Struktural.*" Jakarta.

Jenkins, David. 1994. "Hear the People, Not Soeharto, Says Lawyer." *Sydney Morning Herald*, 16 September 1994.

Kakiailatu, Toeti. 2007. "Media in Indonesia: Forum for Political Change and Critical Assessment." *Asia Pacific Viewpoint* 48 (1): 60–71. https://doi.org/10.1111/j.1467-8373.2007.00330.x.

Kelman, Mark G. 1987. *A Guide to Critical Legal Studies*. Cambridge, Massachusetts and London, England: Harvard University Press.

Khor, Martin, and Li Lin Lim, eds. 2002. *Good Practices and Innovative Experiences in the South*: Volume 3. 1st Edition. London; New York: Zed Books.

Kohno, Takeshi. 2010. *The Emergence of the Legal Aid Institute in Authoritarian Indonesia: How a Human Rights Organization Survived the Suharto Regime and Became a Cornerstone for Civil Society in Indonesia*. Saarbrücken: VDM Verlag Dr. Müller.

Kusumah, Mulyana W. 1990. "Perkembangan Bantuan Hukum." *Hukum Dan Pembangunan*, April.

Kusumah, Mulyana W. 1995. "*Siaran Pers: 25 Tahun Lembaga Bantuan Hukum*." Yayasan LBH Indonesia.

Kusumah, Mulyana W., and Abdul Hakim Garuda Nusantara, eds. 1981. *Beberapa Pemikiran Mengenai Bantuan Hukum: Ke Arah Bantuan Hukum Struktural*. Bandung: Alumni.

LBH Jakarta, ed. 2015. *Rentang Jejak LBH Jakarta: Kisah-Kisah Penanganan Kasus*. Jakarta: LBH Jakarta.

Legal Aid New South Wales. 2020. "*Our History*." Legal Aid New South Wales. 26 May 2020. https://www.legalaid.nsw.gov.au/about-us/who-we-are/our-history.

Legge, JD. 2010. *Intellectuals and Nationalism in Indonesia: A Study of the Following Recruited by Sutan Sjahrir in Occupied Jakarta*. Singapore: Equinox Publishing. https://doi.org/10.2307/2760306.

Leifer, Michael. 2001. *Dictionary of the Modern Politics of South-East Asia*. New York: Routledge. https://doi.org/10.4324/9780203198599.

Lev, Daniel. 1987. *Legal Aid in Indonesia*. Working Papers (Monash University. Centre of Southeast Asian Studies) 44. Clayton, Victoria: Monash University.

Lev, Daniel. 1998. "Lawyers' Causes in Indonesia and Malaysia." In *Cause Lawyering: Political Commitments and Professional Responsibilities*, edited by Austin Sarat and Stuart A. Scheingold, 431–452. New York; Oxford: Oxford University Press. https://doi.org/10.1093/oso/9780195113198.003.0013.

Lev, Daniel. 2000. *Legal Evolution and Political Authority in Indonesia: Selected Essays*. London-Leiden Series on Law, Administration and Development: Volume 4. The Hague; London; Boston: Kluwer Law International. https://doi.org/10.1163/9789004478701.

Lev, Daniel. 2008. "Between State and Society: Professional Lawyers and Reform in Indonesia." In *Indonesia: Law and Society*, edited by Tim Lindsey, 2nd Edition, 48–67. Sydney, Australia: The Federation Press.

Lindsey, Tim, and Mas Achmad Santosa. 2008. "The Trajectory of Law Reform in Indonesia: A Short Overview of Legal Systems and Change in Indonesia." In *Indonesia: Law and Society*, edited by Tim Lindsey, 2nd Edition, 2–22. Sydney, Australia: The Federation Press.

Linnan, David K. 1999. "Indonesian Law Reform, or Once More unto the Breach: A Brief Institutional History." *Australian Journal of Asian Law* 1 (1): 1–33.

Lubis, Todung Mulya. 1985a. "Legal Aid in the Future (A Development Strategy for Indonesia)." *Third World Legal Studies* 4 (1): 133–148.

Lubis, Todung Mulya. 1985b. "Legal Aid: Some Reflections." In *Access to Justice: Human Rights Struggles in South East Asia*, edited by Laurie S. Wiseberg and Harry M. Scoble, 40–45. London: Zed Books.

Lubis, Todung Mulya. 1986. *Bantuan Hukum dan Kemiskinan Struktural*. Jakarta: Lembaga Penelitian, Pendidikan dan Penerangan Ekonomi dan Sosial (LP3ES).

McCann, Michael. 2006. "Law and Social Movements: Contemporary Perspectives." *Annual Review of Law and Social Science* 2 (1): 17–38. https://doi.org/10.1146/annurev.lawsocsci.2.081805.105917.

McEvoy, Kieran, and Anna Bryson. 2022. "Boycott, Resistance and the Law: Cause Lawyering in Conflict and Authoritarianism." *The Modern Law Review* 85 (1): 69–104. https://doi.org/10.1111/1468-2230.12671.

Melvin, Jess. 2018. *The Army and the Indonesian Genocide: Mechanics of Mass Murder*. Rethinking Southeast Asia: 15. London and New York: Routledge. https://doi.org/10.4324/9781351273329.

Moustafa, Tamir. 2007. "Mobilising the Law in an Authoritarian State: The Legal Complex in Contemporary Egypt." In *Fighting for Political Freedom: Comparative Studies of the Legal Complex and Political Liberalism*, edited by Terence C. Halliday, Lucien Karpik, and Malcolm M. Feeley, 193–218. Oxford: Hart Publishing. https://doi.org/10.5040/9781472560179.ch-006.

Mulyono, Agus Joko. 2020. "Demo Tolak UU Cipta Kerja di Semarang, Ini 5 Tuntutan Massa." *Tagar.id*, 7 October 2020. https://www.tagar.id/demo-tolak-uu-cipta-kerja-di-semarang-ini-5-tuntutan-massa.

Munger, Frank, Scott L. Cummings, and Louise Trubek. 2013. "Mobilising Law for Justice in Asia: A Comparative Approach." *Wisconsin International Law Journal* 31 (3): 353–420.

Nasution, Adnan Buyung. 1981. *Bantuan Hukum Di Indonesia*. Jakarta: Lembaga Penelitian, Pendidikan dan Penerangan Ekonomi dan Sosial (LP3ES).

Nasution, Adnan Buyung. 1985. "The Legal Aid Movement in Indonesia: Towards the Implementation of the Structural Legal Aid Concept." In *Access to Justice: Human Rights Struggles in South East Asia*, edited by Laurie S.Wiseberg and Harry M.Scoble, 31–39. London: Zed Books.

Nasution, Adnan Buyung. 1994. "Defending Human Rights in Indonesia." *Journal of Democracy* 5 (3): 114–123. https://doi.org/10.1353/jod.1994.0048.

Nasution, Adnan Buyung. 2004. *Pergulatan Tanpa Henti: Menabur Benih Reformasi*. Jakarta: Aksara Karunia.

Nasution, Adnan Buyung. 2011. "*Towards Constitutional Democracy in Indonesia.*" 1. Papers on Southeast Asian Constitutionalism. Melbourne: Asian Law Centre, Melbourne Law School.

Palmier, Leslie. 1973. "Guided Democracy." *Modern Asian Studies* 7 (2): 296–305. https://doi.org/10.1017/s0026749x00004637.

Pils, Eva. 2014. *China's Human Rights Lawyers: Advocacy and Resistance*. Routledge Research in Human Rights Law. London and New York: Routledge. https://doi.org/10.4324/9780203769061.

Pompe, Sebastiaan. 2005. *The Indonesian Supreme Court: A Study of Institutional Collapse*. Ithaca, New York: Cornell University Press. https://doi.org/10.7591/9781501718861.

Pratama, Aswab Nanda. 2019. "Hari Ini Dalam Sejarah: Mengenang Tragedi Tanjung Priok." *Kompas.com*, 12 September 2019. https://nasional.kompas.com/read/2018/09/12/11453751/hari-ini-dalam-sejarah-mengenang-tragedi-tanjung-priok.

Riana, Friski. 2016. "Berikut Perubahan Mendasar Pemilihan Ketua YLBHI 2017–2021." *Tempo.co*, 6 December 2016. https://nasional.tempo.co/read/825909/berikut-perubahan-mendasar-pemilihan-ketua-ylbhi-2017-2021.

Rich, Roland. 2004. "Applying Conditionality to Development Assistance." *Agenda* 11 (4): 321–334. https://doi.org/10.22459/ag.11.04.2004.03.

Ricklefs, MC. 2001. *A History of Modern Indonesia Since c.1200*. 3rd Edition. Stanford, California: Stanford University Press. https://doi.org/10.5040/9781350394582.

Romano, Angela. 1996. "The Open Wound: Keterbukaan and Press Freedom in Indonesia." *Australian Journal of International Affairs* 50 (2): 157–169. https://doi.org/10.1080/10357719608445177.

Roosa, John. 2006. *Pretext for Mass Murder: The September 30th Movement and Suharto's Coup d'Etat in Indonesia*. Madison, Wisconsin: University of Wisconsin Press.

Routledge, Paul. 1985. "Dissident Former General Uses Trial to Attack Suharto; Indonesia's President Publicly Criticised by Ali Sadikin for Not Governing Constitutionally." *The Times*, 1 November 1985.

Rush, James R. 2014. "Sukarno: Anticipating an Asian Century." In *Makers of Modern Asia*, edited by Ramachandra Guha, 172–198. Cambridge, Massachusetts and London, England: Harvard University Press. https://doi.org/10.2307/j.ctt7zswrr.10.

Sadikin, Ali. 2012. *Ali Sadikin, Membenahi Jakarta Menjadi Kota Yang Manusiawi*. Jakarta: Ufuk Press.

Saptono, Irawan. 2012. "Kisah Panjang Gerakan Bantuan Hukum." In *Verboden Voor Honden En Inlanders dan Lahirlah LBH: Catatan 40 Tahun Pasang Surut Keadilan*, edited by Irawan Saptono and Tedjabayu, 1–112. Jakarta: YLBHI.

Schwarz, Adam. 1992. "Indonesia: NGOs Knocked." *Far Eastern Economic Review*, 14 May 1992.

Sitompul, Martin. 2018. "HR Dharsono, Jenderal Terpidana." *Historia*, 8 January 2018. http://historia.id/persona/articles/h-r-dharsono-jenderal-terpidana-vZ5qo.

Soekarno. 1951. *Indonesia Menggugat: Pidato Pembelaan Bung Karno di Depan Pengadilan Kolonial Bandung, 1930*. Djakarta: Seno.

Tapol. 1995. "*Tempo Defeats Harmoko in Court*," June 1995.

Tapol. 1996. "*Supreme Court in Crisis*," August 1996.

Tempo. 2013. *Wiji Thukul*. Seri Buku Tempo: Prahara Orde Baru. Kepustakaan Populer Gramedia.

Tempo. 2014. "Cerita Janggalnya Peradilan Malari." *Tempo.co*, 15 January 2014. https://nasional.tempo.co/read/544907/cerita-janggalnya-peradilan-malari.

The Jakarta Post. 1996. "Bintang Gets 34 Months for Insulting the President." *The Jakarta Post*, 9 May 1996.

The Times. 1985. "Ex-Asean Chief May Be Tried." *The Times*, 5 August 1985.

The Times. 1986. "Ex-General Jailed on Subversion Charges; HR Dharsono Pleads Innocence at Indonesian Trial." *The Times*, 9 January 1986.

Trubek, David, and Marc Galanter. 1974. "Scholars in Self-Estrangement: Some Reflections on the Crisis in Law and Development Studies in the United States Law and Society." *Wisconsin Law Review*, no. 4: 1062.

Uhlin, Anders. 1993. "Transnational Democratic Diffusion and Indonesian Democracy Discourses." *Third World Quarterly* 14 (3): 517–544. https://doi.org/10.1080/01436599308420340.

UPI. 1996. "Prominent Indonesian Dissident Dies." *UPI*, 5 June 1996. https://www.upi.com/Archives/1996/06/05/Prominent-Indonesian-dissident-dies/8507833947200/.

US Department of State. 1997. "*Country Reports on Human Rights Practices for 1996*." Report Submitted to the Committee on International Relations, US House of Representatives, and Committee on Foreign Relations, US Senate. U.S. Government Printing Office.

Vatikiotis, Michael. 1990. "Influential Inmate." *Far Eastern Economic Review*, 10 May 1990.

Winarta, Frans Hendra. 2013. *Pro Bono Publico*. Jakarta: Gramedia Pustaka Utama.

Woodward, Mark. 2007. "Religious Conflict and the Globalization of Knowledge in Indonesian History." In *Religion and Conflict in South and Southeast Asia: Disrupting Violence*, edited by Linnel E.Cady and Sheldon W.Simon, 85–104. London and New York: Routledge. https://doi.org/10.4324/9780203967485-15.

Zen, A. Patra M. 2004. "Indonesian Legal Aid Foundation: Struggling for Democracy and Its Own Sustainability." *Focus* 38 (December 2004). https://www.hurights.or.jp/archives/focus/section2/2004/12/indonesian-legal-aid-foundation struggling-for-democracy-and-its-own-sustainability.html#note4.

Zifcak, Spencer. 1999. "'But a Shadow of Justice': Political Trials in Indonesia." In *Indonesia: Law and Society*, edited by Tim Lindsey, 1st Edition. Sydney, Australia: Federation Press.

Part Two

Cause lawyering in a time of democratic reform and regression

Democratic reform and regression

Introduction to Part Two

To understand how cause lawyering and LBH cause lawyers have been affected by democratic reform and regression, it is important to have an awareness of the dramatic political and institutional changes that have occurred since the authoritarian New Order collapsed. This brief introductory section to Part Two highlights key aspects of Indonesia's democratic transition, which established many of the structural factors considered important in providing opportunities for cause lawyering. I also sketch the outlines of recent democratic regression in the country, which has further affected cause lawyers' stance toward these structural conditions.

Although the New Order faced considerable challenges to its authority during the 1990s, few thought there was any real chance of Soeharto stepping aside until the Asian financial crisis struck in 1997 (Lindsey and Santosa 2008, 11). When Soeharto was reappointed for another five-year term as president in March 1998, student protests broke out in major cities across the country (Hadiwinata 2003, 77). Rising food and fuel prices led to demonstrations and violence. Calls for *reformasi* intensified. On 13–14 May, widespread rioting and looting broke out across Jakarta, much of it targeting ethnic Chinese-owned businesses. Some reports estimated more than 1,000 people were killed, many burned to death in shopping malls that had been set on fire (Purdey 2008, 518). A combination of economic collapse, student protests, rioting by the urban poor, splits within the military, and the loss of support from domestic cronies and then former Cold War ally the US eventually spelled the end for Soeharto (Davidson 2018, 1–2; Lindsey and Santosa 2008, 12). On 21 May 1998, Soeharto announced his resignation, marking the end of the New Order and the beginning of the *reformasi* period, setting in motion major changes for Indonesian state and society. Soeharto's vice president, Bacharuddin Jusuf Habibie, replaced him and enthusiastically embraced reform.

Habibie had a close relationship with Soeharto and was previously minister of research and technology under the New Order, so few imagined he would have much of an appetite for instituting sweeping change. But after the corruption and nepotism of the final years of Soeharto's rule, the government desperately needed to regain political legitimacy. There was a genuine desire

DOI: 10.4324/9781003486978-4

for institutional and regulatory reform and "little open resistance from within the state to notions of democracy, accountability, supremacy of law, even civil society itself" (Aspinall 2004, 84). Although the Habibie government was only in power for about 18 months, it instituted major and lasting changes. Among its most significant achievements were the release of dozens of political prisoners,[1] the lifting of restrictions on freedom of the press, the revocation of the 1963 anti-subversion law, the ratification of ILO Convention No. 87 on Freedom of Association, the cancelling of the requirement for all organisations to adopt Pancasila as their 'sole foundation', reforms allowing new political parties to be established, and the holding of the first democratic elections since 1955 (Anwar 2010; Lindsey and Santosa 2008, 12; Beittinger-Lee 2013, 73–74). Responding to the New Order's highly centralised rule, Habibie began a "radical" decentralisation programme, involving the devolution of political and fiscal power to regional (district and municipal) administrations, which was implemented from 2001 (Anwar 2010, 109). In the legal realm, the vital 'One Roof' policy was introduced, with the passage of Law No. 35 of 1999 on Judicial Power, which strengthened judicial independence by removing the Ministry of Judicial Affairs' control over the judiciary (Lindsey and Santosa 2008, 13).[2] Crucially, under Habibie, constraints on constitutional amendment were removed, initiating a transformative constitutional reform process that was eventually completed in 2002 (Horowitz 2013). The incremental reforms to the Constitution introduced expansive new human rights protections, stronger separation of powers between the executive and the legislature, limitations on presidential terms to two five-year periods, and provisions establishing an independent Constitutional Court, among other changes. These vital constitutional amendments provided the basis for transition to a liberal democratic system, even if the result was "a long way from being satisfactory" (Butt and Lindsey 2012, 24).

Important reforms continued under the two presidents following Habibie, Abdurrahman Wahid (commonly known as Gus Dur) (1999–2001) and Megawati Soekarnoputri (2001–4). Under Wahid, the police were formally separated from the military,[3] and significant strides were made in judicial reform (Lindsey and Santosa 2008, 15; Yon and Hearn 2016). Under Chief Justice Bagir Manan (2001–8), the Supreme Court began a comprehensive reform process, and in 2003 published five 'blueprints' for reform, prepared in collaboration with CSOs (Yon and Hearn 2016, 9).[4] Wahid is also remembered fondly among human rights activists for dismantling Soeharto-era restrictions on Chinese language and culture, and permitting the celebration of Chinese New Year (Davidson 2018, 22). Megawati, meanwhile, presided over the establishment of the Corruption Eradication Commission (KPK), a new anti-money laundering agency – the Financial Transaction Reports and Analysis Centre (PPATK), and an independent Judicial Commission (Lindsey and Santosa 2008, 18). Of considerable significance to cause lawyers was the establishment of the Constitutional Court under Megawati's presidency, as mandated by the third constitutional amendment of 2001. Despite these major achievements, Megawati's presidency was also marked by her closeness to the military and a deteriorating human rights situation (Beittinger-Lee 2013, 79).

Susilo Bambang Yudhoyono assumed the presidency in 2004 as the first directly elected president in Indonesian history.[5] Most scholars now consider the decade of the Yudhoyono presidency a wasted opportunity (Aspinall, Mietzner, and Tomsa 2015). Although he maintained political stability, he did little to advance the quality of Indonesia's democracy and "slowed the pace of reform to a halt" (Davidson 2018, 7). There is no universally agreed upon definition for the period of *reformasi*, but most scholars agree it was over by the end of Yudhoyono's first term. Mietzner, for example, states that Indonesia reached its democratic peak between 2004 and 2008 (Mietzner 2021a, 11–16). One of the defining features of the Yudhoyono era was the rise of conservative Islam. Attacks on religious minorities and restrictions on their places of worship became more common (Lindsey and Pausacker 2016) and, as such, the defence of religious minorities became an increasingly important component of LBH's work (see Chapter 4). Although the Yudhoyono years are often described as a missed opportunity, his government was open to collaboration with civil society (particularly in its first term), and some progressive reforms were passed under his watch. These included Law No. 14 of 2008 on Freedom of Information and Law No. 16 of 2011 on Legal Aid, both of which were formulated with the active involvement of CSOs, including LBH in the case of the Legal Aid Law. Moreover the KPK, established in 2003, began pursuing corruption suspects with gusto under Yudhoyono, and claimed several high-profile scalps, including senior members of Yudhoyono's own party (Butt and Lindsey 2018, 290–91).

President Joko "Jokowi" Widodo came to power in 2014 on the back of a deeply polarised and at times poisonous election in which he defeated Prabowo Subianto, a New Order-era general and one-time son-in-law of Soeharto. In Indonesia and among foreign observers, the election was cast as a choice between ongoing democratisation under Jokowi versus a reversion to New Order-style authoritarianism under Prabowo (Aspinall and Mietzner 2014). Prominent human rights, religious freedom and anti-corruption activists, including some former LBH staff, publicly backed Jokowi's campaign. Conservative Islamists, meanwhile, rallied behind Prabowo, inflaming a longstanding ideological divide between pluralists and Islamists in the country (Mietzner 2021b, 166–67). Jokowi was the first president with no direct links to the New Order elite, and had a reputation as a no-frills and corruption-free mayor of Solo (Central Java) and governor of Jakarta, so there were enormous expectations that he would reinvigorate Indonesia's moribund reform project. It was not long, however, before the high hopes progressive civil society had for Jokowi were dashed (Muhtadi 2015, 349–50).

As his presidency progressed, disquiet about democratic regression in Indonesia continued to grow. Seemingly spooked by massive Islamist protests in late 2016 and early 2017 against the ethnic Chinese governor of Jakarta

Basuki "Ahok" Tjahaja Purnama (a one-time ally of Jokowi), Jokowi responded with a mix of accommodation and criminalisation, selecting a conservative Islamic figure as his running mate in the 2019 elections at the same time as outlawing hard-line Islamist groups (Mietzner 2020b; 2018). The 2013 Law on Mass Organisations was revised to provide the government with the authority to outlaw CSOs without needing judicial approval, and the government has since banned the conservative Hizbut Tahrir Indonesia (HTI) and the Islamic Defenders Front (*Front Pembela Islam*, FPI).[6] A defining feature of Jokowi's presidency has been a dramatic shrinking of civic space. In addition to the government imposing new restrictions on freedom of association, politicians and other powerful figures have increasingly used prosecution under defamation, hate speech and blasphemy provisions to silence their opponents, activists and citizens (Warburton and Aspinall 2019, 260–61; Setiawan 2020). According to Amnesty International Indonesia, between January 2019 and May 2022 more than 300 people were charged with alleged offences under the notorious Electronic Information and Transactions Law (Amnesty International Indonesia 2022, 4).[7] Sometimes simply the threat of prosecution is used to constrain freedom of expression. In 2021, individuals who had painted murals criticising the government's response to the Covid-19 pandemic were briefly detained or forced to make public apologies. While none were ultimately charged, the incidents contributed to an environment in which the public has become increasingly wary about expressing their views in public (Setiawan 2022, 284–85).

For more than a decade, anti-democratic elites in the legislature have sought to wind back key reforms of the post-authoritarian era (Mietzner 2021b, 162). Under Jokowi, these efforts met with some success. One of the most blatant assaults occurred in 2019, when the president and legislature agreed on revisions to the 2002 Law on the KPK that effectively gutted the body (Butt 2019).[8] This event, along with attempts to pass a repressive new criminal code, triggered the *Reformasi Dikorupsi* protests discussed at the opening of this book. One year later, as Covid-19 spread across the nation and restrictions on public gatherings were introduced, the government rushed through a massive 'Omnibus' Law on Job Creation that combined 79 existing laws and critics said weakened environmental protections, as well as threatening workers' and Indigenous Peoples' rights. A hasty and opaque drafting process suggested that the government was more focused on the needs of business elites than the public (Argama 2020). Mass protests again erupted across Indonesia, and police again responded excessively, arresting thousands of protesters (BBC Indonesia 2020). Since 2020, the government and legislature have also taken increasingly shameless steps to undermine the independence of the Constitutional Court,[9] long considered one of the more respected and dependable institutions established in the democratic era (Butt and Lindsey 2018, 100). Further, after their unsuccessful attempt to pass a highly problematic Criminal Code in 2019, legislators and the government managed to pass a modified, yet still regressive, version in late 2022.

These and other developments have resulted in increasingly pessimistic academic assessments of Indonesia's democratic quality. Scholars have drawn attention to growing efforts among the political elite to wind back democratic reforms (seen clearly in revisions to the Law on the KPK, the Omnibus Law on Job Creation and the revised Criminal Code), attacks on minority rights (especially religious minorities and LGBTIQ+ Indonesians), shrinking civic space, worsening political polarisation along Islamist and pluralist lines, and the growing role of the military and police in social and political life, among other concerns.[10] Scholarly accounts of the erosion of Indonesian democracy are also mirrored in international measures of democratic quality. Freedom House now considers Indonesia only 'partly free' after ranking the country 'free' from 2006 to 2013 (Freedom House 2013; 2023). Similarly, the V-Dem Institute, which considers Indonesia only an 'electoral democracy', noted in 2023 that the country had seen a significant trend toward autocratisation over the past decade (Papada et al. 2023).

Although there is no question that anxiety about democratic regression in Indonesia has grown over recent years, scholars have long raised concerns about the extent of Indonesia's democratic transition. These concerns have generally focused on the ability of New Order-era elites to survive the transition and capture the state's new democratic institutions (Aspinall 2013, 102). Oligarchy scholars such as Jeffrey Winters (Winters 2011), and Richard Robison and Vedi Hadiz (Robison and Hadiz 2004), have been particularly influential in advancing aspects of this argument. While there are differences in these scholars' conceptualisations of oligarchy, they all generally contend that "democratisation has changed the form of Indonesian politics without eliminating oligarchic rule" (Ford and Pepinsky 2013, 3–7). Or in other words, the social and political elites that emerged under the New Order have readily adapted to new democratic structures and shaped policy formulation and implementation to their benefit (Aspinall 2013, 102; Winters 2013, 18–19). Winters' comments on the rule of law are particularly relevant for understanding the predicament faced by Indonesian cause lawyers in the post-Soeharto period:

> There is an expectation in the scholarly literature that as democratic consolidation progresses, gains will be made in the rule of law (typically portrayed as a 'quality of democracy' matter). But there is no inherent reason for this to be so. A democracy thoroughly captured by oligarchs has no strong inherent incentives to impose independent and punishing legal constraints on itself. Nor is a democracy necessarily unstable or vulnerable to crippling illegitimacy just because the legal regime is feeble at the top. Evidence from a broad sample of transitions to democracy globally suggest that 'democracy without law' can persist for decades. It is a scenario Indonesia has been playing out since 1998.
>
> (Winters 2013, 33).

A result of this situation is that "on matters of property, wealth, economy, corruption, and criminality of all kinds, the law bends to individual oligarchs and elites rather than the reverse" (Winters 2013, 19). This may be an overly pessimistic view, but it is a view to which many LBH lawyers subscribe. LBH lawyers now view the state as completely in the grip of oligarchs.[11]

As this short overview of political and institutional developments since 1998 has shown, many of the reforms implemented in the period immediately after Soeharto fell were those considered supportive of legal mobilisation – constitutional rights protections, the establishment of a constitutional court, judicial independence, political openness, and reforms allowing the emergence of a vibrant civil society sector and free media. But the last decade has been characterised by serious democratic backsliding. This decline in democratic quality has had a major impact on the form of cause lawyering practiced by LBH.

Notes

1 Including labour leader Muchtar Pakpahan, discussed in Chapter 2.

2 This amended Law No. 14 of 1970 on Judicial Power, under which the Ministry of Judicial Affairs had authority over the general and administrative courts (the Ministry of Religious Affairs oversaw the religious courts and the Ministry of Defence the military courts). The revised 1999 Law stated that the Supreme Court was to have exclusive authority over the organisational, administrative and financial affairs of the judiciary. The Ministry of Judicial Affairs was required to transfer this authority within five years, which it did in April 2004.

3 Through the passage of two decrees of the People's Consultative Assembly (*Majelis Permusyawaratan Rakyat*, MPR): MPR Decree No. VI/MPR/2000 and Decree No. VII/MPR/2000.

4 Notably, LBH elected not to engage in this reform process (see discussion in Chapter 3).

5 Direct presidential elections were introduced in the third amendment to the Constitution, in November 2001 (Article 6A).

6 Through the passage of Government Regulation in Lieu of Law (Perppu) No. 2 of 2017 on Revision of Law No. 17 of 2013 on Mass Organisations. The revised 2017 Law on Mass Organisations provided the minister of law and human rights with the authority to dissolve organisations, a power used to ban HTI in July 2017 and FPI in 2020.

7 Law No. 11 of 2008 on Electronic Information and Transactions (as amended by Law No. 19 of 2016 and Law No. 1 of 2024).

8 Among the most controversial amendments were those requiring the KPK to seek approval from a new 'Supervisory Board', appointed by the president, before it can conduct wiretapping, searches and seizures (Butt 2019).

9 See discussion in Chapter 4.

10 See, for example, Power 2018; Diprose, McRae, and Hadiz 2019; Schäfer 2019; Marta, Agustino, and Wicaksono 2019; Aspinall and Mietzner 2019; Warburton and Aspinall 2019; Mietzner 2018; 2020a; 2021b; 2021a; Setiawan 2020; Mujani and Liddle 2021; Sambhi 2021.

11 A simple demonstration of this is that YLBHI's annual reports for 2019 and 2020 were titled, respectively, "Reform Corrupted by Oligarchs" (*Reformasi Dikorupsi Oligarki*) and "Authoritarians and Oligarchs Agitate in the Midst of the Pandemic" (*Otoritarian dan Oligarki Membuncah di Tengah Pandemi*).

References

Amnesty International Indonesia. 2022. "*Silencing Voices, Suppressing Criticism: The Decline in Indonesia's Civil Liberties.*" Jakarta: Amnesty International Indonesia. https://www.amnesty.org/en/documents/asa21/6013/2022/en/.

Anwar, Dewi Fortuna. 2010. "The Habibie Presidency: Catapulting Towards Reform." In *Soeharto's New Order and Its Legacy: Essays in Honour of Harold Crouch*, edited by Edward Aspinall and Greg Fealy, 99–117. Asian Studies Series Monograph 2. Canberra: ANU Press. https://doi.org/10.22459/SNOL.08.2010.07.

Argama, Rizky. 2020. "Major Procedural Flaws Mar the Omnibus Law." *Indonesia at Melbourne*, 9 October 2020. https://indonesiaatmelbourne.unimelb.edu.au/major-procedural-flaws-mar-the-omnibus-law/.

Aspinall, Edward. 2004. "Indonesia: Transformation of Civil Society and Democratic Breakthrough." In *Civil Society and Political Change in Asia: Expanding and Contracting Democratic Space*, edited by Muthiah Alagappa, 61–96. Stanford, California: Stanford University Press. https://doi.org/10.1515/9780804767545-008.

Aspinall, Edward. 2013. "Popular Agency and Interests in Indonesia's Democratic Transition and Consolidation." *Indonesia*, no. 96 (October): 101–121. https://doi.org/10.5728/indonesia.96.0011.

Aspinall, Edward, and Marcus Mietzner. 2014. "Indonesian Politics in 2014: Democracy's Close Call." *Bulletin of Indonesian Economic Studies* 50 (3): 347–369. https://doi.org/10.1080/00074918.2014.980375.

Aspinall, Edward, and Marcus Mietzner. 2019. "Indonesia's Democratic Paradox: Competitive Elections Amidst Rising Illiberalism." *Bulletin of Indonesian Economic Studies* 55 (3): 295–317. https://doi.org/10.1080/00074918.2019.1690412.

Aspinall, Edward, Marcus Mietzner, and Dirk Tomsa, eds. 2015. *The Yudhoyono Presidency: Indonesia's Decade of Stability and Stagnation.* Indonesia Update Series. Singapore: ISEAS Publishing. https://doi.org/10.1355/9789814620727.

BBC Indonesia. 2020. "Demo Tolak Omnibus Law di 18 Provinsi Diwarnai Kekerasan, YLBHI: 'Polisi Melakukan Pelanggaran.'" *BBC Indonesia*, 9 October 2020. https://www.bbc.com/indonesia/indonesia-54469444.

Beittinger-Lee, Verena. 2013. *(Un)Civil Society and Political Change in Indonesia: A Contested Arena.* Routledge Studies on Civil Society in Asia. London and New York: Routledge. https://doi.org/10.4324/9780203868799.

Butt, Simon. 2019. "Amendments Spell Disaster for the KPK." *Indonesia at Melbourne*, 18 September 2019. https://indonesiaatmelbourne.unimelb.edu.au/amendments-spell-disaster-for-the-kpk/.

Butt, Simon, and Tim Lindsey. 2018. *Indonesian Law.* Oxford; New York: Oxford University Press. https://doi.org/10.1093/oso/9780199677740.001.0001.

Butt, Simon, and Tim Lindsey. 2012. *The Constitution of Indonesia: A Contextual Analysis.* Constitutional Systems of the World. Oxford; Portland: Hart Publishing. https://doi.org/10.5040/9781509955732.

Davidson, Jamie S. 2018. *Indonesia: Twenty Years of Democracy.* Elements in Politics and Society in Southeast Asia. Cambridge: Cambridge University Press. https://doi.org/10.1017/9781108686518.

Diprose, Rachael, Dave McRae, and Vedi R. Hadiz. 2019. "Two Decades of Reformasi in Indonesia: Its Illiberal Turn." *Journal of Contemporary Asia* 49 (5): 691–712. https://doi.org/10.1080/00472336.2019.1637922.

Ford, Michele, and Thomas B. Pepinsky. 2013. "Beyond Oligarchy? Critical Exchanges on Political Power and Material Inequality in Indonesia." *Indonesia*, no. 96 (October): 1–9. https://doi.org/10.7591/9781501719158.

Freedom House. 2013. "*Freedom in the World 2013: Democratic Breakthroughs in the Balance.*" Freedom House. https://freedomhouse.org/report/freedom-world/2013/democratic-breakthroughs-balance.

Freedom House. 2023. "*Indonesia.*" Freedom House. 2023. https://freedomhouse.org/country/indonesia/freedom-world/2023.

Hadiwinata, Bob S. 2003. *The Politics of NGOs in Indonesia: Developing Democracy and Managing a Movement*. London and New York: RoutledgeCurzon. https://doi.org/10.4324/9780203200124.

Horowitz, Donald L. 2013. *Constitutional Change and Democracy in Indonesia*. New York: Cambridge University Press. https://doi.org/10.1017/cbo9781139225724.

Lindsey, Tim, and Helen Pausacker, eds. 2016. *Religion, Law and Intolerance in Indonesia*. London and New York: Routledge. https://doi.org/10.4324/9781315657356.

Lindsey, Tim, and Mas Achmad Santosa. 2008. "The Trajectory of Law Reform in Indonesia: A Short Overview of Legal Systems and Change in Indonesia." In *Indonesia: Law and Society*, edited by Tim Lindsey, 2nd Edition, 2–22. Sydney, Australia: The Federation Press.

Marta, Auradian, Leo Agustino, and Baskoro Wicaksono. 2019. "Democracy in Crisis: Civic Freedom in Contemporary Indonesia." In *International Conference on Democratisation in Southeast Asia (ICDeSA 2019)*, 255–257. Atlantis Press. https://doi.org/10.2991/icdesa-19.2019.52.

Mietzner, Marcus. 2018. "Fighting Illiberalism with Illiberalism: Islamist Populism and Democratic Deconsolidation in Indonesia." *Pacific Affairs* 91 (2): 261–282. https://doi.org/10.5509/2018912261.

Mietzner, Marcus. 2020a. "Authoritarian Innovations in Indonesia: Electoral Narrowing, Identity Politics and Executive Illiberalism." *Democratization* 27 (6): 1021–1036. https://doi.org/10.1080/13510347.2019.1704266.

Mietzner, Marcus. 2020b. "Populist Anti-Scientism, Religious Polarisation, and Institutionalised Corruption: How Indonesia's Democratic Decline Shaped Its Covid-19 Response." *Journal of Current Southeast Asian Affairs* 39 (2): 227–249. https://doi.org/10.1177/1868103420935561.

Mietzner, Marcus. 2021a. *Democratic Deconsolidation in Southeast Asia*. Elements in Politics and Society in Southeast Asia. Cambridge: Cambridge University Press. https://doi.org/10.1017/9781108677080.

Mietzner, Marcus. 2021b. "Sources of Resistance to Democratic Decline: Indonesian Civil Society and Its Trials." *Democratization* 28 (1): 161–178. https://doi.org/10.4324/9781003346395-9.

Muhtadi, Burhanuddin. 2015. "Jokowi's First Year: A Weak President Caught Between Reform and Oligarchic Politics." *Bulletin of Indonesian Economic Studies* 51 (3): 349–368. https://doi.org/10.1080/00074918.2015.1110684.

Mujani, Saiful, and R. William Liddle. 2021. "Indonesia: Jokowi Sidelines Democracy." *Journal of Democracy* 32 (4): 72–86. https://doi.org/10.1353/jod.2021.0053.

Papada, Evie, David Altman, FabioAngiolillo, LisaGastaldi, TamaraKöhler, MartinLundstedt, NataliaNatsika, et al. 2023. "*Democracy Report 2023: Defiance in the Face of Autocratization.*" University of Gothenburg: Varieties of Democracy Institute (V-Dem Institute).

Power, Thomas. 2018. "Jokowi's Authoritarian Turn and Indonesia's Democratic Decline." *Bulletin of Indonesian Economic Studies* 54 (3): 307–338. https://doi.org/10.1080/00074918.2018.1549918.

Purdey, Jemma. 2008. "Legal Responses to Violence in Post-Soeharto Indonesia." In *Indonesia: Law and Society*, edited by Tim Lindsey, 2nd Edition, 515–531. Sydney, Australia: The Federation Press.

Robison, Richard, and Vedi R. Hadiz. 2004. *Reorganising Power in Indonesia: The Politics of Oligarchy in an Age of Markets.* London and New York: Routledge-Curzon. https://doi.org/10.4324/9780203401453.

Sambhi, Natalie. 2021. "Generals Gaining Ground: Civil-Military Relations and Democracy in Indonesia." *Brookings Institution*, 22 January 2021. https://www.brookings.edu/articles/generals-gaining-ground-civil-military-relations-and-democracy-in-indonesia/.

Schäfer, Saskia. 2019. "Democratic Decline in Indonesia: The Role of Religious Authorities." *Pacific Affairs* 92 (2): 235–255.

Setiawan, Ken. 2020. "A State of Surveillance? Freedom of Expression Under the Jokowi Presidency." In *Democracy in Indonesia: From Stagnation to Regression?*, edited by Thomas Power and Eve Warburton, 254–274. Singapore: ISEAS Publishing. https://doi.org/10.1355/9789814881524-018.

Setiawan, Ken. 2022. "Vulnerable but Resilient: Indonesia in an Age of Democratic Decline." *Bulletin of Indonesian Economic Studies* 58 (2): 273–295. https://doi.org/10.1080/00074918.2022.2139168.

Warburton, Eve, and Edward Aspinall. 2019. "Explaining Indonesia's Democratic Regression: Structure, Agency and Popular Opinion." *Contemporary Southeast Asia* 41 (2): 255–285. https://doi.org/10.1355/cs41-2k.

Winters, Jeffrey A. 2011. *Oligarchy.* Cambridge: Cambridge University Press. https://doi.org/10.1017/cbo9780511793806.

Winters, Jeffrey A. 2013. "Oligarchy and Democracy in Indonesia." *Indonesia*, no. 96 (October): 11–33. https://doi.org/10.5728/indonesia.96.0099.

Yon, Kwan Men, and Simon Hearn. 2016. "*Laying the Foundations of Good Governance in Indonesia's Judiciary: A Case Study as Part of an Evaluation of the Australia Indonesia Partnership for Justice.*" London: Overseas Development Institute.

3 Transitions and troubles

Organisational challenges after Soeharto

The New Order cast a long shadow over LBH. The repressive Soeharto regime was a central factor in LBH's establishment in 1970, and had a pronounced influence over its approach to cause lawyering. The momentous collapse of this regime was therefore bound to have a major impact on LBH and its approach to cause lawyering. In the decade following the demise of Soeharto, Indonesia transformed its institutions, establishing the structural conditions that should have allowed organisations like LBH to thrive. In contrast to expectations, however, LBH struggled following the democratic transition. There was a period when LBH appeared weaker under democracy than it was under an authoritarian regime. This chapter examines the reasons behind LBH's failure to flourish in the post-Soeharto era, attributing its struggles largely to internal organisational factors.

There is considerable literature on how organisational factors affect the work of CSOs, including in the Indonesian context (Lewis 2014; Fowler 2000; Ahmed 2021; Antlöv, Brinkerhoff, and Rapp 2010; Antlöv, Ibrahim, and van Tuijl 2006; Davis 2015; McGlynn Scanlon and Alawiyah 2015; Lassa and Li 2015). This work covers considerations of funding, leadership, organisational culture, operational systems, human resources, organisational learning, networking, relationship management and more. Studies of cause lawyering are mostly silent on these factors, apart from the crucial role of funding (Epp 1998; Ellmann 1998; Cummings 2008; Meili 2009). Epp's influential 1998 work, *The Rights Revolution*, does emphasise rights-advocacy organisations as a key element of the support structure essential for legal mobilisation, but internal organisational factors such as leadership are only covered in passing (see, for example, Epp 1998, 98–99). This chapter shines a light on these organisational factors, demonstrating how they have had a major influence on LBH and its approach to cause lawyering in the period after Soeharto fell. Most, but not all, of these organisational factors were also affected by democratic change.

The chapter begins by reflecting on how the demise of the New Order affected LBH. With the fall of Soeharto, LBH was forced to think more deeply about its identity and the strategies of cause lawyering it wished to prioritise now that the state was no longer authoritarian. For some time, it struggled to define its place in a dramatically different political context. I then

DOI: 10.4324/9781003486978-6

examine leadership and organisational management issues. Following the leadership turbulence of the 1990s, major leadership squabbles continued in the democratic era, and most reflected underlying tensions about the most appropriate strategy of cause lawyering for the time.

Another change precipitated by the democratic transition was that it prompted LBH's international donors to rethink their approach to development assistance. There were new opportunities to support the institutions of the transitional state and promote democratic consolidation, and the vigorous human rights and democracy activism conducted by LBH fell out of favour. Donors' changing priorities had significant implications for LBH, which experienced serious financial stress. Finally, the chapter briefly covers other critical organisational aspects that are less directly connected to democratic change, such as staff recruitment and training, gender equality, and research and knowledge management. The chapter reflects on how these aspects of institutional capacity have shaped LBH's cause lawyering work.

Reckonings with organisational identity post-Soeharto

Cause lawyers at LBH had long defined themselves in opposition to the New Order state. For much of civil society, and LBH too, the state was viewed "as the primary problem of political life and its constraint as a foremost political goal" (Aspinall 2004, 91). Indeed, LBH's ideology of legal aid – structural legal aid – could be said to have been focused almost entirely on bringing about changes to the authoritarian system of the New Order.

This was a relatively uncomplicated position to maintain for the first couple of years after Soeharto fell. Although Soeharto's successor and long-time protégé, BJ Habibie, surprised many observers with his attitude to reform, civil society still considered him a continuation of the old regime. In these early years of the democratic transition, LBH remained a powerful voice calling out state violence and pushing for democratic reform. For example, YLBHI led vocal campaigns over disappeared activists and students killed by security forces (The Jakarta Post 1998e; 1998c; 1998a; 1998b) and state-sponsored violence in Aceh (The Jakarta Post 1999a; 1998d). YLBHI, LBH Jakarta and the National Consortium for Law Reform (KRHN), the law reform organisation established by YLBHI, were members of the Civil Society Coalition for a New Constitution (*Koalisi Ornop untuk Konstitusi Baru*), which advocated for constitutional reform (Yadav and Mukherjee 2014, 295; Butt 2015, 25). But with the collapse of the New Order, the locus of power had shifted, and patterns of relations quickly became more complex. Although figures hostile to democratic change remained in government, there were far greater opportunities for civil society to contribute to reform.

Following the collapse of the New Order, LBH was suddenly confronted with questions over its raison d'être. Among civil society circles in Jakarta, it is common to hear that LBH suffered an identity crisis. Former YLBHI and LBH staff themselves acknowledge as much. Luhut Pangaribuan, a senior

YLBHI lawyer during the Soeharto era, said the change from a closed authoritarian system to a more open system led to LBH becoming "dis-oriented" (Saptono 2012, 85). LBH had become accustomed to a political, oppositional form of cause lawyering less suited to a more open regime. Structural legal aid, with its strong focus on changing the structure of the authoritarian New Order state, soon dipped in relevance, or was neglected. As YLBHI's director of organisational development from 2017–21, Febi Yonesta (commonly known as Mayong), explained: "With the passing of time, the structural legal aid concept was interpreted differently, and there were even some LBH offices that left it behind completely."[1]

At the same time as this was occurring, the lifting of New Order restrictions on associational life led to an explosion in CSOs. One account suggested the number of Indonesian CSOs had swelled to 70,000 by 2000 (Hadiwinata 2003, 113). Under the New Order, YLBHI had acted as a 'generalist' organisation. It was comfortable in its role as the "*rumah demokrasi*" (home of democracy) (Walhi 2017), a place that welcomed a broad range of social movement groups and activists. As Chapter 2 discussed, LBH worked politically on issues as diverse as labour rights, agrarian reform, freedom of expression, democratic reform, corruption, state violence, criminal law and procedure, and more. As former LBH lawyer Mas Achmad Santosa said: "At that time, YLBHI stood alone as a CSO that was brave. Now that everything is more open, LBH no longer dominates the civil society movement as it did under Soeharto" (Saptono 2012, 90). New organisations with sharply focused issues of concern and sound technical skills emerged, some of which had similar mandates to LBH. These included CSOs that are still strong today, such as the Indonesian Centre for Law and Policy Studies (*Pusat Studi Hukum dan Kebijakan Indonesia,* PSHK), the Indonesian Institute for an Independent Judiciary (*Lembaga Kajian dan Advokasi untuk Independensi Peradilan*, LeIP), the Indonesian Court Monitoring Society (*Masyarakat Pemantau Peradilan Indonesia*, MaPPI), the Association for Elections and Democracy (Perludem), the Indonesian Forum for Budget Transparency (*Forum Indonesia untuk Transparansi Anggaran*, Fitra), and the Institute for Criminal Justice Reform (ICJR). Meanwhile, LBH remained an all-rounder. Herlambang P. Wiratraman, formerly of LBH Surabaya, explains:

> The crisis at LBH was related to the development of civil society. CSOs thrived in the *reformasi* period, and other organisations in the civil society network started to fill the roles that were once filled by LBH… LBH experienced a phase where it couldn't position itself correctly.[2]

Donors encouraged and facilitated these new organisations to play a role in strengthening Indonesia's new democratic institutions. For example, LeIP and PSHK prepared the landmark Blueprint for Reform of the Supreme Court, which was then adopted by the reform-minded Chief Justice Bagir Manan in 2003 (Cox, Duituturaga, and Sholikin 2012, 31; Yon and Hearn 2016, 9). In

2004, Bagir Manan even established a 'Judicial Reform Team Office' (which still exists today), comprising Supreme Court officials and ministerial and civil society representatives to implement the blueprint. Not only were there new opportunities for civil society to contribute to the drafting of legislation but there was now even room for cause lawyers to enter the state and actively shape the reform process.

On the whole, however, LBH was uninterested in strengthening state institutions. Most LBH staff were not inclined to build the legitimacy of a state that they considered contained too many elements of the former authoritarian regime they had spent decades opposing and that they believed had little interest in improving social welfare. LBH offices in regional areas, especially, struggled to move beyond the oppositional role they had established over many years. This was understandable, given regional LBH offices were often on the frontlines of advocacy against the Soeharto government's major development projects. YLBHI's Mayong summarises these dynamics well:

> LBH was involved in the pushing for the birth of independent institutions, and organisations that were born from LBH did that sectorally, and were engaged in policy reform.[3] Meanwhile, LBH offices [in the regions] didn't do that. They were very sceptical of efforts to collaborate with the government. The institutional reform agenda was viewed by many LBH offices as a futile effort that would only expend a great deal of funds, and only end up strengthening the state, which is capitalist in nature, and not focused on improving the prosperity of the people. This was a common discourse among LBH offices when we had national level meetings.[4]

This tension between maintaining an adversarial role and engaging with the state on policy reform is a challenge common to cause lawyers in newly democratic states (Klug 2001, 265; Meili 2009, 62) and is a significant focus of Chapter 6. As Mayong's comment indicates, many of LBH's reflections in the years following the fall of Soeharto focused on how to use the law and for what cause. His comment also touches on the anti-capitalist sentiment that had infiltrated some LBH offices in the late New Order years (Aspinall 2005, 108). Within LBH, ideals of distributive justice may often exist alongside, and be expressed at the same time as, arguments emphasising individual civil and political rights. This highlights the deeply complex and sometimes divisive nature of these kinds of questions.

For some of the older guard of LBH, the collapse of the New Order meant that it was time for the organisation to return to providing conventional legal aid, and abandon the highly political form of cause lawyering that had developed under Soeharto. Consistent with his proceduralist inclinations, LBH founder Adnan Buyung Nasution tried to push the organisation in a more professional lawyerly direction. "LBH should return to its true nature as a professional institution. LBH is not a CSO that just does demonstrations," Buyung said in 2001 (Hukum Online 2001).[5] Similar views were also expressed to me by former YLBHI figure Hendardi, the more activist inclined lawyer who founded PBHI following the 1996 leadership dispute in YLBHI.

> After *reformasi*, I always said, 'we have to reorientate ourselves'. Although this country is not fully democratic, the transition is progressing, as small as that progress may be... LBH should return to providing more conventional legal aid. There are still so many poor and oppressed people who do not get representation. We can still prioritise communal cases, but there is no need to be so political, to play a political role.[6]

In the most extreme demonstration of this belief in a return to 'conventional' legal aid, in 1999 LBH founder Buyung chose to lead the defence team for former Armed Forces Commander General Wiranto and other senior military figures accused of human rights abuses in East Timor during the turbulent period surrounding its vote for independence from Indonesia in August 1999 (England 1999). The decision to represent Wiranto put Buyung in direct conflict with LBH. Respected former YLBHI staff Todung Mulya Lubis, Nursyahbani Katjasungkana, and Munir Said Thalib had been appointed to a fact-finding team that reported that Wiranto, as Armed Forces commander, should bear responsibility for crimes against humanity in East Timor (Commission to Investigate Human Rights Violations in East Timor (KPP-HAM) 2000). While Buyung was no longer part of the YLBHI leadership team and did not represent Wiranto as a YLBHI lawyer, he still sat on its advisory board, and in the eyes of the public (and YLBHI's donors) he was forever associated with YLBHI. Buyung's colleagues at YLBHI-LBH were livid, and there was almost immediately a move to dismiss him from the YLBHI board (The Jakarta Post 1999b).

Buyung's decision to represent Wiranto was difficult for many to understand, given Buyung had often squared off against the military under the New Order. But Buyung was "unrepentant" and said he was motivated by attempts to promote a professional legal culture (England 1999). In deciding to defend Wiranto, Buyung had returned to a purely proceduralist position. In line with this proceduralist understanding, Buyung's main cause was the rule of law and strengthening the legitimacy of the legal system, an important goal in a newly democratic state. Buyung had moved away from the classic cause lawyering position of identifying with the cause of his client, and toward more mainstream professional ideals of non-partisanship (Hilbink 2004, 665). This was a fundamental shift compared to Buyung's highly political approach in the HR Dharsono case, but it is a shift that was only made possible by the democratic transition.

Buyung's defence of Wiranto was not a one-off either. Buyung went on to defend controversial figures such as accused Bali Bombing mastermind Abu Bakar Ba'asyir, for terrorism related offences, and former Democratic Party chair Anas Urbaningrum and tax official Gayus Tambunan in corruption cases. He repeatedly asserted that, despite the charges against them, they had a right to a fair trial. But Buyung moved too fast for his colleagues at YLBHI, who still considered the legal system to be deeply defective. Buyung's defence of Wiranto, even if he was doing it in a professional capacity (BBC 1999), along with his closeness with the military, continued to resonate and cause problems for LBH well into the future, as I explain below.[7]

It is natural that YLBHI would reassess its organisational identity following such a fundamental change to the nature of the state. However, the tension between returning to a more conventional legal aid role and the grassroots style of politically motivated cause lawyering associated with structural legal aid continued for several years. But as I demonstrate in subsequent chapters, YLBHI has been invigorated by the democratic regression that has occurred in Indonesia over the past decade, and seems to have developed a renewed clarity about its role. These discussions are a major feature of subsequent chapters.

Organisational structure and governance

Before reflecting more closely on the organisational factors that have affected cause lawyers at YLBHI, it is important to recap YLBHI's organisational structure and governance. In accordance with the Indonesian Law on Foundations (Law No. 16 of 2001 as amended by Law No. 28 of 2004), YLBHI is comprised of an advisory board (*Dewan Pembina*), a supervisory board (*Dewan Pengawas*) and an executive board or leadership team (*Dewan Pengurus*), which is responsible for the day-to-day management of the organisation. The roles of each of the bodies are laid out in YLBHI's foundation document (*akta pendirian*) and articles of association or constitution (*anggaran dasar*). Briefly, the unelected and unpaid advisory board consists of senior statespeople with legal expertise and commitment to YLBHI's organisational priorities, and it is responsible for appointing members of the supervisory board and executive board. The advisory board meets at least twice annually but does not have a major role in the daily running of the organisation. The supervisory board, meanwhile, consists of at least three people (preferably with backgrounds in law, accounting and management), and is tasked with providing oversight and advice to the executive. It can temporarily suspend members of the executive if they have breached the articles of association, but the advisory board makes the decision on whether a member will be dismissed or retained. LBH lawyers are also guided by organisational bylaws (*anggaran rumah tangga*) and a 'Guide to YLBHI Struggle Values and Code of Ethics for Legal Aid Lawyers'.[8]

In the media and among the public, YLBHI and LBH Jakarta are often equated with one another, especially because they are both housed in the same building. However, it's important to remember that YLBHI is the 'Foundation of Legal Aid Institutes of Indonesia'. As Figure 3.1 below demonstrates, it serves as the umbrella body for 18 regional Legal Aid Institutes (LBH).[9] The divisions within YLBHI may vary depending on the priorities of the director at the time, but generally cover advocacy, campaigns, networking, organisational development, research and knowledge management, and finance. Although YLBHI does not involve itself in direct service provision to the same extent as the regional offices, it does take on cases from

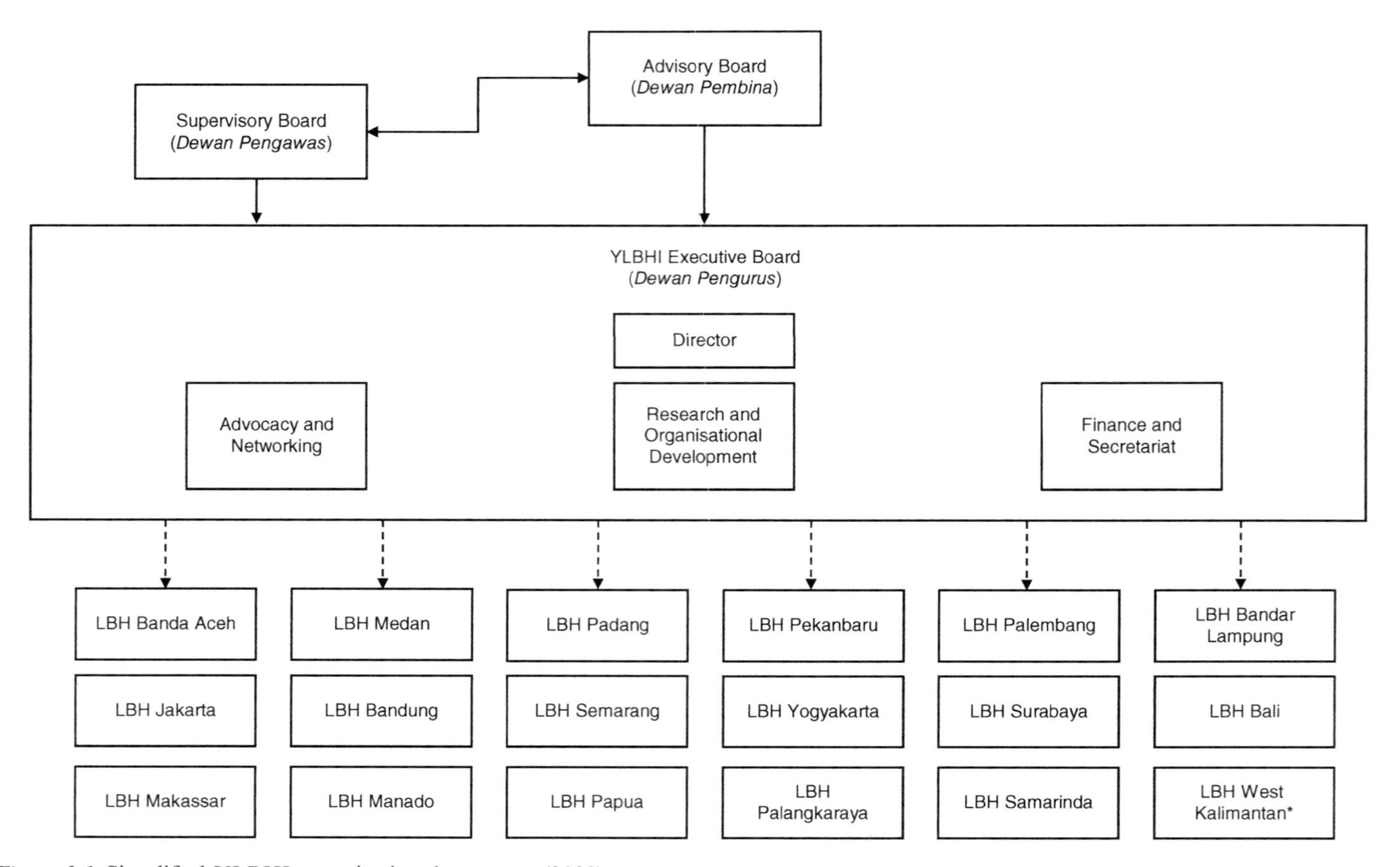

Figure 3.1 Simplified YLBHI organisational structure (2023)

time to time, usually those of national importance, as is indicated by the existence of its advocacy division.

Leadership and organisational management

Leadership is recognised as a key factor for the organisational effectiveness of CSOs (Fowler 1997; Hailey and James 2004; Amagoh 2015; Smillie and Hailey 2001). It is not hard to see how leadership can influence approaches to cause lawyering practiced from CSOs. Strong leadership can play an important role in setting an organisation's direction and inculcating shared values (Lewis 2014, 135; Amagoh 2015, 229). Accounts of LBH after 1998 commonly state that the organisation failed to thrive after the democratic transition because of leadership turmoil and internal conflict (Antlöv, Ibrahim, and van Tuijl 2006, 156; Ford Foundation 2003, 80). It is too simplistic to depict leadership turmoil as the primary cause of LBH's struggles, but it is true that repeated leadership friction has been a defining feature of YLBHI in the post-authoritarian period. Almost every transition between YLBHI's leaders (see Figure 3.2) has resulted in significant organisational turmoil. These leadership disputes have almost always reflected conflict over underlying organisational priorities and debates about the most appropriate form of cause lawyering for the political context.

Figure 3.2 YLBHI directors of the post-Soeharto era

Following the major conflict in 1996 between the proceduralist and grassroots arms of the organisation discussed in Chapter 2, the next leadership crisis occurred in late 2001. This conflict was reportedly sparked by efforts from YLBHI director Bambang Widjojanto and other senior figures, including YLBHI deputy Munir Said Thalib, Dadang Trisasongko and Poengky Indarti, to reform the organisation in order to bring it closer to the people it represented (Simanjuntak 2001). Those in favour of reform proposed changing the structure of the organisation from a foundation (*yayasan*) to an association or federation, allowing academics and members of YLBHI and LBH's target groups, such as labourers, farmers and the urban poor, to sit on the advisory board and have a role in determining the direction of the organisation.[10] This was, therefore, partly a conflict over reforming the somewhat undemocratic *yayasan* structure to an organisational form more appropriate for the (democratic) times. Those advocating for change also reportedly pushed for greater financial transparency and more power for LBH offices in the regions, and were strongly critical of the close relationships Buyung and YLBHI board member Victor Sibarani maintained with military figures (The Jakarta Post 2001; Hukum Online 2001).

The advisory board and Buyung rejected the proposed reforms, reportedly because of the foundation's "historical values" (Simanjuntak 2001). Although Buyung was no longer formally a member of the board, as founder he remained highly influential. The conflict came to a head with the resignations of the respected director Widjojanto and deputy Munir. Widjojanto went on to work in private practice, while continuing to be involved in legal reform and governance issues, eventually serving as a respected deputy head of the KPK (Kompas.com 2015). Munir, meanwhile, helped found and played leading roles in human rights organisations KontraS and Imparsial, but was murdered by arsenic poisoning in 2004 on a Garuda Indonesia flight to Amsterdam, in a case that has become one of the most notorious human rights abuses of the democratic era in Indonesia (Human Rights Watch 2017).[11]

The 2001 leadership conflict was widely covered in the media, with legal portal Hukum Online lamenting that the "locomotive of democracy" had "lost its democratic spirit" (Hukum Online 2001). Widjojanto's comments on exiting the organisation are significant. The Jakarta Post quoted him as saying:

> I have chosen to quit… I will work together with other people to promote democracy. I refuse to work with people who love playing in the grey area. I respect my seniors, but none of you here is good enough to be my role model.

His comments were clearly directed at Buyung, but they also reflected the broader organisational tensions that emerged following the transition to democracy. Some LBH lawyers were open to closer engagement with government figures, while others continued to favour an uncompromising, oppositional approach.

Leadership turmoil again returned to the organisation in 2006. In September 2002, the advisory board had appointed Munarman as YLBHI director, following nine months during which the organisation had been led by a temporary caretaker team. Munarman was previously head of YLBHI's civil and political rights division and former coordinator of KontraS in Aceh (Simanjuntak 2002). Munarman proved to be a highly controversial figure. Lindsey and Crouch have charted his evolution from a cause lawyer concerned with the promotion of democracy, human rights and the rule of law, to an Islamist openly hostile to the rights of religious minorities (Lindsey and Crouch 2013, 637–42). They suggest Munarman was deeply affected by the death of his friend and colleague Munir, leading him to lose faith in the legal system as a force for change (Lindsey and Crouch 2013, 638).

Munarman became increasingly close to Hizbut Tahrir Indonesia (HTI),[12] the Indonesian arm of the Islamist organisation that seeks to establish an Islamic caliphate. He increasingly brought his religiously conservative beliefs to bear on his work with YLBHI (Fitzpatrick 2008) and became quite erratic and difficult to deal with. He was openly combative with LBH's foreign donors and at one point spoke at an HTI protest outside the US Embassy in Jakarta (Antara 2006). This was too great a deviation from LBH's core values for Buyung, who had returned to chair YLBHI's advisory board in 2002, and he asked Munarman to resign voluntarily (Siboro 2006a). Although Munarman was asked to leave because of his growing relationship with HTI, other factors were also at play in the 2006 leadership turmoil. Munarman's time in power coincided with a significant decline in organisational funding (partly of his own making), discussed further below. There was once again significant tension between members of the leadership team and the advisory board, and Buyung as its dominant figure (Detik 2006). This time, conflict was sparked by the advisory board's decision to accept a Rp 200 million (about US $28,000 in 2024 terms) donation from oligarch Tomy Winata to renovate the YLBHI building (Ardiansyah 2006).[13] In May 2006, ten senior YLBHI staff left the organisation, including Munarman and prominent figures like deputy Robertus Robet, director of research Daniel Hutagalung and director of advocacy Syamsul Bahri Radjam, claiming that accepting the donation conflicted with YLBHI internal rules banning donations from environmentally destructive organisations or individuals (Siboro 2006b).

Munarman was replaced by Patra M. Zen, an LBH lawyer from Palembang, South Sumatra, who had also worked in Aceh, East Nusa Tenggara and Papua (Musthofid 2006). Patra was considered intelligent and was well liked by international donors, in part because he was receptive to collaborating with the government on legal reform, and because he went out of his way to heal the rifts with donors that Munarman had created.[14] This was significant given the political context of the time, during Indonesia's "democratic peak" between 2004 and 2008 (Mietzner 2021), a time when donors were enthusiastically promoting governance programming in collaboration with state institutions. But despite improved relations with donors, Patra was not strong on internal organisational management. Just a year into his leadership, YLBHI was again struck by crisis. Eighteen staff signed a motion of no confidence in Patra's leadership. Staff were highly critical of his authoritarian leadership style, lack of transparency and behaviour towards women (Hukum Online 2007).[15] The 18 staff took their concerns to the advisory board and threatened to leave unless Patra was replaced. The board was unmoved and said Patra had not broken any internal rules (Hukum Online 2007). The 18 left. YLBHI was decimated, and lost many talented young staff who cared deeply about using the law for social justice. Taufik ('Tobas') Basari, who had represented the 18 staff in discussions with the board, as well as YLBHI staff Ricky Gunawan and Yasmin Purba, went on to co-found LBH Masyarakat (Community Legal Aid Institute) with another lawyer, Dhoho Ali Sastro.[16]

Patra remained director of YLBHI for another three years but ended his term early, in 2010. Erna Ratnaningsih was selected to serve out the remaining portion of his term as leader (Detik 2010). Despite his managerial failings, Patra played a key role in policy advocacy and helping to formulate the 2011 Law on Legal Aid (KUBAH 2010). Patra and YLBHI worked closely with Deputy Law and Human Rights Minister Denny Indrayana to prepare the draft bill, with eager backing from donors.[17] Yet few within YLBHI look back on this period with much fondness. In fact, while YLBHI staff acknowledge that collaborating with the legislature and government on legal reform can be a viable cause lawyering strategy, many felt that Patra had been too accommodative. In any case, the organisational turmoil and loss of staff eventually discouraged donors, and by the end of Patra's time as director, YLBHI was in a difficult financial position.

Alvon Kurnia Palma was selected by the advisory board as the next director for the 2011–15 period, and his term was extended through to 2016. This covered the final years of the Yudhoyono presidency and part of Jokowi's first term. Alvon was a former director of LBH Padang, and the only candidate who stood for the YLBHI leadership. There is a broad consensus from YLBHI's donors and its staff that the organisation stagnated under Alvon. While Alvon is regarded as an individual of integrity and strongly committed to the legal aid cause, he was widely considered to be a weak leader.[18] He reportedly had a limited understanding of the donor landscape and was not adept at communicating his vision for the organisation to regional offices. As a result, he lacked the support of regional offices and was largely ignored by donors.[19] While in the past, YLBHI had looked for donor-supported projects and channelled funds to regional LBH offices, since the mid-2000s LBH offices were largely left to fundraise on their own.[20] Though some regional LBH offices thrived under this arrangement, others barely managed to stay afloat. As a result, some LBH offices ended up acting as organisational 'shells' for the pro bono work of directors.[21] Given the weaknesses in YLBHI, many donors began looking to the stronger regional LBH offices like LBH Jakarta,

LBH Bandung and LBH Surabaya, funding them directly. While Alvon was at the helm of YLBHI, LBH Jakarta rose in prominence, and to some degree displaced YLBHI. In the words of one donor, LBH Jakarta "ended up being the peak body for international projects on legal aid".[22] Under Alvon's leadership, YLBHI continued efforts to support the implementation of the 2011 Law on Legal Aid and coordinated a prominent but unsuccessful civil society challenge to the 2013 Law on Mass Organisations at the Constitutional Court. However, after five years in which organisational management took a back seat, donors had begun to lose faith in YLBHI.[23]

An example of the discord between YLBHI and regional offices emerged in 2013, when YLBHI and LBH Jakarta took different positions on reforming Indonesia's outdated Criminal Procedure Code (KUHAP). LBH Jakarta was one of the leading organisations in the Civil Society Committee for Criminal Procedure Code Reform (*Komite Masyarakat Sipil Untuk Pembaharuan KUHAP*), which conducted public campaigns and engaged in discussions with the legislature on key priorities for reform (Civil Society Committee for KUHAP Reform 2014). With its focus on police arrest, detention and investigation, and the rights of suspects and defendants, including to legal aid, the KUHAP is a policy area where YLBHI-LBH has a direct interest. No other civil society organisation in Indonesia would have the same first-hand experience of the operationalisation of the code – and its almost routine contravention by police (Butt and Lindsey 2018, 210). Yet YLBHI elected not to be involved in reform discussions, to the bafflement of other members of the reform coalition.[24] The incident was a further illustration of the fraught and sometimes divisive nature of deliberations over the most appropriate strategies of cause lawyering for a more open regime.

Things changed dramatically following the election of Asfinawati (Asfin) as YLBHI leader in early 2017. Asfin was elected under new leadership rules that gave the directors of regional LBH offices a vote alongside the members of the advisory board (Riana 2016), a change that seemed designed to strengthen the legitimacy of the new director among regional LBH offices and overcome some of the divisions that had emerged. The new rules also allowed former LBH and YLBHI staff to put themselves forward for the role (provided they were under 40),[25] a tacit acknowledgement of the fact that the organisation no longer had a deep pool of talent from which to draw. Asfin had previously served as head of LBH Jakarta and was known and respected among broader civil society, especially among organisations working on the rights of religious minorities.[26] She assembled a strong team around her, including former LBH Jakarta director Febi Yonesta (Mayong), former LBH Jakarta senior lawyer Muhamad Isnur, former LBH Semarang director Siti Rakhma Mary Herwati (Rakhma), and former LBH Bandung director Arip Yogiawan (Yogi). Although Yogi was Asfin's main rival for the leadership position, they worked well together. Asfin proved to be a passionate, consultative and egalitarian leader. These qualities were repeated often by her colleagues at LBH and in broader civil society.[27] Unlike many other civil society leaders, she was not particularly focused on building a public profile, although she certainly ended up gaining one by the time her term was over. When Asfin assumed the leadership, YLBHI was several hundred million rupiah in debt,[28] but she turned the organisation around. YLBHI has regained the trust of international donors and is now in the best financial position it has been in for many years.

Asfin and her team recognised that the YLBHI network had faced almost a decade of neglect and dedicated significant energy to promoting strengthened systems and procedures, improved knowledge and data management, a shared understanding of structural legal aid, and gender and minority rights among the LBH regional offices. Donors have supported them in these revitalisation efforts. Soon after Asfin took over the leadership position, YLBHI conducted a strategic planning process, updated its standard operating procedures, assessed the organisational capacity of regional LBH offices (in terms of funding, internal management and application of structural legal aid) and supported leadership transitions in the weaker offices.[29] Two new regional offices were opened, in Palangkaraya, Central Kalimantan, and Samarinda, East Kalimantan, both significant sites of conflict between local communities and plantation and mining companies. Organisational strengthening efforts have resulted in an improved division of roles and responsibilities between YLBHI as the central umbrella body and the LBH offices in the regions. Asfin also made an effort to visit regional LBH offices, some of which had reportedly not had visits from the YLBHI director for several years.[30] There has been a renewed focus on promoting structural legal aid across the LBH network (Asfinawati 2017), with a strong emphasis on the community organising and community legal empowerment aspects of the approach. A new internal "Guide to Structural Legal Aid" has been published, covering the concept, its history and its practical application (LBH-YLBHI 2020).

YLBHI has also tried to strengthen the quality of its public reports and research products, such as fact sheets and policy papers. Having visited the YLBHI office several times around 2012–14, and then again in 2019, it was remarkable to see the extent of changes to the library and knowledge management more broadly. Asfin also sought to re-establish YLBHI as a 'hub' for civil society, opening YLBHI's doors for members of its network who wish to hold discussions there, and hosting meetings with civil society groups to plan joint advocacy activities. This would not have been possible, of course, if Asfin had not been so liked and respected across a broad range of groups in Indonesian civil society.[31] One former LBH Jakarta lawyer offered a particularly blunt assessment of why YLBHI is now in a stronger position: "Before Asfin there was no work being done. After Bambang Widjojanto, YLBHI experienced a decline. YLBHI is stronger now because Asfin is there, and she is doing the work!"[32] It is worth noting that Asfin was the first leader elected following the death of YLBHI's charismatic founder Buyung in 2015. Given that so many CSOs falter following the loss of founder-leaders (Lewis 2014,

135–36), YLBHI's improved position is a testament to Asfin and her leadership team, as well as the organisation's maturity.

In late 2021, YLBHI selected a new leader for the 2022–26 period to replace Asfin after her leadership term expired. Asfin and her team were acutely aware of the organisational turmoil that had accompanied previous leadership transitions, and held the selection process earlier than in previous years, to allow for a longer transition period. The leadership team was also aware that succession planning has been a weakness in the past, and spent time investing in and preparing strong regional LBH directors as potential future leaders, such as the former director of LBH Padang, Era Purnama Sari, and the former director of LBH Semarang, Zainal Arifin. In the end, there were two contenders for the top job: Era and Muhamad Isnur, head of advocacy under Asfin. Although only regional LBH leaders and advisory board members can vote, in a sign of how YLBHI sees its leadership role in Indonesian civil society, the organisation held a public debate, accessible online, between the two candidates (YLBHI 2021b).[33] Isnur eventually won the contest and commenced as leader in January 2022. He has worked hard to ensure continuity, retaining influential YLBHI staff Asfin, Siti Rakhma Mary Herwati and Arip Yogiawan in a 'Knowledge Management Team'. Although these experienced staff are not formally part of the leadership structure, they are providing mentorship to the new YLBHI senior staff.

Not all the leadership tensions described above were directly connected to democratic transition, but most were. As noted, following democratic transition, when the government was more open, LBH experienced a period of disorientation where it experimented with more accommodative versions of cause lawyering involving closer collaboration with the government, for example in the formulation of the 2011 Legal Aid Law. Some within LBH even suggested reverting to more proceduralist or conventional forms of legal aid. These moves toward more accommodative or proceduralist versions of cause lawyering were not always well accepted by cause lawyers within the organisation who continued to favour more political, oppositional approaches. However, cause lawyering strategies are not influenced by leadership orientation alone.

Funding and donor relations

Funding is another critical organisational factor influencing the shape of cause lawyering, and is particularly relevant in Indonesia, where cause lawyering is mainly conducted from CSOs that are highly dependent on donor funding (Antlöv, Brinkerhoff, and Rapp 2010, 427).[34] There is no doubt that repeated leadership turmoil and weak organisational management affected YLBHI's financial position, particularly in the decade from 2006. But the democratic transition, and the political and institutional developments that accompanied it, altered the way donors viewed their approach to civil society support, and this also had implications for LBH and the forms of cause lawyering donors wanted to back.

In the immediate aftermath of the fall of the New Order, international donor funds for Indonesian CSOs were plentiful. Donors poured money into civil society to encourage the consolidation of democracy (Davis 2015, 6; Aspinall 2010, 2–3). Yet after the initial explosion in funds for democracy assistance, donor funds for civil society fell sharply, particularly from the early 2000s (Davis 2015, 6). Many donors considered the hard work of democratisation largely over and began to view Indonesia as a democratic "success story" (Aspinall 2010, 6). Also influential was Indonesia's designation as a lower middle-income country in 2006, which prompted many donors to reduce their commitments to Indonesia and shift their focus to lower income countries (Davis 2015, 3). Indonesia's growing economic prominence altered donor attitudes, leading major donors like Australia to place greater priority on building strong government-to-government relationships, rather than supporting CSOs that were likely to be critical of the Indonesian government (Aspinall 2010, 7–8). A trend tied to these developments was that, since the early 2000s, there has been a growing preference among international donors for governance programming and strengthening government institutions (Aspinall 2010, 6). These factors combined had negative implications for YLBHI's financial situation.

Under Soeharto, YLBHI had enjoyed a charmed relationship with its main donor Novib, the Netherlands development organisation that has been an affiliate of Oxfam since 1994. It was flush with organisational funds, and was allowed considerable freedom in determining its programme priorities. In the early 2000s, however, Oxfam Novib decided to conduct an evaluation of its assistance to YLBHI, a decision that was apparently prompted by Buyung's defence of General Wiranto.[35] The evaluation concluded YLBHI had not developed sufficiently detailed plans to ensure organisational sustainability and recommended discontinuing support for the organisation.[36] YLBHI leader Munarman and deputy Robertus Robet travelled to the Netherlands to plead YLBHI's case but were only able to secure "bridging funds" through to the end of 2003.[37] Mainly because of the loss of Novib, but also some of its other donors (Saraswati 2003), YLBHI saw its funding plunge from Rp 16.2 billion in 1999, immediately after the fall of Soeharto, to Rp 4.9 billion in 2003, as Table 3.1 shows (Munarman et al. 2004, 53).[38]

In its 2003 annual report, YLBHI reflected on these dynamics. It recognised that repeated leadership skirmishes had undoubtedly dampened donors' enthusiasm for supporting YLBHI, but it also noted that the shifting priorities of donors described above had played a role. Donors "are increasingly focusing on consolidation of democracy, institutional governance and decentralisation", the report stated (Munarman et al. 2004, 56). While strengthening civil society was a focus under the New Order, YLBHI noted, after 1998 donors were more concerned with good governance. Therefore, it was not that donors did not want to back cause lawyering, but rather that donor funding was available for a certain version of cause lawyering. Donors were more interested in funding organisations like LeIP,

Table 3.1 YLBHI income 1996–2003[39]

Year	*Total funds (Rp)*
1996	2,371,090,732
1997	3,484,909,401
1998	7,750,405,841
1999	16,217,881,388
2000	9,407,839,088
2001	10,588,007,686
2002	7,776,198,566
2003	4,970,843,795

MaPPI and PSHK to support, for example, reform efforts in the Supreme Court. There was a preference for the state-centred, elite/vanguard model of cause lawyering over the more explicitly activist grassroots model typically favoured by LBH. Tellingly, around this time, prominent former YLBHI staff and board member Frans Hendra Winarta publicly argued for a more politically engaged, accommodative version of cause lawyering. Following loss of donor support, he said, the organisation should shift its focus to drafting legislation (Saraswati 2003).

These trends have largely continued. Donors remain more likely to encourage collaboration between civil society and state institutions on policy reform, strategies more typically associated with elite/vanguard models of cause lawyering. LBH staff comment that there is comparatively far less funding available for case handling or representation (litigation) and community organising, activities LBH offices see as their core responsibilities.[40] Under the New Order, when funds and human resources were plentiful, LBH staff would spend weeks living with communities, gathering data and conducting community organising work.[41] LBH simply does not have sufficient resources or staff for this kind of strategy anymore. Civil society organisations (including LBH) also often complain that it has become rare for donors to provide core funding for institutional support and development (Antlöv, Brinkerhoff, and Rapp 2010, 428).[42] A recent example of these dynamics can be seen in the approach of the Australian Department of Foreign Affairs and Trade (DFAT), one of the largest bilateral donors to Indonesia (Prizzon, Rogerson, and Jalles d'Orey 2017), which dropped the civil society strengthening component of the Australia Indonesia Partnership for Justice (AIPJ) when its second phase was launched in 2017.

One of the consequences of dependence on foreign donors is that organisations like YLBHI are vulnerable to the project-based and short-term funding cycles of international development. Studies on the CSO sector in Indonesia have noted that many Indonesian CSOs are generalist in nature, and adapt their missions according to funding availability (Antlöv, Ibrahim, and van Tuijl 2006, 156; Davis 2015, 5; McGlynn Scanlon 2012, 15). When meeting with LBH around 2012–14 in the course of my previous work, I occasionally heard LBH staff grumble about having to implement programmes that were not in line with their core expertise or organisational priorities because of financial considerations. Yet in my discussions with LBH lawyers, they repeatedly asserted that they did not accept donor projects that were not consistent with their organisational priorities. For example, LBH Jakarta director Arif Maulana said:

> LBH Jakarta is always strict on this matter. If donors are not able to align themselves with our vision and mission, we will not accept them… That is what caused the crisis [LBH Jakarta suffered a funding crisis in 2011]. There were many [donor] programmes at the time. In fact, the conditions are even worse now. But LBH is very strict about this.[43]

Indeed, LBH Jakarta once rejected an offer from the World Bank to collaborate on its access to justice-focused 'Justice for the Poor' programme. This was partly motivated by LBH's longstanding ideological antipathy toward the World Bank, which burgeoned under the New Order, through cases like the Kedung Ombo reservoir conflict. But it also reflected a deeper distrust about the World Bank's support for access to justice programming at the same time as it was backing neoliberal initiatives on privatisation and labour market flexibility (Wiratraman 2006, 65–66). Access to justice programmes have been criticised for individualising communal problems and ignoring the underlying causes of inequality and injustice (Bedner and Vel 2010, 15). LBH lawyers made similar comments when speaking to me, suggesting that access to justice programmes are often too focused on professionalising the paralegal class, or that they only end up legitimising the corrupt Indonesian justice sector, placing insufficient emphasis on legal empowerment at the grassroots level. Again, these types of critiques are indicative of LBH's aversion to proceduralist forms of cause lawyering that they believe do less to address the substantive problems of the poor and marginalised.

A consequence of the donor affinity for policy advocacy is that many LBH offices have created a dichotomy between 'programme work' focused on policy advocacy (for donors) and legal aid, strategic litigation and community organising. In some offices, this division is even institutionalised in staffing arrangements, with certain staff members assigned to implementing donor programmes and others focused on case management. To some extent this is understandable, given the fact that proposal writing, engaging donors and report writing requires particular skills, but it can have the effect of demarcating donor work as separate from other organisational activities. Mayong, the former head of LBH Jakarta and head of organisational development at

YLBHI under Asfin, has a reputation for being more open to collaboration with government than many of his colleagues. He reflected on these concerns:

> In the eyes of donors, case management is not a sustainable use of their funds, whereas policy reform is sustainable. Most LBH offices work on case management and community organising, and are reluctant to work on policy advocacy, unless they are forced to do so to implement programmes. I have criticised this way of viewing things in this dichotomy. The core of case representation is policy change. We can't always just work through legal channels to change policy. It is not reliable. Except if the Indonesian legal system were strong enough that we could change policy through legal decisions. Can we do this? No way! The reality is not like this.[44]

Mayong's comments demonstrate that despite what LBH lawyers say, there does seem to be some influence from donors in pushing LBH toward governance programming.

Whether a result of Asfin's strong leadership, strengthened internal management or changing donor priorities – or, more likely, a combination of these factors – YLBHI's financial situation has improved dramatically. In 2020, it received almost Rp 14 billion in funds (see Figure 3.3), with its main donors including the International Law Development Organisation (IDLO) (managing funds for the Indonesia-Netherlands Rule of Law Fund of the Embassy of the Kingdom of the Netherlands in Jakarta), The Asia Foundation (which implements the Empowering Access to Justice (MAJU) programme for USAID), and Tifa Foundation (an Indonesian affiliate of the Open Society Foundations).

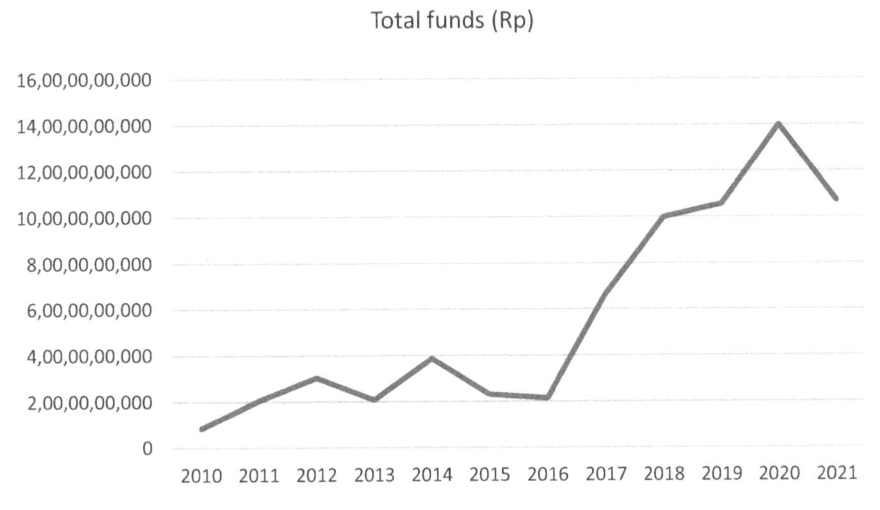

Figure 3.3 YLBHI income 2010–21[45]

Although YLBHI has not always embraced the kind of policy advocacy donors typically favour, it has been willing to work on governance programming in areas in which it has a direct interest, or in which it believes the government is genuinely committed to reform, such as strengthening the national legal aid system (discussed below). This has played a part in YLBHI's improved financial position. But developments in the donor landscape have also played a role. Recent years have seen growing international donor support for legal empowerment programming.[46] LBH can quite conveniently fit its pro-democracy work within the broader framework of legal empowerment (such as with the 2020 protests against the Omnibus Law on Job Creation, discussed in Chapter 7). Of greater significance, however, is a belated donor appreciation of the extent of democratic regression taking place in Indonesia. Civil society activists I spoke to mentioned that since 2019, donors have been much more willing to provide institutional funds to strengthen civil society as source of resistance to further democratic decline.[47]

> They understand that the situation is like what was faced under the New Order. You can no longer achieve major change by working on policy reform. Donors realise that the best chance that they have [to prevent further democratic backsliding] is to strengthen the community, with support from civil society, and if there is a problem, provide them with funds.[48]

In a sign of just how bad restrictions on dissent have become, some donors are now providing CSOs with 'emergency funds', to promote a quick response or allow for the evacuation of community members and activists at risk, much as they did decades ago under the New Order.[49]

Other sources of funding

While the past few years have seen YLBHI's financial position improve, less dependence on foreign funding would provide the organisation with even further autonomy over the form of cause lawyering it practices. However, like many other Indonesian CSOs, LBH has faced a persistent challenge identifying viable sources of funding other than foreign donors. Private sector donations are minimal and, in any case, LBH would have difficulty accepting donations from most major corporations in Indonesia for ethical reasons, such as involvement in environmental destruction or violations of labour rights. Similarly, there is little culture of philanthropy in Indonesia that could support the work of LBH. YLBHI and regional LBH offices have therefore regularly leant on former staff who have gone on to more lucrative private legal practice, asking them to donate small sums to support organisational stability. There has also occasionally been talk about trying to set up an endowment fund to support YLBHI's work, but these discussions have never progressed far.[50] At various points throughout its history, LBH Jakarta has experimented with public fundraising. For example, during a funding crisis in 2011, LBH Jakarta raised more than Rp 242 million, effectively saving the organisation (Tempo Magazine 2011). This is a strategy that LBH Jakarta turned to again when approaching its 50th birthday in 2021, calling on its supporters to donate the small sum of Rp 10,000 each (about US$0.6).[51] While important, these initiatives have never come close to matching the funding available through foreign donors. A promising recent development, however, has been the introduction of government-funded legal aid.

In addition to affecting donor funding priorities, the democratic transition also provided a new space for the formal recognition of legal assistance provided by CSOs, and new opportunities for accessing government funds. Law No. 16 of 2011 on Legal Aid introduced a robust legal framework for government-supported legal aid (Butt and Lindsey 2018, 217). As stated, YLBHI, with the backing of donors like AusAID (now DFAT) and the Tifa Foundation, was heavily involved in drafting the law (Yon and Hearn 2016, 9–10). This law recognised that the state had a responsibility to provide free legal aid to poor citizens. Rather than establishing a new central legal aid commission or agency to deliver legal aid, as occurred in other countries such as South Africa, Indonesia elected to accredit existing legal aid providers in civil society and reimburse them for services provided. This decision recognised the tradition of legal aid established by LBH offices and other CSOs over several decades (Yon and Hearn 2016, 10). YLBHI also favoured this model because it did not feel it could trust any government provider to deliver legal aid at an appropriate standard and without corruption (Rizaldi 2019).[52]

The National Law Development Agency (*Badan Pembinaan Hukum Nasional,* BPHN), a body under the Ministry of Law and Human Rights, was tasked with implementing the system. When the scheme was launched in 2013, there was an energetic debate within LBH about the degree to which it should participate in a government-managed system. Many LBH staff were concerned government-supported legal aid could distract them from their core mission of structural legal aid, with its emphasis on communal cases and addressing unbalanced power relationships. As Mayong explained, "We were worried that if we became dependent on the government legal aid scheme, then structural legal aid would be abandoned. We would end up busy dealing with the minor everyday cases of poor people."[53]

Some LBH offices complained they did not want to become "administrators for the government" and worried their autonomy could be threatened, or that they would be less willing to challenge the government if it were also providing them with funds.[54] In fact, experienced providers like LBH Bandung, LBH Surabaya and LBH Manado initially failed the verification and accreditation process, reportedly because of communication and administration issues (Yon and Hearn 2016, 16). But speaking to the directors of LBH Bandung and LBH Surabaya at the time, I remember that there was not much enthusiasm for accessing the scheme. Several years later, however, many LBH offices have now been accredited, and regularly access BPHN funds to support their work.[55] Febi Yonesta explains the reasoning for this change of heart:

> We realised that if we didn't get involved, we would end up allowing the government's legal aid scheme to strengthen organisations that harm the community, that are enemies of justice and YLBHI. The strategic choice we took was to try to dominate those resources and change the direction of policy gradually so that it was more in line with the vision and mission of structural legal aid... Now even the Islamic Defenders Front (FPI) has a legal aid arm, political parties have set up legal aid offices. If we allow them to thrive, they will control these resources.[56]

Some LBH offices have had more success than others in accessing government funds. For a few years, LBH Makassar, in South Sulawesi, had no international donor funds to support its work. Yet it was able to support its institutional costs and survive because of the funds it received through the BPHN.[57] And despite fears about government interference or LBH moderating its criticism of the government, LBH Makassar has continued to bring cases against the regional government and BPHN has been willing to offer reimbursement in such cases.[58]

The government allocated Rp 53 billion (US$3.6 million) to the scheme for the 2019–21 period (Elnizar 2019), a small amount given Indonesia's population and needs for legal services (World Bank 2020).[59] While state funding may eventually become an important source of support for LBH, several other fundamental aspects of the system mean it has yet to work in LBH's favour. Chief among these is that the system is heavily weighted in favour of litigation. This disadvantages organisations like LBH, which tend to take on cases that can extend for months – sometimes years – and involve a considerable amount of nonlitigation activity. LBH could quite easily send a lawyer to a prison, pick up 'simple' cases requiring litigation and minimal work outside the courtroom, and be reimbursed for the litigation costs. But it does not do so, because this is not in line with a 'structural legal aid' approach to case selection. Further, there are other nonlitigation aspects of LBH's work not covered under the government scheme, such as community legal empowerment, strategic litigation at the Constitutional Court, and reporting cases to independent institutions like the Indonesian Ombudsman (*Ombudsman Republik Indonesia*) or Komnas HAM (Asfinawati, Azhar, and Munti 2019, 22). Although the scheme does cover some nonlitigation legal aid activities, the amounts available are small. For example, the total amount available for research is just Rp 2.0 million ($US135) per project.[60]

The government-funded legal aid system has provided LBH with a new important stream of funding, but it is yet to have much influence on LBH's strategies of cause lawyering. The new Indonesian system, unsurprisingly, is focused on expanding access to justice for individuals, and offers comparatively less support for political-activist versions of cause lawyering. If LBH were to have its work supported by government funding, it would need to practice a largely proceduralist version of cause lawyering. For most LBH cause lawyers, this is unacceptable. As long as LBH still has access to foreign donor funds, which provide it with enough freedom to practice grassroots versions of cause lawyering, it is likely to continue to see the government system as complementary at best.

Staffing and training

This chapter has so far focused on how the democratic transition affected key organisational factors at LBH, and the impact these factors have had on the forms of cause lawyering practiced. But there are other critical organisational factors influencing forms of cause lawyering that are only loosely connected to democratic change, such as the ability to recruit, train and retain capable staff. The recruitment process at LBH offices is conducted through an intensive process known as Kalabahu (*Karya Latihan Bantuan Hukum*, or Legal Aid Training). Kalabahu was affected by democratic change to the extent that funding challenges and management weaknesses in YLBHI – precipitated partly by the democratic transition – meant some regional offices did not implement Kalabahu for several years.[61] This has changed since Asfin was installed as leader. While it is true that YLBHI's improved financial position has allowed the organisation to pay greater attention to these institutional issues, strong leadership has also played a role. Under Asfin, YLBHI published a standardised Kalabahu curriculum to be implemented across the LBH network, the first time such an effort had been made in the organisation's 50-year history.[62]

The Kalabahu process varies slightly at each regional office, but generally, prospective participants must submit an application, sit an exam, submit a writing sample or pass an interview before they are selected as trainees (YLBHI 2018). This process can be extremely competitive. LBH Jakarta reportedly can have more than 300 applicants for the 50 positions it offers.[63] Throughout 2018–19, before the Covid-19 pandemic interfered with in-person activities, all of LBH's then 16 offices held Kalabahu training events, involving nearly 500 participants (YLBHI 2020, 1). Kalabahu typically runs for at least two weeks (in some offices it is more than a month), and involves 'in-class' and 'out-class' components. The in-class component involves training on LBH values, human rights, structural poverty and structural legal aid, social analysis, and aspects related to target communities, such as land and environmental law, religious freedom, gender and sexual orientation. The out-class component involves activities such as participatory action research, visits to communities and, usually, a period of living with a community (YLBHI 2018). The live-in component is designed primarily to encourage humility and solidarity with the poor and marginalised people LBH typically represents, but it also serves as an introduction to community organising. Eti Oktaviani, from LBH Semarang, explains the purpose of this live-in process:

> Sometimes there are LBH lawyers who think, 'I am a lawyer, I am on a higher level than the community'. We ask people like that to live in more often. We won't let them go to court. They have to be brought down to earth, to learn how to listen to community complaints. These are the principles we learn when we first start at LBH.[64]

The process aims to break down perceptions of the lawyer as a professional offering their services as a form of charity, and to promote less hierarchical interactions between lawyers and the community. LBH documents describe the importance of Kalabahu participants developing *keberpihakan* (literally partisanship, but in this case more accurately translated as 'concern for the rights of…'). This choice of language is illustrative. One of the key defining features of cause lawyering is identification with the causes of clients (Scheingold and Sarat 2004, 8–9). This live-in process is therefore designed in part to encourage participants to understand and identify with the causes of their clients. It is a key part of the process for them to become cause lawyers. Further, given the emphasis on working closely alongside clients and among social movements, the process encourages them to become cause lawyers of a particular type: grassroots cause lawyers.

Following the successful completion of Kalabahu, participants may then be invited to join the selection process to become an LBH volunteer. Typically, only a handful of Kalabahu participants are accepted as volunteers. Volunteers then work at their respective LBH office, usually spending up to a year rotating through the various divisions (labour, land, minority rights, women and children, research and documentation, and so on, depending on the office). If a volunteer's progress is deemed satisfactory, he or she may then be promoted to 'assistant legal aid lawyer' (*asisten pengabdi bantuan hukum*). As mentioned in Chapter 1, LBH uses the term *pengabdi bantuan hukum* (legal aid servant) to describe its legal aid lawyers. This is another method used to convey LBH values of humanitarianism or volunteerism to staff, and to differentiate LBH from commercial firms (Febry 2019; Isnur 2017). As LBH staff acknowledge, it is also an implicit recognition of the modest remuneration available to LBH lawyers.

The Kalabahu process therefore selects for, and encourages the development of, staff who are inclined toward social justice activism over conventional legal aid or policy oriented legal research. The lengthy training process (in Kalabahu and on the job) is an important strategy in inculcating shared values, and in introducing trainees to social movement strategies. To some extent, this process is designed to rectify the lack of a public interest perspective in most Indonesian law schools. And it is effective – there is little question of LBH lawyers' commitment to social justice once they have completed this extended training. Cause lawyering in Indonesia would look very different were it not for the longstanding influence of LBH's Kalabahu programme.

Another aspect of staffing to consider is LBH's ability to attract the right staff. Since the democratic transition, LBH is no longer one of the only outlets for progressive lawyers interested in social justice and policy change. For graduates of the top law schools, like University of Indonesia (UI), Gadja Mada University (UGM) and Padjadjaran University (Unpad), LBH is not the drawcard that it was under the New Order.[65] Graduates interested in using their legal skills to contribute to social change now have other options, like PSHK, LeIP, MaPPI, ICJR, or the Indonesian Centre for Environmental Law (ICEL), which are focused on legal research and policy advocacy. It is not hard to imagine how the middle- and upper-class graduates of the top law schools would find the more business-like environment in organisations like PSHK more appealing than the tired offices of LBH.

> These organisations seem to be more attractive for young people compared to shouting, demonstrating on the roads, or arguing in court [at LBH]. They can make a meaningful contribution to policy [in these other organisations] but the offices are cleaner, and the salary would be better too.[66]

The democratic transition and the expansion of civil society therefore provided different opportunities for cause lawyering. Although legal graduates who go on to careers in organisations like PSHK or LeIP would rarely spend time in court (unlike their colleagues at LBH), they are still practicing a form of cause lawyering. Yet these organisations are more focused on an elite/vanguard version of cause lawyering, one that views legal reform as capable of bringing about social change (Hilbink 2004, 673). These organisations have typically been more willing to collaborate with or work in closer proximity to state institutions. This contrasts with the oppositional, grassroots style of cause lawyering favoured by LBH. These discussions are a major focus of Chapter 6.

Women's representation and awareness of gender and sexuality

Another organisational factor that has been only partly affected by democratic change but remains a crucial element in consideration of the form of cause lawyering practiced by LBH is women's representation and gender equality. One of LBH's longstanding weaknesses is its masculine organisational culture, a problem that LBH staff and its donors recognise.[67] Indeed, structural legal aid's lack of attention to women's rights and gender equality was one of the reasons that led Nursyahbani Katjasungkana to leave LBH and set up LBH Apik in the 1990s.[68] In 2019, when the bulk of research for this book was conducted, there was only one other female director among the then-16 regional LBH offices, Ni Kadek Vany Primaliraning, the director of LBH Bali.[69] In 2020, across the entire YLBHI-LBH network, there were only 57 women among the 189 lawyers appointed to permanent staff status.[70] LBH offices strive for gender parity when conducting Kalabahu but they often struggle to retain female staff. When I visited LBH Surabaya, for example, the only female lawyer was an intern, who had yet to be appointed to permanent staff status (this has also since improved). Difficulties in retaining female staff may reflect low levels of female participation in the labour market in Indonesia. In 2013, 51.4% of Indonesian women over 15 participated in the labour market (barely up from 50.2% in 1990) – low by global standards. By contrast, the male labour participation rate was 84.2% (Cameron and Contreras Suarez 2017, 7). Yet other factors are likely at play in LBH's case. Its historical association with direct action tactics like street-level demonstrations and the also typically masculine labour and farmers' movements, as well as an organisational culture that has been tolerant of sexist language and jokes were mentioned as factors that may be behind low levels of female representation.

When Asfin and her leadership team developed a standardised Kalabahu curriculum, one of their key changes was to include compulsory modules on gender, sexuality and minority rights, in part with support from donors.[71] Senior YLBHI staff are personally committed to improving attention to gender equality and minority rights in the organisation. They also recognise that if LBH is to continue to attract donor funds and effectively criticise the government, it must not only respond to the "call of the times" but also practice what it preaches.[72] As Asfin explained:

> Gender, minority rights, it must be a perspective. We realise that LBH is very masculine. We have no legitimacy to speak to the state on these matters if we don't pay attention to these matters ourselves. If we are corrupt, how can we shout at the government about corruption? It is the same thing with gender and minority rights.[73]

Although there has been more attention to gender equality over recent years, for some LBH staff these efforts were still falling short. As Rezky Pratiwi, then based at LBH Makassar, explained:

> Students [in Kalabahu] don't get enough information on mainstreaming gender, gender equality, what kinds of behaviours are discriminatory, or how to make programmes gender responsive. These issues are very rarely discussed. The emphasis is on formal equality, equal rights to work and education, for example. But the special needs of women, sexist language, how to make situations safe, or discussions that are safe and friendly toward women, these issues are almost never discussed.[74]

YLBHI staff also spoke of an incident in which a regional LBH office was reluctant to take on a case involving the persecution of sexual and gender minorities, because of prejudice among local LBH staff. YLBHI senior staff reportedly had to visit the office twice and remind it of its responsibilities before it finally agreed to provide legal assistance. The staff members who handled the case were then teased by their colleagues for doing so.[75] Recent efforts to strengthen attention to gender and sexuality will take some time to have an impact. It is clear, however, that this aspect of organisational culture affects the way in which LBH cause lawyers approach their work. Historically, many LBH offices referred gender-based violence or women's rights cases on to other organisations outside the YLBHI network, such as LBH Apik, stating that such cases were not 'structural' in nature.[76] Similarly, the presence of a 'women's and children's rights' division in some LBH offices can mean that gender concerns are siloed in this division. As such, other divisions may fail to consider how forced evictions or land conflict, for example, affect men and women differently. Lack of sufficient gender awareness can also result in the exploitation of women on the frontline of protests, as they are considered less likely to be treated violently by law enforcement officials (see Chapter 5). Despite these problems, the increased focus on gender and sexuality has resulted in some noticeable changes. In 2020, for example, YLBHI hosted a high-profile three-day 'People's Forum' (*Sidang Rakyat*) on the importance of passing the bill on the crime of sexual violence (RUU TPKS) (CNN Indonesia 2020). This was a rare example of YLBHI attempting to take a leading role on a gender issue of national importance.

Research capacity and knowledge management

A final organisational challenge that LBH has faced since the democratic transition relates to research. Research products are a key part of cause lawyering organisations' policy advocacy work, and are important marketing tools for an organisation to shape its profile and legitimacy among donors, the government, broader civil society and the public. There is a common perception that YLBHI has experienced an "intellectual decline" since the late Soeharto period,[77] when scholar-activists such as Adnan Buyung Nasution and Todung Mulya Lubis were helping to shape national discourse on the rule of law and democratisation (Saptono 2012, 35–36). Given that YLBHI was one of the few outlets for progressive thought during the repressive New Order period, it is unfair to expect YLBHI to be driving discourse around democracy and human rights in a similar fashion now – times have changed. Nevertheless, there is a perception that YLBHI's research capacity is weaker than it once was, and that the organisation is not attracting or producing prominent thinkers in the same way.[78]

This decline can be partly explained by the matters of funding and donor priorities discussed above. Former YLBHI deputy Robertus Robet, now a prominent academic, complained that donors no longer wanted to provide funding for publishing books, contributing to a decline in the academic tradition at YLBHI over recent years.[79] Indeed, given the funding models and outcome orientation of international donor supported programmes, donors rarely want to support research unless it has a clear policy goal, and often do not want to fund the organisational overhead costs associated with producing research (Suryadarma, Pomeroy, and Tanuwidjaja 2011, 25).

In addition to funding concerns, the degree to which the organisational leadership prioritises research is also influential, and this is connected to the leader's vision of organisational identity. Former YLBHI leader Alvon (2012–16) reportedly firmly believed that YLBHI was not a research organisation.[80] YLBHI still produced a few research products under Patra M. Zen (2006–10), particularly related to the development of the 2011 Legal Aid Law, but there is little doubt YLBHI's research output has declined significantly since the mid-2000s. Alvon's position is completely understandable. YLBHI should not have to 'compete' with research and advocacy focused organisations in civil society like Elsam or PSHK. Organisations like these are staffed by dedicated researchers, and do not have the additional responsibility of service provision. Nonetheless, there are areas in which YLBHI could be making a much stronger contribution. YLBHI has an enormous wealth of case data, but it has struggled to compile and analyse it. Few other civil society organisations possess the same extent of first-hand knowledge of police and prosecutorial misconduct during arrest, detention, investigation and trial procedures. Yet this data remains underutilised in public advocacy efforts.

YLBHI's production of research and knowledge products received a significant boost with the appointment of former head of LBH Semarang and part-time lecturer Rakhma as Knowledge Management division head under Asfin. Over recent years, with Rakhma's encouragement, YLBHI has improved the quality of its annual reports, and published thematic reports and fact sheets on key human rights concerns. It has actively tried to gather data from regional offices, and play a larger role in synthesising and analysing data from the regions.[81] For example, LBH regional offices have collected data, mainly from the media, on local violations of freedom of expression and association which, as discussed, have been a major element of anxieties about declining democratic quality in Indonesia over the past five years. The public release of this data by YLBHI made a major splash in the media (Aditya 2021). While things are improving, the accessibility of the writing, formatting, graphic design and overall finish of YLBHI's knowledge products is still behind that of its peers in civil society. In any case, it would be overstating things to suggest that YLBHI's recent emphasis on research and knowledge production was somehow linked to democratic regression, or renewed donor support for research activities. Rather, it has been a conscious effort from the leadership team to address a weakness of the post-Soeharto era.

Conclusions

The collapse of the authoritarian New Order regime and transition to democracy should have provided the conditions for LBH to thrive, but it did not flourish after Soeharto stood down. There were four main reasons why LBH struggled in a more democratic Indonesia. Most of these factors (though not all) were directly connected to democratic change.

First, the transition to democracy forced LBH to reckon with its organisational identity. Much of its cause lawyering work, directed through the framework of 'structural legal aid', was focused on changing the structure of Soeharto's New Order state to a more democratic structure. With Soeharto and the New Order gone, LBH had to rethink its mandate in relation to the state and legal system. Many within LBH were not comfortable with more accommodative versions of cause lawyering focused on strengthening the state and, for some time, LBH struggled to define its role. The second critical factor was funding. During the authoritarian Soeharto years, international donors supported LBH's highly oppositional, grassroots approach to cause lawyering. But the democratic transition ushered in major changes, and donor funds flowed toward programmes that were geared toward strengthening Indonesia's new democratic institutions. LBH found there was less donor appetite for supporting the kind of oppositional pro-democracy advocacy that had become its trademark under Soeharto. A third, related factor was that the opening up of civil society also meant LBH no longer dominated civil society. New, specialised organisations emerged, some of which filled the advocacy space formerly taken up by LBH. Many of them were more prepared to engage directly with the government on reform. Crucially, these new organisations did not face the same competing demands of service delivery faced by LBH and could also dedicate more time to research and policy advocacy. Finally, repeated leadership tensions and organisational management concerns, related partly to the factors described above, resulted in losses of staff and further discouraged donors, particularly in the central umbrella organisation YLBHI. Funding and management challenges also affected LBH's ability to recruit and train new staff and conduct research to supplement or strengthen other cause lawyering strategies.

The experience of LBH in the post-New Order period highlights the critical role of cause lawyering organisations' internal organisational characteristics and capacity in supporting cause lawyering work. LBH's struggles following the fall of Soeharto illustrate the importance of looking beyond formal legal structures when considering processes of legal mobilisation and democratic change, and of paying closer attention to these internal organisational factors. It is true that rights advocacy organisations are recognised as an essential element of the "support structure" critical for legal mobilisation (Epp 1998), yet their internal organisational workings are seldom, if ever, discussed in scholarly accounts of cause lawyering.

The organisational challenges outlined in this chapter resulted in a weaker YLBHI, with subsequent impacts on its strategies of cause lawyering. It is too simplistic, however, to depict the story of LBH in the post-New Order period as a straightforward story of decline. The transition to democracy resulted in stronger legal protections of rights, new legal forums in which to claim these rights, and improved judicial independence. These developments provided new opportunities for legal mobilisation, and LBH has explored these new legal opportunities, with varied success. These strategies are the focus of the next chapter.

Notes

1 Interview, Febi Yonesta, September 2019.
2 Interview, Herlambang P. Wiratraman, November 2019.
3 Here he is referring to organisations like ICW and KRHN.
4 Interview, Febi Yonesta, September 2019.
5 A year later, he made similar calls for LBH to stop being so "radical" (Hukum Online 2002).
6 Interview, Hendardi, October 2019.
7 In line with his commitment to promoting professionalism, in his later years Buyung tried and failed to establish a united and professional bar association, which he long believed was a major missing piece in the struggle for the rule of law in Indonesia (Lindsey 2015). Indonesia's legal profession remains divided, with repeated tension and conflict between the country's multiple bar associations, hampering professionalism and contributing to public distrust (Butt and Lindsey 2018, 110).
8 *Pedoman Pokok Nilai-Nilai Perjuangan YLBHI dan Kode Etik Pengabdi Bantuan Hukum.*
9 At the time of writing, LBH West Kalimantan had been in operation for less than one year, and was still considered a 'project base'. It was expected to achieve formal status following a year of satisfactory operations.
10 Interview, Poengky Indarti, October 2019.
11 State Intelligence Agency (*Badan Intelijen Negara*, BIN) officials were implicated in planning his murder but were never convicted. A Garuda Indonesia pilot and suspected BIN agent, Pollycarpus Budhi Priyanto, was eventually imprisoned for his role in the case (Wahyuningroem 2014; Tempo 2014).
12 HTI was banned by the government in 2017 (see discussion in Introduction to Part Two).
13 Tomy Winata is a notorious businessman with longstanding links to the Indonesian military and political elite. His massive Artha Graha Network has interests in property, plantations (including oil palm), hotels, banking and more (Braithwaite, Charlesworth, and Soares 2012, 239). Winata is also alleged to be a major figure in organised crime in Indonesia, and often faces rumours of involvement in activities including prostitution, gambling and drugs (Chua 2008, 131).
14 Interviews, confidential, 2019; 2020.
15 Interviews, confidential, 2019; 2020.
16 Despite its name, LBH Masyarakat is not a member of the YLBHI network.
17 Interviews, Yasmin Purba, December 2019; Tim Lindsey, July 2020.
18 Interviews, confidential, 2019; 2020.
19 Interviews, confidential, 2019.
20 Interviews, confidential, 2019.
21 LBH internal regulations allow LBH senior staff to take on paid work separate to LBH for the survival of the organisation. Some directors took on this paid work, in addition to pro bono work at LBH, to ensure their offices could survive.
22 Interview, confidential, 2020.
23 Interviews, confidential, 2019; 2020.
24 Interviews, confidential, 2023.

25 Previously, candidates were required to be existing senior staff.
26 Asfin is well known for representing Lia Eden (Lia Aminuddin), the leader of a new religious movement, who was convicted for blasphemy in 2006 and 2009 (Makin 2016, 17) and Tajul Muluk, a Shi'a Muslim convicted of blasphemy in 2012 (BBC Indonesia 2012).
27 Interviews, Sandra Hamid, June 2019; Mohammad Doddy Kusadrianto, July 2019; Nurkholis Hidayat, September 2019; Ricky Gunawan, December 2019; Robertus Robet, April 2020.
28 Interviews, Nursyahbani Katjasungkana, October 2019; Siti Rakhma Mary Herwati, August 2019.
29 Interviews, Asfinawati, December 2019; August 2021; Renata Arianingtyas, August 2019; August 2021.
30 Interview, Asfinawati, December 2019.
31 I look further at this convening role of LBH in Chapter 7.
32 Interview, confidential, 2019.
33 It was also a move that suggested an awareness of past leadership turmoil, under the New Order, when broader civil society demanded a say in the selection of the YLBHI leader.
34 Some studies suggest Indonesian CSOs derive about 85–90% of their funding from foreign donors (Davis 2015, 4).
35 Interview, confidential, 2019; Saraswati 2003. At the same time, however, Oxfam Novib was facing its own financial constraints, and increasing pressure in the Netherlands to demonstrate results for expenditures abroad (Scheffer and Benning 2018, 11).
36 Interviews, confidential, 2019.
37 Interview, Robertus Robet, April 2020.
38 Adjusted for inflation, Rp 16.2 billion in 1999 is equivalent to about Rp 66.1 billion (US$4.1 million) in 2024 and Rp 4.97 billion in 2003 is equivalent to about Rp 14.7 billion in 2024 (US$910,500).
39 Figures from Munarman et al. (2004, 53).
40 Interviews, Febi Yonesta, August 2019; Lasma Natalia, September 2019; Eti Oktaviani, November 2019.
41 Interview, Hendardi, October 2019.
42 One recent exception is the USAID-funded MADANI project (2019–24), which includes a major civil society strengthening element. Notably, it also has a strong focus on facilitating collaboration between CSOs and local governments on local policy issues (USAID MADANI 2021).
43 Interview, Arif Maulana, December 2019.
44 Interview, Febi Yonesta, September 2019.
45 Interview, Fanti Yusnita, July 2020; YLBHI annual reports.
46 Interview, Asfinawati, 2021. LBH's legal empowerment work is discussed further in Chapter 5. Broadly, it encompasses efforts to empower poor or marginalised individuals and communities to use the law to assert their rights or interests (Domingo and O'Neil 2014, 4).
47 Interviews, confidential, 2021; 2023.
48 Interview, confidential, 2023.
49 Interviews, confidential, 2023.
50 Interview, Febi Yonesta, August 2021.
51 LBH Jakarta (@LBH Jakarta) "Menjelang #50tahunLBHJakarta mari kita terlibat dalam #10000untukKeadilan utk layanan bantuan hukum gratis bagi masyarakat miskin, buta hukum dan tertindas…" [Approaching #LBHJakarta50-Years Get Involved in #10000forJustice to Support Free Legal Services for the Poor, Legally Ignorant and Oppressed…] 15 March 2021, 18:20. Tweet.
52 Interview, Patrick Burgess, May 2020.

53 Interview, Febi Yonesta, December 2019.
54 Interviews, Arip Yogiawan, October 2019; Eti Oktaviani, November 2019; Haswandy Andy Mas, November 2019; Mohamad Soleh, November 2019.
55 Interviews, Febi Yonesta, December 2019; Haswandy Andy Mas, November 2019.
56 Interview, Febi Yonesta, December 2019.
57 Interviews, Mohammad Doddy Kusadrianto, July 2019; Febi Yonesta, December 2019.
58 Interview, Haswandy Andy Mas, Salman Azis, November 2019.
59 In 2020, the World Bank estimated there were about 25.1 million Indonesians living below the poverty line. Of course, not all these people would have needed legal services over the 2019–21 period, but a budget of Rp 53 billion is equivalent to just Rp 2,100 (AU$0.20) for each poor person for the three-year period. Some other government ministries have budget items for legal aid and advocacy, but it is difficult to get accurate details on amounts and expenditure.
60 See Law and Human Rights Ministry Regulation No. M.HH-01.HN.03.03 of 2021 on Payment Amounts for Litigation and Nonlitigation Legal Aid.
61 Interviews, Renata Arianingtyas, August 2019; Eti Oktaviani, November 2019.
62 The basic Kalabahu syllabus is divided into 17 units: 1) structural poverty; 2) militarism; 3) identity politics and impacts on injustice; 4) LBH history and LBH lawyers; 5) rule of law and democracy; 6) structural legal aid; 7) introduction to human rights; 8) social analysis; 9) community organising; 10) advocacy strategies; 11) participatory action research; 12) economic, social and cultural rights; 13) agrarian law; 14) Indigenous Peoples' rights; 15) the right to religious freedom; 16) gender equality, sexual orientation and gender identity; and 17) campaigning (Putri et al., n.d.).
63 Interview, Alghiffari Aqsa, November 2019.
64 Interview, Eti Oktaviani, November 2019.
65 Interviews, Ajeng Wahyuni, August 2019; Windu Kisworo, September 2019.
66 Interview, Windu Kisworo, September 2019.
67 Interviews, Renata Arianingtyas, August 2019; Ajeng Wahyuni, August 2019; Rezky Pratiwi, November 2019; Asfinawati, December 2019.
68 Interview, Nursyahbani Katjasungkana, October 2019. Recall that despite the similar names, LBH Apik is not connected to the YLBHI-LBH network.
69 This situation has since changed. In early 2020, Lasma Natalia, a woman, was selected as director of LBH Bandung for the 2020–24 period, and later that year, Eti Oktaviani, another woman, was selected as the new director of LBH Semarang for the 2020–24 period.
70 This includes legal aid lawyers (*pengabdi bantuan hukum*) and assistant legal aid lawyers (*asisten pengabdi bantuan hukum*). The proportion is higher among volunteers/interns (*volunter/pemagang*), with women comprising 37 of 78 (47%) volunteers/interns (YLBHI 2021a, 10).
71 Interviews, Renata Arianingtyas, Ajeng Wahyuni, August 2019.
72 Interview, Asfinawati, December 2019.
73 Interview, Asfinawati, December 2019.
74 Interview, Rezky Pratiwi, November 2019.
75 Interviews, confidential, 2019.
76 Interview, Renata Arianingtyas, August 2019.
77 Interview, Indro Sugianto, September 2019.
78 Interview, Abdul Fatah, November 2019.
79 Interview, Robertus Robet, April 2020.
80 Interviews, Siti Rakhma Mary Herwati, August 2019; Carolina S. Martha, September 2019.
81 Interviews, Siti Rakhma Mary Herwati, August 2019; Nurkholis Hidayat, September 2019.

References

Aditya, Nicholas Ryan. 2021. "Catatan YLBHI, 351 Kasus Pelanggaran Hak dan Kebebasan Sipil Terjadi Selama 2020." *Kompas.com*, 26 January 2021. https://nasional.kompas.com/read/2021/01/26/19193351/catatan-ylbhi-351-kasus-pelanggaran-hak-dan-kebebasan-sipil-terjadi-selama.

Ahmed, Shamima. 2021. *Effective Nonprofit Management: Context, Concepts, and Competencies.* 2nd Edition. New York: Routledge. https://doi.org/10.4324/9781003240150-8.

Amagoh, Francis. 2015. "Improving the Credibility and Effectiveness of Non-Governmental Organizations." *Progress in Development Studies* 15 (3): 221–239. https://doi.org/10.1177/1464993415578979.

Antara. 2006. "Hizbut Tahrir Indonesia Stages Rally Outside US Embassy." *Antara*, 5 March 2006.

Antlöv, Hans, Derick W. Brinkerhoff, and Elke Rapp. 2010. "Civil Society Capacity Building for Democratic Reform: Experience and Lessons from Indonesia." *VOLUNTAS: International Journal of Voluntary and Nonprofit Organizations* 21 (3): 417–439. https://doi.org/10.1007/s11266-010-9140-x.

Antlöv, Hans, Rustam Ibrahim, and Peter van Tuijl. 2006. "NGO Governance and Accountability in Indonesia: Challenges in a Newly Democratizing Country." In *NGO Accountability: Politics, Principles and Innovations*, edited by Lisa Jordan, Peter van Tuijl, and Michael Edwards, 147–163. London: Earthscan.

Ardiansyah, Arif. 2006. "Lembaga Bantuan Hukum Daerah Persoalkan Sumbangan Tomi Winata." *Tempo.co*, 15 January 2006. https://nasional.tempo.co/read/72318/lembaga-bantuan-hukum-daerah-persoalkan-sumbangan-tomi-winata.

Asfinawati. 2017. "Bantuan Hukum Struktural: Sejarah, Teori, dan Pembaruan." *Buletin Bantuan Hukum* 2: 69–99.

Asfinawati, Ajeng Larasati, Dio Azhar, and Ratna Batara Munti. 2019. "*Perluasan Akses Keadilan Melalui Optimalisasi Layanan Bantuan Hukum yang Berkualitas.*" Rekomendasi Konferensi Nasional Bantuan Hukum. Jakarta: YLBHI, ILRC, Asosiasi LBH Apik Indonesia, LBH Jakarta, LBH Masyarakat, LBH Apik Jakarta, MaPPI, PBHI. https://ylbhi.or.id/bibliografi/konferensi-nasional-bantuan-hukum-perluasan-akses-keadilan-melalui-optimalisasi-layanan-bantuan-hukum-yang-berkualitas/.

Aspinall, Edward. 2004. "Indonesia: Transformation of Civil Society and Democratic Breakthrough." In *Civil Society and Political Change in Asia: Expanding and Contracting Democratic Space*, edited by Muthiah Alagappa, 61–96. Stanford, California: Stanford University Press. https://doi.org/10.1515/9780804767545-008.

Aspinall, Edward. 2005. *Opposing Suharto: Compromise, Resistance, and Regime Change in Indonesia.* East-West Center Series on Contemporary Issues in Asia and the Pacific. Stanford, California: Stanford University Press. https://doi.org/10.1515/9780804767316.

Aspinall, Edward. 2010. "*Assessing Democracy Assistance: Indonesia.*" FRIDE Project Report: Assessing Democracy Assistance. FRIDE.

BBC. 1999. "Former Army Chief's Jakarta Meeting Under Scrutiny." *BBC Monitoring Service: Asia-Pacific*, 14 December 1999.

BBC Indonesia. 2012. "Hukuman Tajul Muluk Menjadi Empat Tahun Penjara." *BBC Indonesia*, 21 September 2012. https://www.bbc.com/indonesia/berita_indonesia/2012/09/120921_vonistajul.

Bedner, Adriaan, and Jacqueline AC Vel. 2010. "An Analytical Framework for Empirical Research on Access to Justice." *Law, Social Justice and Global Development Journal* 15: 1–29.

Braithwaite, John, Hilary Charlesworth, and Adérito Soares. 2012. *Networked Governance of Freedom and Tyranny: Peace in Timor-Leste*. Canberrra: ANU Press. https://doi.org/10.22459/NGFT.03.2012.

Butt, Simon. 2015. *The Constitutional Court and Democracy in Indonesia*. Leiden; Boston: Brill Nijhoff. https://doi.org/10.1163/9789004250598.

Butt, Simon, and Tim Lindsey. 2018. *Indonesian Law*. Oxford; New York: Oxford University Press. https://doi.org/10.1093/oso/9780199677740.001.0001.

Cameron, Lisa, and Diana Contreras Suarez. 2017. "*Women's Economic Participation in Indonesia: A Study of Gender Inequality in Employment, Entrepreneurship, and Key Enablers for Change*." The Australia-Indonesia Partnership for Economic Governance and the Australian Department of Foreign Affairs and Trade.

Chua, Christian. 2008. *Chinese Big Business in Indonesia: The State of Capital*. Routledge Contemporary Southeast Asia Series. London and New York: Routledge. https://doi.org/10.4324/9780203931097.

Civil Society Committee for KUHAP Reform. 2014. "*Press Release: Perubahan KUHAP: Jalan Baru Mencegah Praktik Penyiksaan*," 26 June 2014. https://bantuanhukum.or.id/perubahan-kuhap-jalan-baru-mencegah-praktik-penyiksaan/.

CNN Indonesia. 2020. "LBH Gelar Sidang Rakyat, Desak Pemerintah Sahkan RUU PKS." *CNN Indonesia*, 2 October 2020. https://www.cnnindonesia.com/nasional/20201002185038-32-553836/lbh-gelar-sidang-rakyat-desak-pemerintah-sahkan-ruu-pks.

Commission to Investigate Human Rights Violations in East Timor (KPP-HAM). 2000. "*Executive Summary Report on the Investigation of Human Rights Violations in East Timor*." Executive Summary. Jakarta: Commission to Investigate Human Rights Violations in East Timor.

Cox, Marcus, Emele Duituturaga, and Nur Sholikin. 2012. "*Indonesia Case Study: Evaluation of Australian Law and Justice Assistance*." Canberra, Australia: Australian Agency for International Development (AusAID), Office of Development Effectiveness.

Cummings, Scott L. 2008. "The Internationalization of Public Interest Law." *Duke Law Journal* 57 (4): 891–1036.

Davis, Ben. 2015. "*Financial Sustainability and Funding Diversification: The Challenge for Indonesian NGOs*." Cardno, prepared for the Department of Foreign Affairs and Trade.

Detik. 2006. "10 Personel BP YLBHI Mundur." *Detik*, 30 May 2006. https://news.detik.com/berita/d-605301/10-personel-bp-ylbhi-mundur.

Detik. 2010. "Erna Ratnaningsih Jadi Ketua YLBHI Gantikan Patra M Zein [sic]." *Detik*, 19 July 2010. https://news.detik.com/berita/d-1402193/erna-ratnaningsih-jadi-ketua-ylbhi-gantikan-patra-m-zein.

Domingo, Pilar, and Tam O'Neil. 2014. *The Politics of Legal Empowerment: Legal Mobilisation Strategies and Implications for Development*. London: Overseas Development Institute.

Ellmann, Stephen. 1998. "Cause Lawyering in the Third World." In *Cause Lawyering: Political Commitments and Professional Responsibilities*, edited by Austin Sarat and Stuart A.Scheingold, 349–430. New York; Oxford: Oxford University Press. https://doi.org/10.1093/oso/9780195113198.003.0012.

Elnizar, Norman Edwin. 2019. "Pemerintah Sediakan 53 Miliar untuk Bantuan Hukum Masyarakat Marginal 2019–21." *Hukum Online*, 7 January 2019. https://www.hukumonline.com/berita/baca/lt5c33123bb3929/pemerintah-sediakan-53-miliar-untuk-bantuan-hukum-masyarakat-marginal-2019-2021/.

England, Vaudine. 1999. "Rights Lawyer Heads Defence of Generals." *South China Morning Post*, 18 December 1999.

Epp, Charles R. 1998. *The Rights Revolution: Lawyers, Activists, and Supreme Courts in Comparative Perspective*. Chicago: University of Chicago Press. https://doi.org/10.7208/chicago/9780226772424.001.0001.

Febry, Pratiwi. 2019. "Rekrutmen Asisten Pengabdi Bantuan Hukum LBH Jakarta." *LBH Jakarta*, 8 May 2019. https://www.bantuanhukum.or.id/web/rekrutmen-asisten-pengabdi-bantuan-hukum-lbh-jakarta/.

Fitzpatrick, Stephen. 2008. "Rebel Lawyer Turns Himself In." *The Australian*, 11 June 2008.

Ford Foundation. 2003. *Celebrating Indonesia: Fifty Years with the Ford Foundation, 1953–2003*. Jakarta: Ford Foundation.

Fowler, Alan. 1997. *Striking a Balance: A Guide to Enhancing the Effectiveness of Non-Governmental Organisations in International Development*. London and New York: Earthscan. https://doi.org/10.4324/9781315070735.

Fowler, Alan. 2000. *The Virtuous Spiral: A Guide to Sustainability for NGOs in International Development*. London, United Kingdom; Sterling, USA: Earthscan. https://doi.org/10.4324/9781315071374.

Hadiwinata, Bob S. 2003. *The Politics of NGOs in Indonesia: Developing Democracy and Managing a Movement*. London and New York: RoutledgeCurzon. https://doi.org/10.4324/9780203200124.

Hailey, John, and Rick James. 2004. "'Trees Die from the Top': International Perspectives on NGO Leadership Development." *VOLUNTAS: International Journal of Voluntary and Nonprofit Organizations* 15 (4): 343–353. https://doi.org/10.1007/s11266-004-1236-8.

Hilbink, Thomas M. 2004. "You Know the Type: Categories of Cause Lawyering." *Law & Social Inquiry* 29 (3): 657–698. https://doi.org/10.1086/430155.

Hukum Online. 2001. "YLBHI, Lokomotif Yang Kehilangan Roh Demokrasi." *Hukum Online*, 6 December 2001. https://www.hukumonline.com/berita/baca/hol4417/ylbhi-lokomotif-yang-kehilangan-roh-demokrasi—.

Hukum Online. 2002. "Perwira TNI Mulai Disidang, Adnan Buyung Mundur dari Tim." *Hukum Online*, 19 March 2002. https://hukumonline.com/berita/a/font-size1-colorff0000bsidang-pengadilan-hambfontbrperwira-tni-mulai-disidang-adnan-buyung-mundur-dari-tim-hol5122/.

Hukum Online. 2007. "Air Mata Perpisahan di YLBHI." *Hukum Online*, 16 November 2007. https://www.hukumonline.com/berita/baca/hol18000/air-mata-perpisahan-di-ylbhi/.

Human Rights Watch. 2017. "*Justice Denied for Slain Indonesian Rights Activist Munir*," 3 September 2017. https://www.hrw.org/news/2017/09/03/justice-denied-slain-indonesian-rights-activist-munir.

Isnur, Muhamad. 2017. "Kami Menyebutnya 'Pengabdian.'" *Bantuan Hukum* 2: 15–23.

Klug, Heinz. 2001. "Local Advocacy, Global Engagement: The Impact of Land Claims Advocacy on the Recognition of Property Rights in the South African Constitution." In *Cause Lawyering and the State in a Global Era*, edited by Austin

Sarat and Stuart A. Scheingold, 264–286. New York: Oxford University Press. https://doi.org/10.1093/0195141172.003.0010.

Kompas.com. 2015. “Siapa Bambang Widjojanto?” *Kompas.com*, 23 January 2015. https://nasional.kompas.com/read/2015/01/23/10303131/Siapa.Bambang.Widjojanto.

KUBAH. 2010. *Bantuan Hukum Dan Pembentukan Undang-Undang Bantuan Hukum: Pertanyaan Dan Jawaban*. Jakarta: Koalisi Masyarakat Sipil Untuk Undang-Undang Bantuan Hukum (KUBAH).

Lassa, Jonathan, and Dominggus Elcid Li. 2015. “*NGO Networks and the Future of NGO Sustainability in Indonesia*.” Cardno, prepared for the Department of Foreign Affairs and Trade.

LBH-YLBHI. 2020. *Panduan Bantuan Hukum Struktural* I. Jakarta: YLBHI.

Lewis, David. 2014. *Non-Governmental Organizations, Management and Development*. Third Edition. London and New York: Routledge. https://doi.org/10.4324/9780203591185.

Lindsey, Tim. 2015. “Farewell Adnan Buyung Nasution.” *Indonesia at Melbourne*, 5 October 2015. https://indonesiaatmelbourne.unimelb.edu.au/farewell-adnan-buyung-nasution-2/.

Lindsey, Tim, and Melissa Crouch. 2013. “Cause Lawyers in Indonesia: A House Divided.” *Wisconsin International Law Journal* 31 (3): 620–645.

Makin, Al. 2016. *Challenging Islamic Orthodoxy: Accounts of Lia Eden and Other Prophets in Indonesia*. Switzerland: Springer International Publishing. https://doi.org/10.1007/978-3-319-38978-3.

McGlynn Scanlon, Megan. 2012. “*NGO Sector Review: Phase I Findings*.” Findings Report: Final Draft. STATT, prepared for AusAID’s Knowledge Sector Unit. https://dfat.gov.au/about-us/publications/Documents/indo-ks15-ngo-sector-review-phase1.pdf.

McGlynn Scanlon, Megan, and Tuti Alawiyah. 2015. “*The NGO Sector in Indonesia: Context, Concepts and an Updated Profile*.” Cardno, prepared for the Department of Foreign Affairs and Trade.

Meili, Stephen. 2009. “Staying Alive: Public Interest Law in Contemporary Latin America.” *International Review of Constitutionalism* 9 (1): 43–71.

Mietzner, Marcus. 2021. *Democratic Deconsolidation in Southeast Asia*. Elements in Politics and Society in Southeast Asia. Cambridge: Cambridge University Press. https://doi.org/10.1017/9781108677080.

Musthofid. 2006. “Patra M. Zen: Facing a Test of His Independence.” *The Jakarta Post*, 19 September 2006.

Prizzon, Annalisa, Andrew Rogerson, and Maria Ana Jalles d’Orey. 2017. “*Moving Away From Aid? The Case of Indonesia*.” London: Overseas Development Institute.

Putri, April Pattiselanno, Arip Yogiawan, Asfinawati, Era Purnama Sari, Febi Yonesta, Muhamad Isnur, Niccolo Attar, and Siti Rakhma Mary Herwati. n.d. “*Modul Kalabahu Dasar LBH Indonesia*.” YLBHI.

Riana, Friski. 2016. “Berikut Perubahan Mendasar Pemilihan Ketua YLBHI 2017–2021.” *Tempo.co*, 6 December 2016. https://nasional.tempo.co/read/825909/berikut-perubahan-mendasar-pemilihan-ketua-ylbhi-2017-2021.

Rizaldi, Muhammad. 2019. “Why Legal Aid Is Not Working in Indonesia.” *Indonesia at Melbourne*, 15 May 2019. https://indonesiaatmelbourne.unimelb.edu.au/why-legal-aid-is-not-working-in-indonesia/.

Saptono, Irawan. 2012. “Kisah Panjang Gerakan Bantuan Hukum.” In *Verboden Voor Honden En Inlanders dan Lahirlah LBH: Catatan 40 Tahun Pasang Surut Keadilan*, edited by Irawan Saptono and Tedjabayu, 1–112. Jakarta: YLBHI.

Saraswati, Muninggar Sri. 2003. "Alumni Urge YLBHI to Introspect and Change." *The Jakarta Post*, 7 July 2003.

Scheffer, Rudolf, and Esther Benning. 2018. "*Oxfam Novib and Partnerships: A Historical Perspective.*" Oxfam Research Reports. Oxfam Novib.

Scheingold, Stuart A., and Austin Sarat. 2004. *Something to Believe In: Politics, Professionalism, and Cause Lawyering*. Stanford, California: Stanford University Press.

Siboro, Tiarma. 2006a. "Leading Legal Aid Group Head Ousted Over Internal Dispute." *The Jakarta Post*, 11 April 2006.

Siboro, Tiarma. 2006b. "Resignations Leave Legal Aid Body on Brink of Collapse." *The Jakarta Post*, 31 May 2006.

Simanjuntak, Tertiani ZB. 2001. "'Loco of Democracy' Shuns Democracy." *The Jakarta Post*, 7 December 2001.

Simanjuntak, Tertiani ZB. 2002. "Munarman Elected YLBHI Top Executive." *The Jakarta Post*, 25 September 2002.

Smillie, Ian, and John Hailey. 2001. *Managing for Change: Leadership, Strategy and Management in Asian NGOs*. London and New York: Earthscan. https://doi.org/10.4324/9781315071749.

Suryadarma, Daniel, Jacqueline Pomeroy, and Sunny Tanuwidjaja. 2011. "*Economic Factors Underpinning Constraints in Indonesia's Knowledge Sector.*" Jakarta: Australian Agency for International Development (AusAID).

Tempo. 2014. "Pollycarpus Never Dismisses Allegations of Being BIN Agent." *Tempo.co*, 12 December 2014. https://en.tempo.co/read/628048/pollycarpus-never-dismisses-allegations-of-being-bin-agent.

Tempo Magazine. 2011. "Menyelamatkan LBH Jakarta." *Tempo Magazine*, 11 July 2011. https://majalah.tempo.co/read/pokok-dan-tokoh/137228/farhan-dan-najwa-shihabmenyelamatkan-lbh-jakarta.

The Jakarta Post. 1998a. "Activist Gives Public Account of Abduction." *The Jakarta Post*, 5 May 1998.

The Jakarta Post. 1998b. "Activist Gives Grisly Details of Torture." *The Jakarta Post*, 13 May 1998.

The Jakarta Post. 1998c. "The Gallery of Thirteen Activists Who Are Still Missing." *The Jakarta Post*, 2 August 1998.

The Jakarta Post. 1998d. "Activists Call for Excavation of All Suspected Mass Graves." *The Jakarta Post*, 31 August 1998.

The Jakarta Post. 1998e. "Parents of Dead Students Seek Justice." *The Jakarta Post*, 1 December 1998.

The Jakarta Post. 1999a. "Aceh Death Toll Hits 38." *The Jakarta Post*, 6 May 1999.

The Jakarta Post. 1999b. "Buyung Faces Dismissal from YLBHI's Board of Trustees." *The Jakarta Post*, 21 December 1999.

The Jakarta Post. 2001. "LBH to Lose Munir, Widjojanto." *The Jakarta Post*, 5 December 2001.

USAID MADANI. 2021. "*Strengthening Civil Society Organizations (Fact Sheet).*" USAID.

Wahyuningroem, Sri Lestari. 2014. "Solving Munir's Murder Case, a Test for Indonesia's President-Elect." *The Conversation*, 9 September 2014. http://theconversation.com/solving-munirs-murder-case-a-test-for-indonesias-president-elect-31293.

Walhi. 2017. "*Press Release: Darurat Demokrasi! Penyerangan Terhadap YLBHI-LBH Jakarta, Ancaman Serius Bagi Demokrasi Di Indonesia,*" 16 September 2017.

http://www.walhi.or.id/darurat-demokrasi-penyerangan-terhadap-ylbhi-lbh-jakarta-ancaman-serius-bagi-demokrasi-di-indonesia.

Wiratraman, Herlambang. 2006. "*Good Governance and Legal Reform in Indonesia.*" Master of Arts, Thailand: Mahidol University.

World Bank. 2020. "*The World Bank in Indonesia: Overview.*" World Bank. 7 April 2020. https://www.worldbank.org/en/country/indonesia/overview.

Yadav, Vineeta, and Bumba Mukherjee. 2014. *Democracy, Electoral Systems, and Judicial Empowerment in Developing Countries.* Ann Arbor, USA: University of Michigan Press. https://doi.org/10.3998/mpub.5037026.

YLBHI. 2018. *Panduan Penyelenggaraan Kalabahu LBH-YLBHI.* YLBHI.

YLBHI. 2020. *Reformasi Dikorupsi Oligarki: Laporan Hukum dan HAM YLBHI Tahun 2019.* Jakarta: YLBHI. https://ylbhi.or.id/bibliografi/laporan-tahunan/laporan-hukum-dan-ham-ylbhi-tahun-2019-reformasi-dikorupsi-oligarki/.

YLBHI. 2021a. *Otoritarian Dan Oligarki Membuncah Di Tengah Pandemi: Laporan Hukum Dan Hak Asasi Manusia Yayasan LBH Indonesia 2020.* Jakarta: YLBHI.

YLBHI. dir. 2021b. *Debat Terbuka Calon Ketua Umum Pengurus YLBHI 2022–2026.* https://www.youtube.com/watch?v=BlCQ5cvTfpk.

Yon, Kwan Men, and Simon Hearn. 2016. "*Laying the Foundations of Good Governance in Indonesia's Judiciary: A Case Study as Part of an Evaluation of the Australia Indonesia Partnership for Justice.*" London: Overseas Development Institute.

Zen, A. Patra M., Robertus Robet, Daniel Hutagalung, Daniel Pandjaitan, Ikravany Hilman, and Rita Novella. 2004. "*Public Report: One Year Period Yayasan LBH Indonesia the Board of Director (October 2002-December 2003).*" Annual Report. Jakarta: Yayasan Lembaga Bantuan Hukum Indonesia.

4 Mobilising the law

New opportunities, new strategies

Cause lawyers from LBH have a complicated relationship with the formal legal system. They are dependent on it and attempt to use it strategically, but they also are wary of its deficiencies – of which there are many, despite the changes of the post-Soeharto years. This chapter focuses on the litigation strategies used by LBH to advance their causes, looking at how LBH cause lawyers' approach to formal legal strategies has been affected by democratic reform and regression.

Under the New Order, the courts were subservient to the executive. Weaknesses in the rule of law and state repression led to an extremely pessimistic attitude among LBH lawyers about the possibility of achieving change through the courts, prompting LBH to develop a deeply oppositional approach to cause lawyering involving grassroots organising and direct action tactics. On the occasions when litigation strategies were pursued, the courts were used as a public, and mostly safe, site to critique New Order rule. The democratic transition that began in 1998 involved structural reforms that provided opportunities for new and different strategies of cause lawyering. As discussed, judicial independence was vastly improved, a broad array of rights protections were introduced into the amended constitution, and a Constitutional Court was established, which readily granted standing to civil society activists. These changes provided LBH with new legal structures for experimentation with strategic impact litigation and the deployment of tactics similar to those commonly associated with cause lawyers in liberal democratic contexts.

This chapter comprises five case studies examining different legal strategies undertaken by LBH in the post-authoritarian era. After providing a brief overview of the main types of cases seen across the LBH offices, I will examine an example of individual case defence. LBH's defence of leaders of the minority religious community Gafatar is a cause lawyering strategy that has changed little since the New Order years.[1] I then examine four examples of strategic impact litigation, an approach that was only possible following the fall of Soeharto, and one commonly associated with elite/vanguard cause lawyering. I describe LBH's challenges to book banning and blasphemy provisions at the Constitutional Court, its use of the administrative courts to

DOI: 10.4324/9781003486978-7

challenge an internet shutdown in Papua, and its citizen lawsuit against air pollution in Jakarta. The chapter also examines recent attempts by the government to undermine the integrity of the Constitutional Court, and reflects on how these and other attacks on Indonesia's democracy have affected LBH's approach to litigation strategies. The cases examined in this chapter serve to demonstrate how the form that cause lawyering takes is closely related to the quality of democracy. While there was great enthusiasm among LBH lawyers for engaging in strategic impact litigation in the years when Indonesian democracy was at its peak, democratic decline has led to much more ambivalent attitudes toward courtroom litigation.

Legal strategies: An overview

At the fall of the New Order, the Supreme Court was regarded as corrupt, poorly skilled and compliant (Pompe 2005, 471–72). Decades of marginalisation, poor educational standards, low wages, limited institutional support and an absence of effective performance evaluation processes contributed to a judiciary that was not only lacking in status but also largely incompetent (Pompe 2005, 472–73; Lindsey 2004, 21). The Supreme Court therefore had much to gain from democratisation, but in the first few years of *reformasi* it was highly resistant to reform and refused to acknowledge responsibility for its weaknesses (Pompe 2005, 472).

Reformasi did result in several important changes. The 'One Roof' reforms, which transferred authority over the judicial system from the government to the Supreme Court, have been considered largely effective (Butt and Lindsey 2018, 83). The recruitment of non-career judges, in particular Chief Justice Bagir Manan (2001–08), helped to drive reform, including through the publishing of the reform 'blueprints' with civil society and the establishment of the Judicial Reform Team Office (Pompe 2005, 475). A new detailed blueprint was published in 2010, with greater Supreme Court ownership of the process, covering the 2010–35 period (Yon and Hearn 2016, 10). There have been notable improvements in transparency, with millions of court decisions uploaded to a publicly accessible database, and the courts have been much more willing to assert their independence and make decisions against the government (Assegaf 2018; Butt and Lindsey 2018, 83). On the whole, however, the judicial system remains deeply dysfunctional.

One of the problems is that judges have been highly resistant to efforts to impose oversight and increase accountability (Crouch 2017b). Pushback from judges neutered the oversight function of the Judicial Commission that was founded in 2004,[2] and the Supreme Court escaped any meaningful supervision. This ensured that in the democratic era, as Nadirsyah Hosen succinctly puts it, Indonesian courts "moved from non-independence to independence without accountability" (Hosen 2017, 187). Consequently, judicial corruption has barely been touched by post-Soeharto reforms – in fact, it may have become worse. Judges, court administrators, prosecutors and police collude in patronage networks commonly described as "the judicial mafia" (*mafia hukum* or *mafia peradilan*). Bribery is widespread. It is even said that judges sometimes "auction" off decisions to the party willing to pay the largest bribe (Butt and Lindsey 2018, 299–303). This is the system in which LBH cause lawyers – who refuse to engage in corrupt practices – are forced to operate.

Before looking in detail at the types of legal strategies deployed by LBH in the post-authoritarian era, it is helpful to provide a broad overview of the numbers and types of cases handled. In 2019, the last year before the Covid-19 pandemic affected figures, the LBH network received 4,174 complaints, up from 3,455 in 2018 (YLBHI 2020, 16). LBH Jakarta received just over a third of the complaints in 2019, with 1,496. Figure 4.1 demonstrates the distribution of complaints across 16 offices, including the newly established office in Palangkaraya, Central Kalimantan.[3]

This book is largely concerned with the strategies deployed by LBH to advance social change, but the importance of this daily 'grunt work' of providing legal counsel for people who would otherwise struggle to access it should not go understated. LBH plays a vital role in holding law enforcement officials to account, limiting incidents of torture and abuses of power in detention, and helping to reduce the time suspects and the accused are detained for. Former YLBHI senior staff member Robertus Robet summarises this role well:

> Having been questioned by police, I know first-hand that if we are accompanied by a lawyer, we feel so much more secure. This feeling cannot be replaced. This is one of LBH's main advantages. People know that LBH provides free legal aid and does so sincerely. They know LBH wants to protect our fate.[4]

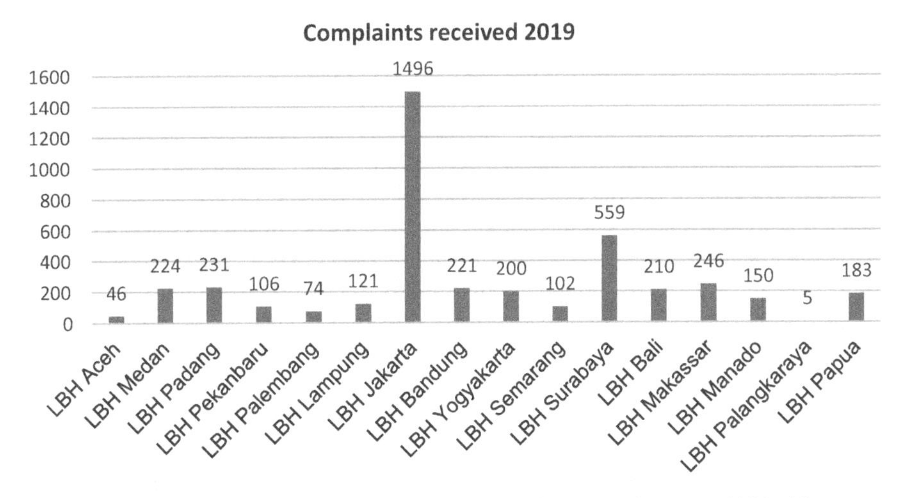

Figure 4.1 Complaints received by LBH offices during 2019 (YLBHI 2020, 16)

Resource constraints and organisational priorities mean that not all the complaints listed above are handled by LBH – for example, after an initial consultation, some cases may be referred on to other providers. It is worth noting too that not all of these cases end up in court. If a complaint is followed up by LBH, it might be addressed through litigation or nonlitigation means, or both. In terms of priorities, LBH continues to focus on the kinds of social and economic rights cases that made it such a prominent organisation under the New Order: those involving land and natural resource conflicts, labour rights, and cases of the urban poor facing forced eviction. But it has also made a name for itself defending unpopular minorities who would struggle to secure representation elsewhere, including religious minorities, people accused of communism, gender and sexual minorities, and Indigenous and traditional belief communities. LBH states its priority issues are as follows:

1 Land rights and agrarian conflict
2 Labour rights
3 Rights of the urban poor, including forced evictions and the rights of informal street traders (*pedagang kaki lima*)
4 Environmental rights
5 The rights of minority and vulnerable groups, focusing mainly on religious freedom and tolerance (but also the rights of other minority groups and vulnerable people, such as sexual and gender minorities, women and children)
6 The right to a fair trial, covering issues such as false arrests, mistrials and torture (YLBHI 2020, 8–9)

Many of the regional LBH offices structure their offices around these issues, or their associated target groups of focus (for example farmers, labourers and the urban poor), depending on local dynamics. Others, however, structure their offices more broadly, for example into civil and political rights and economic, social and cultural rights divisions (YLBHI 2020, 9).

As discussed in Chapter 2, LBH's ideology of structural legal aid involves an approach to case selection that emphasises communal cases over those involving individuals. This also plays a role in determining whether a complaint will be followed up. But individual cases are still regularly taken on when they can be connected to a broader cause. Notably, one of the LBH's priority causes is 'the right to a fair trial', a proceduralist (or rule of law) goal, where the focus is on improving the functioning of the system. This underscores the fluid nature of categories of cause lawyering, and the fact that cause lawyers may adopt multiple approaches at the same time or, indeed, change approaches over time.

Individual criminal case defence

LBH's core function is to provide legal assistance to poor and marginalised Indonesians, so there will always be an important role for the defence of criminal cases in its work. Despite the often individual nature of criminal case defence, however, LBH tries to connect these cases to higher causes and broader principles, in particular human rights and democracy. Since its establishment, LBH has developed a reputation for being one of the few organisations in Indonesia willing to take on the most politically sensitive, unpopular and hopeless causes. Knowing that social and political circumstances are often not in its favour, LBH nevertheless uses individual case defence to expose illegitimate state action to public scrutiny, and demand that constitutional rights do not just exist on paper but are actually implemented in practice.

For example, YLBHI and LBH Jakarta were among the few CSOs that came to the defence of more than 140 gay men who were arrested in a police raid on a Jakarta sauna in 2017 (Advocacy Coalition for Violence Against Gender and Sexual Minority Groups 2017). Police denied many of the men access to legal counsel, and reportedly leaked photos of them naked or partially clothed, which then went viral (Taher 2017). In the end, only 10 of the men were sentenced for violating the Pornography Law (Law No. 44 of 2008) (Pausacker 2021, 450).[5] At the time of their arrest, LBH Jakarta spoke out strongly in defence of the men's rights, describing the arrests as a "bad precedent" for the rights of sexual and gender minorities in the country that could be "treated as a reference [justification] for violent actions by the public" (Maulia 2017). It also stressed the accused men's right to privacy and criticised the police for degrading their humanity (Riana 2017). The raid came just months after a national moral panic over LGBTIQ+ rights, when multiple senior public officials condemned the gay community and 93% of the population said "homosexuality should be rejected" (Pew Research Center 2013). There is therefore little to be gained in terms of public or political support by defending such cases, but LBH consistently takes them on as part of its commitment to defending the poor and marginalised and to articulate values of democracy and human rights.

One of the defining human rights challenges of the post-authoritarian era in Indonesia has been escalating intolerance and the persecution of minority religious groups. Significant scholarly attention has been devoted to documenting and understanding the so-called "conservative turn" in Indonesian Islam since the fall of Soeharto (Bruinessen 2013; Lindsey and Pausacker 2016b), and increasing legal and physical attacks against minority and unorthodox groups (Lindsey and Pausacker 2016a; Human Rights Watch 2013). Almost all the major legal cases targeting religious minorities in the democratic era have involved LBH lawyers. For example, LBH lawyers defended religious movement leader Lia Eden,[6] a member of the unorthodox Muslim minority Ahmadiyah sect that was attacked in Cikeusik, Banten,[7] Shi'a community leader Tajul Muluk from Sampang, East Java,[8] and Alexander Aan from West Sumatra, who was convicted after publicly expressing his atheism in a Facebook group.[9] Another prominent case involved the new religious movement Gafatar, which was attacked and its leaders were charged with blasphemy and treason in 2016. While LBH's approach to individual case defence has changed little over the years, taking on these cases is a key means by which LBH defends individual freedoms and challenges government abuses of power amid democratic decline.

Gafatar blasphemy case

Escalating attacks on religious minorities is at odds with the stronger protections for religious freedom in the amended Indonesian Constitution. Article 29(2) of the original constitution guaranteed "all persons the freedom to embrace their religion and to worship according to their own religion or belief". The constitutional reforms implemented after 1998 reinforced this right. Article 28E of the amended Constitution states that: "(1) Each person is free to embrace their religion and to worship according to their religion… and (2) Each person has the freedom to possess beliefs, and to express their thoughts and attitudes according to their conscience". Article 28I of the amended constitution even states that the right to a religion cannot be limited under any circumstances.[10] It is significant that the Constitution makes a clear distinction between 'religions' and 'beliefs'. Indonesia officially recognises only six religions: Islam, Protestantism, Catholicism, Hinduism, Buddhism and Confucianism.[11] Yet it is also home to hundreds of beliefs, commonly called *aliran kepercayaan* or *aliran kebatinan*, encompassing traditional or Indigenous beliefs, as well as emerging new religious movements.

Many followers of unorthodox religious groups have been charged under the so-called Blasphemy Law, which was enacted in 1965, as Law No. 1/PNPS/1965 on the Prevention of the Misuse and or Insulting of a Religion.[12] The Blasphemy Law only recognises the six official religions and in fact describes beliefs as a danger to the recognised religions.[13] It inserted an article into the Criminal Code (KUHP) stating that "Any person who intentionally and publicly expresses sentiments or conducts acts that (a) involve enmity, abuse or insulting of a religion adhered to in Indonesia; or (b) are intended to stop a person from believing in a religion based on an Almighty God" faces up to five years in prison.[14] The Law was used only a handful of times under the New Order, but since the fall of Soeharto, at least 130 people have been convicted (Crouch 2016b, 106). The leaders of Gafatar are among them.

The 1997 Asian financial crisis, and the political turmoil associated with the collapse of the Soeharto regime in 1998, saw the emergence of a range of new 'prophets' who claimed that they could save Indonesians from the troubles they faced (Makin 2016, 5). One of those who claimed to be a prophet was Ahmad Mushaddeq (also known as Abdussalam), who, with Mahful Muis, founded Qiyadah Islamiyah in 2001, which later became Komar (*Komunitas Milah Abaraham*, Abrahamic Religion Community), and eventually Gafatar. The history of the movement is convoluted and the teachings of Mushaddeq and Mahful evolved over time, but controversially, since about

2007, they advanced the idea of a return to the religion of Abraham (which they called *Milah Abraham*), and read the Qur'an alongside the Old and New Testaments (Makin 2019, 92–94). After establishing Gafatar in 2011, Mushaddeq began urging his followers to migrate to regional areas of Indonesia and purchase land, where they would establish a new centre of world civilisation (Makin 2019, 91, 96). The movement raised suspicions among both conservative Muslims who considered its teachings a threat to Islamic orthodoxy,[15] and the government, which was worried about it challenging the authority of the state (Makin 2019, 96). In response to increasing pressure, in August 2015 the group agreed to disband and form a new organisation: *Negeri Karunia Tuan Semesta Alam Nusantara* (NKTSAN), which translates roughly as Archipelagic Nation of the Grace of the Lord of the Universe. This new organisation had a strong focus on food security and self-sufficiency, and encouraged its members to migrate to Kalimantan, where they purchased land to conduct their activities. From August 2015 to January 2016, more than 8,000 former Gafatar members moved and established a settlement in Pasir village, Mempawah district, West Kalimantan (Rahman 2018, 7).

Following escalating media coverage of the group (Syaifullah 2016a; 2016b; 2016c; The Jakarta Post 2016; Susanto and Adi 2016), mobs attacked two Gafatar communities in Mempawah on 18 January 2016, destroying their settlements and crops, as police and military officials reportedly watched on (Human Rights Watch 2016). Military officials then evacuated hundreds of Gafatar followers from West Kalimantan to government shelters in Java (Reuters 2016). A Gafatar spokesperson said that almost 8,000 followers were forcibly evicted to unofficial detention centres in Java during January and February (Human Rights Watch 2016). Not long after the attack, on 3 February 2016, the Indonesian Council of Ulama (*Majelis Ulama Indonesia*, MUI) issued a fatwa declaring Gafatar to be a deviant sect,[16] and then on 29 February, the Religious Affairs Ministry, Attorney General's Office and Home Affairs Ministry issued a joint decision formally banning the organisation from spreading its teachings (Ribka 2016).[17]

In May 2016, three senior figures in the organisation were named suspects for alleged blasphemy and treason and detained by police: Mahful Muis, the former head of Gafatar and deputy of NKTSAN; spiritual leader Ahmad Mushaddeq; and Andry Cahya, Mushaddeq's son and the president of NKTSAN. YLBHI and LBH Jakarta lawyers took on the case. From the beginning, public opinion was firmly against Gafatar, yet LBH repeatedly asserted the group's right to practice their religion, as guaranteed by the Constitution. Then YLBHI director Asfin spoke to the media immediately after they were named suspects: "According to the Constitution, religious freedom and belief is guaranteed by the state. As such, naming them suspects is a form of criminalisation of religious freedom and belief," she said. She cited Article 28E of the Constitution, and described the right to a religion as a non-derogable right (Kompas.com 2016). As discussed below, however, this is not how the government and Constitutional Court have interpreted this right.

The trial heard from 24 witnesses. Several of the witnesses brought by the prosecution contradicted the accusations against the ex-Gafatar leaders, saying that the focus of the community was on food security, that they did not oppose the government or have any inclination towards violence, and even that they had good relations with local officials and the community.[18] The most controversial allegations against the community came from witness statements collected during the investigation stage, which were then read by prosecutors in court.[19] These witnesses were never called to appear in court.

Prosecutors sought sentences of 12 years in prison for Mushaddeq and Mahful, and 10 years for Andry. The panel of judges ultimately handed down sentences of five years for Mushaddeq and Mahful, and three years for Andry. The judges concluded that the group's teachings that prayer, fasting, zakat and the hajj are not compulsory conflicted with Islamic teaching and therefore "it can be concluded that Milah Abraham [referring to the group and their religious beliefs] have offended Muslims and simultaneously desecrated the purity of the teachings of Islam, which are followed by the majority of Indonesians".[20] The court found them not guilty of the charges of treason. LBH lawyers were highly critical of the process, especially the judges' apparent reliance on the written statements of witnesses obtained during the investigation stage over the oral statements of witnesses who provided testimony in court. Their frustrated lawyer, LBH Jakarta's Pratiwi Febry, described the courts as a "killing field" for minorities (BBC Indonesia 2017).

Defending the cause, not just individuals

LBH has never managed to secure an acquittal in a religious freedom case (Lindsey and Crouch 2013, 634). Indeed, until early 2020, no person charged with blasphemy in Indonesia had ever been acquitted.[21] As LBH Jakarta lawyer Muhammad Rasyid Ridha said, pursuing religious freedom cases through the courts is "...the road to hell, because, from the outset, the legal norms are not compatible with human rights, democratic values and justice".[22] Yet LBH still takes on these cases, despite its lack of confidence in the courts. Of course, part of the reason it does so is because it is committed to ensuring persecuted minorities receive representation. Even if many Gafatar followers are middle class, and the Gafatar movement could have paid for private lawyers, few other lawyers outside of the LBH network would be willing to take on such a politically sensitive case.[23]

Another important motivation for taking on these cases is that they are a key means by which LBH advances democratic principles. Indeed, YLBHI leader Asfin acknowledged that in the Gafatar case, LBH was "defending values, not just the person".[24] Typical of a grassroots approach to cause lawyering, the individual case was connected to higher level principles (Hilbink 2004, 681), in particular freedom of religion and freedom of expression. LBH lawyers consistently framed its defence of the Gafatar members in these terms. For example, when responding to the prosecution's sentencing demand,

LBH Jakarta's Pratiwi Febry said: "Do they deserve these charges? … Have they really defamed religion or the existence of the government? As an independent nation, where do we place freedom of thought, belief and expression?" (Kusuma 2017).

Much as was the case under the authoritarian New Order, the courts were transformed into a space to deliver public education, to highlight the problems faced by Indonesia, to criticise and oppose the government, and to strengthen democratic discourse. These trials provide an opportunity for LBH to "seize public space" for the open discussion of politically sensitive issues like blasphemy and minority rights.[25] YLBHI director Asfin explained this further:

> [L]itigation can put us on an even playing field, although in many cases it is far from even because the judges will always pressure us. We can oppose the government and show how it is not 'in order' (*beres*). The courts can be a stage for delivering facts that it would be impossible for us to discuss outside the courtroom without being arrested… Litigation can also mobilise the public, to attend, to listen, to be informed, to see injustice.[26]

Asfin's choice of language is significant. She invokes the idea of political theatre and performance, with the courts used as a "stage" for speaking to a wider audience beyond the courtroom. Her colleague, Pratiwi Febry, reiterated the importance of this safe space, while emphasising its impacts on client empowerment.

> In the current democratic environment, one of our biggest challenges is fighting for civic space. Even though LBH knows that minorities never win, that they always end up in prison, we should not ignore how important it can be for minorities to use the courtroom as a space to declare, 'This is what I believe'. There is nowhere else for them to [safely] do so. And this space is protected by law.[27]

Defending the rights of marginalised individuals and groups like Gafatar against abuses of state power remains one of the most common strategies of LBH cause lawyers in post-authoritarian Indonesia. It is an approach to cause lawyering that has changed little since the democratic transition in 1998. Indeed, as comments from Asfin and Pratiwi suggest, their descriptions of the value of litigation differ little from descriptions of cause lawyering practiced under authoritarian regimes elsewhere, with the courts used as a safe public forum to "speak out legally and politically" (Fu 2011, 355).

Although LBH's approach to criminal case defence might have shifted little since the New Order era, the democratic transition did offer new opportunities for strategic policy-focused litigation, directed at the state, to defend or advance human rights protections. These types of strategies are associated with an elite/vanguard approach to cause lawyering. LBH has had some success with these strategies, as the next section shows, but they never really took root.

Constitutional litigation

The establishment of strong constitutional or legal guarantees of rights and courts where these rights can be claimed are broadly recognised as crucial factors in the development of cause lawyering (Chang 2010; Ginsburg 2007; Goedde 2009; Meili 2009; Rekosh 2008; Tam 2010; Scheingold and Sarat 2004, 130). In Indonesia, reforms to the Constitution and establishing a Constitutional Court provided important new legal opportunities for cause lawyering. Landmark reforms to the Indonesian Constitution between 1999 and 2002 resulted in the insertion of a new chapter (Chapter XA) based largely on the Universal Declaration of Human Rights. Under the original 1945 Constitution, human rights were only covered under Article 28, which stated that "freedom of association and assembly and the like shall be prescribed by law". However, as Lindsey notes, rather than guaranteeing rights, this provision was intended to provide the government with the freedom to restrict or suspend these rights as it wished (Lindsey 2008, 29). The insertion of Chapter XA therefore marked "a radical shift in Indonesia's constitutional philosophy from essentially authoritarian to liberal democratic in nature" (Clarke 2008, 432). It granted some of the strongest protections of human rights in the region, and indeed, beyond those guaranteed in many established western democracies (Lindsey 2008, 29). The constitutional reform process in Indonesia also provided for the establishment of an independent Constitutional Court with the power to review the consistency with the Constitution of statutes (*undang-undang*) enacted by the national legislature (DPR) (Article 24C). The constitutional reforms and 2003 founding of the Constitutional Court provided Indonesian cause lawyers not only with a strong legal institution at which rights claims could be launched, but also helped to boost the persuasiveness of these claims, given these rights were guaranteed by the highest law in the land.

Briefly, the Constitutional Court is comprised of nine justices who can serve for up to 15 years.[28] The president, DPR and Supreme Court nominate three justices each, with their appointment officially confirmed by the president.[29] The court has no appeal function and cannot alter decisions of any other court, but enjoys first and final jurisdiction for the constitutional review of statutes.[30] It has additional responsibility for hearing disputes about the constitutional authority of state institutions, deciding on the dissolution of political parties, hearing disputes over contested election results, and deciding on an opinion of the DPR requesting that the president or vice president be impeached.[31] Notably, the Court only has power to review the constitutionality of statutes – it cannot rule on lower-level government regulations (*peraturan*), which remain the jurisdiction of the Supreme Court.

This means the Court can have little or no say in how statutes are implemented by government or over ministerial regulations and decisions (Hendrianto 2018, 42).

The Court has taken a relaxed approach to the standing of CSOs in submitting petitions for constitutional review. It generally grants standing to CSOs if their organisational charters are related to the issue at hand, while organisations working on the rule of law, human rights, democratisation or the broader public interest are offered standing "almost as a matter of course" (Butt 2015, 49). The first Constitutional Court chief justice, Jimly Asshiddiqie, encouraged LBH to file judicial reviews to "test" the new institution.[32] Jimly recognised that the Court needed to quickly generate its own jurisprudence and therefore wanted it to hear a large number of cases as soon as possible (Butt 2015, 48). Before the emergence of the Constitutional Court, the unreliable and often inconsistent administrative courts, discussed below, were one of two legal options available for cause lawyers to directly challenge state power. As stated, the Supreme Court has the power to review whether lower-level regulations violate statutes, but during the New Order era it was reluctant to exercise this function (Fenwick 2009, 333),[33] a pattern that has continued in the post-authoritarian period (Butt 2021, 536). The relaxed approach the Constitutional Court has taken to standing rules has allowed LBH to help establish a practice of public interest litigation not previously possible under the New Order. Indeed, LBH has developed a reputation among Indonesian civil society as one of the "pioneers" of strategic impact litigation at the Constitutional Court (Kompas 2023).

After it began operations in 2003, the Constitutional Court quickly built a reputation for independence and integrity. It has been willing to take on sensitive political matters, and its judges have demonstrated that they are receptive to rights claims, leading some to describe it as "unusually activist" (Dressel and Inoue 2018, 162). Further, the government has mostly complied with its decisions, even if the DPR has not (Butt and Lindsey 2018, 100–07). These factors encouraged a degree of optimism about the ability of litigation-driven legal reforms to lead to social change, consistent with an elite/vanguard vision of cause lawyering, as indicated by the popularity of the Court among CSOs. The Court's reputation was severely tarnished by the convictions of Chief Justice Akil Mochtar for corruption in 2014 and Justice Patrialis Akbar in 2017 (Kahfi 2017), but it recovered its status quickly and was for several years considered the best regarded of Indonesia's courts (Butt and Lindsey 2018, 108).

Since the establishment of the Court, lawyers from YLBHI and LBH Jakarta have played central roles in coordinating some of the most high-profile civil society challenges to legislation, and have recorded major victories to extend and strengthen rights protections and protect the public interest. Some of the landmark decisions of the Court during the democratic era (not all of which involved LBH lawyers) have included: ruling that former members of the banned Indonesian Communist Party (PKI) should be allowed to vote and stand in elections;[34] cancelling notorious *lèse majesté* and 'hate sowing' provisions in the KUHP used by the Soeharto regime (and Megawati Soekarnoputri's post-authoritarian government) to pursue its critics (Butt and Lindsey 2018, 199, 438);[35] effectively forcing the government to comply with a constitutional requirement for 20% of the state budget to be allocated to education (Butt 2015, 128–29; Hendrianto 2018, 114–16);[36] deciding in favour of Indigenous Peoples' rights over customary forests;[37] and ruling that the minimum age of 16 for women to marry was unconstitutional, and ordering that the national legislature revise the 1974 Marriage Law to increase it to 19, the same as the minimum age for men.[38] This section focuses on two key challenges submitted by LBH lawyers, the Book Banning case and the Blasphemy Law case.

Book Banning case

The 'Book Banning' decision (Decision No. 6–13–20/PUU-VIII/2010) is one of the most prominent victories LBH has had at the Constitutional Court in its short history. The case was sparked by an escalation in book banning by the Attorney General's Office in the mid-2000s. The Attorney General relied on a decades-old presidential regulation, Presidential Decree No. 4/PNPS/1963 on Security of Printed Products That Could Disturb Public Order, which was converted into law by Law No. 5 of 1969.[39] This decree was enacted during the turbulent final years of Soekarno's 'Guided Democracy' regime, but used extensively under Soeharto (Utama 2016). The regulation was not applied in the immediate post-Soeharto period, but from 2006 to 2009, the AGO banned at least 22 books it deemed a threat to public order.[40] In 2009, it banned two books that presented counter-narratives to official accounts of the 1965 anti-communist violence: the Indonesian translation of John Roosa's "Pretext for Mass Murder: The September 30th Movement and Suharto's Coup d'Etat in Indonesia" [*Dalih Pembunuhan Massal: Gerakan 30 September dan Kudeta Suharto*]; and *Lekra Tak Membakar Buku: Suara Senyap Lembar Kebudayaan Harian Rakjat 1950–1965* ["Lekra Didn't Burn Books: The Silenced Voice of the People's Daily 1950–1965"] by Rhoma Dwi Aria Yuliantri and Muhidin M. Dahlan. It also banned a book on Papua and two books on religious pluralism.[41] In early 2010, there was also a public debate about banning another book critical of then-President Susilo Bambang Yudhoyono (Hukum Online 2010).[42] These events catalysed a vigorous civil society campaign against book banning and prompted three simultaneous challenges to Law No. 4/PNPS/1963.[43] YLBHI and LBH lawyers were a major component of the 'Advocacy Team Against Book Banning' that represented the Indonesian Institute of Social History (*Institut Sejarah Sosial Indonesia*, ISSI) and Rhoma Dwi Arya Yuliantri in their challenge (No. 20/PUU-VIII/2010).

In criticising the banning of the books, LBH and ISSI figures repeatedly framed their critiques with references to democracy and human rights. The then YLBHI director Patra Zen, for example, said "banning books is an act of authoritarian and totalitarian regimes" (Kompas.com 2010a). Similarly, then LBH Jakarta director Nurkholis Hidayat said: "Law No. 4/PNPS/1963 violates the principles of the rule of law and democracy that are guaranteed by the Constitution" (Kompas.com 2010b). As he told me:

> Everyone knew the Attorney General's authority to ban books was derived from a problematic legal framework. It was based on an emergency presidential decree issued in a context very different to the democratic system we were in... At LBH at that time, there was a lot of enthusiasm for getting rid of the legal products that were no longer relevant for the [democratic] situation. If there is a problem with a publication, there should be due process, respect for the rule of law. Our argument was as simple as that.[44]

In addition to the legal challenge, other members of the 'Civil Society Coalition for Defence of Freedom of Expression' organised public campaigns to support these legal efforts. A public exhibition was held at the Taman Ismail Marzuki cultural centre, examining the history of book banning in Indonesia (Detik 2010). Murals and posters also promoted the campaign. LBH cause lawyers and their allies in civil society did not view litigation in isolation. There was a recognition of the importance of these broader campaign efforts to accompany the legal process. As Nurkholis explained:

> We are not simply hoping to change the law, we also aim to encourage political support for change, from the public, academics, public figures and so on. When we launch a judicial review, in fact, even before, it is essential to conduct a public campaign.[45]

As Nurkholis's comments indicate, even when LBH has engaged in elite/vanguard strategies of cause lawyering, it has also attempted to incorporate efforts to strengthen rights awareness in the community – a manifestation of its structural legal aid approach.

The Constitutional Court ultimately ruled that Articles 1–9 of Law No. 4/PNPS/1963 were unconstitutional, and as such the entire law was unconstitutional.[46] It stated that in a rule of law state like Indonesia there is due process of law, and as such any decision to remove a book from circulation must be made by the courts. It ruled that providing the Attorney General with the power to ban books without a legal process was "the approach of an authoritarian government, not a rule of law state (*negara hukum*) like Indonesia".[47] The Court's decision also referred to book banning as conflicting with citizens' right to communicate and obtain information (as guaranteed by Article 28F of the Constitution) and the right to freedom of belief and expression (as guaranteed by Article 28E).[48] Much of the language used by the court mirrored the arguments put forward by the civil society coalition.

This was a vitally important decision to strengthen freedom of expression in Indonesia. But raids on leftist books still occur, largely conducted by the military. In early 2019, for example, weeks after the Indonesian military confiscated leftist books from bookshops in West Sumatra and East Java, then Attorney General Muhammad Prasetyo advocated for a "massive" seizure of leftist books (Siddiq 2019). Despite continued government and military efforts to control the public narrative around communism and leftism, the loss of the legal basis for banning books does make it easier for LBH and broader civil society to challenge these kinds of acts or statements.[49] As Muhamad Isnur said, "Confiscating books is now illegal, so the community can resist or criticise these actions... we are on firmer ground".[50] This is one of the few examples where LBH could point to an elite/vanguard method of cause lawyering as having had some impact. Another critically important, but far less successful, example was the challenge to the Blasphemy Law.

Blasphemy Law case

The ultimately unsuccessful challenge to the Blasphemy Law in 2009–10 is one of the highest-profile cases decided by the Constitutional Court since its establishment. The challenge was launched by civil society in recognition of the fact that the Law was increasingly being used to target minority religious communities. But several other threats to religious freedom preceded the challenge to the Law. In 2005, the MUI issued a fatwa condemning secularism, liberalism and pluralism and re-issued an older fatwa condemning the Islamic sect Ahmadiyah as heretical (Menchik 2014, 610). While these fatwas have no legal basis in Indonesian law, they contributed to an environment of growing intolerance of minority religious communities, especially Ahmadiyah, whose teachings were viewed by Islamists as deviations from Islamic orthodoxy. Then, on 1 June 2008, a group of human rights and religious freedom activists held a rally at the National Monument (*Monas*) in support of pluralism and the Ahmadiyah community. They were violently attacked by Islamist protestors, in an event now known as the 'Monas Incident' or the 'Monas Tragedy' (*Insiden Monas* or *Tragedi Monas*). About a week after this incident, the government published a joint ministerial decision that banned the Ahmadiyah community from spreading their beliefs.[51]

With this as the backdrop, a large group of human rights organisations and senior liberal Muslim intellectuals submitted a challenge to the Blasphemy Law at the Constitutional Court. Former President Abdurrahman Wahid ("Gus Dur") was the most prominent of these challengers, and had reportedly encouraged LBH to submit a petition for judicial review as far back as 2005, when Asfin was head of LBH Jakarta.[52] Some 54 lawyers, including Asfin and several other current and former YLBHI and LBH staff, represented the case.

As stated, the Indonesian Constitution provides several protections for religious freedom, under Articles 28E and 29, and Article 28(I) describes the right to religion as a right "that cannot be limited under any circumstances". Nevertheless, Article 28J goes on to state that:

> In carrying out his/her rights and freedoms, every person is obliged to abide by the restrictions in legislation issued to protect the rights and freedoms of others and which accords with moral considerations, religious values, security and public order in a democratic society.

Article 28J is closely based on Article 18(3) of the International Covenant on Civil and Political Rights (ICCPR), although it notably adds "religious values" to the list of permissible reasons for restrictions on religious freedom in the ICCPR. The applicants' arguments have been well described by Melissa Crouch (Crouch 2012a). Briefly, they included that: the Blasphemy Law conflicted with the principle of freedom of religion as guaranteed by Articles 28E and 29 of the Constitution because it only recognised six official religions; although Article 28J(2) of the Constitution permits some limitations on religious freedom, the Blasphemy Law extended too far into internal matters of thought and belief; and the Blasphemy Law was inconsistent with the rule of law and the principle of equal treatment before the law described in the Constitution (Articles 1(3) and 28D(1), respectively) (Crouch 2012a, 18–23).

The court heard from a large number of organisations and expert witnesses, including Muslim organisations and government officials in support of the Blasphemy Law, as well as submissions by religious minorities and 'progressive' Muslims against the Law although, notably, no representatives from the Ahmadiyah community testified, despite reportedly being asked (Crouch 2016a, 5). The case attracted an immense amount of public and media attention, some of which was encouraged by the LBH lawyers themselves. Islamist groups attended the court in large numbers, occasionally disrupting proceedings (Menchik 2014, 612–13). Reflecting on the trial years later, Asfin remarked on their realisation that, if they were going to win, they needed to support litigation with extra-legal activities, such as a "roadshow" to mainstream Muslim organisations Nahdlatul Ulama and Muhammadiyah to get them on side. Ultimately, however, both Nahdlatul Ulama and Muhammadiyah representatives also testified in support of the Law (Menchik 2014, 613–15; Crouch 2012a, 27–28). As Asfin explained:

> Because we suspected we were not going to win, we split up and some handled the nonlitigation aspects, for example by conducting a 'roadshow' to religious institutions. But this aspect did not run as planned. Even the representatives of Confucianism testified in support of the Law![53]

The court ultimately rejected the challenge to the Law, finding that the Law was constitutional. The court stated that "Indonesia is a 'Godly nation' (*bangsa yang bertuhan*)" (Menchik 2014), referring to the first principle of the national ideology, Pancasila (which is 'Belief in One Almighty God'), as well as references to religion in the Constitution and several other statutes.[54] It emphasised the "Indonesianness" of the right to religious freedom in Indonesia, stating that "the practice of religion in Indonesia was different to other countries" (Crouch 2012a, 40).[55] The court specifically said that Indonesia "does not recognise separation of state and religion", and that a fundamental difference between the Indonesian rule of law state (*negara hukum*) and western rule of law states is that in Indonesia, in matters of "governance, formulation of laws, and in the courts, religious teachings and values help to determine which laws are good and which are bad".[56] While the constitutional amendment process might have restructured the state on liberal democratic principles, there was an attempt to define an Indonesian "version" of democracy, one that provided a role for religious values in determining policy, and did not extend to the full enjoyment of religious freedom.

The Court further stated that Indonesia's respect for international legal instruments, including on human rights, "must remain based on the philosophy and Constitution of the Indonesian state".[57] It noted that although Article 28E(1) of the Constitution provides for religious freedom, Article 28J also states that "religious values" are among the factors that may be used to limit human rights.[58] It also acknowledged that Article 28J was different to Article 18 of the ICCPR, which does not include religious values as grounds for limiting religious freedom.[59] The court made a distinction between the private freedom to hold a religious belief (*forum internum*) and the freedom to express or manifest this belief in public (*forum externum*) (Lindsey and Butt 2016, 53–54). The court noted that everyone was free to interpret religious teachings or rules, but such interpretation should use "appropriate methodology" and be based on "relevant holy books".[60] According to the court, "interpretation not based on methodology recognised by followers of a religion or on relevant holy books could threaten security and public order if expressed in public" and, as such, the public expression (*forum externum*) of belief could legitimately be restricted by the state.[61] The court further commented that the Blasphemy Law:

> …did not place limitations on an individual's personal faith (*forum internum*), it only limited publicly expressing statements and attitudes (*forum externum*) that deviated from the fundamental teachings of a religion adhered to in Indonesia, [or] expressing feelings or engaging in behaviour that is hostile, abuses or insults a religion adhered to in Indonesia.[62]

The position of Islam in the state has long been a matter of heated contention in Indonesia, even prior to the birth of the republic (Lindsey and Pausacker 2016a, 30). Yet the Blasphemy Law case prompted one of the most serious and sustained debates on the role of religion and the state in Indonesia for many years. There were deep discussions about the role of law and human rights in regulating this relationship. Ultimately, however, the Constitutional Court came down on the side of maintaining the status quo, as it so far always has in cases on religion (Lindsey and Butt 2016, 48), affirming the government's role in determining religious orthodoxy. Regardless of the protections of religious freedom in the amended constitution, the limits of this formal change to a more liberal democratic system were apparent.

Religious freedom advocates now view the Blasphemy Law case as a "significant setback" for their cause (Crouch 2016a, 12). LBH is much more careful about how it approaches religious freedom cases, preferring to limit public campaigns and advocacy where possible, to avoid attracting the attention of Islamist opponents.[63] This is not its typical approach, however, which tends to combine litigation with other tactics and strategies, as Chapter 5 discusses in more detail.

Legitimising the amended constitution

One advantage of constitutional litigation is that these cases often attract significant media attention, and are sometimes streamed live on television or online (Butt 2015, 294). Even if challenges fail, they can set or influence the political agenda, initiating debate around the need for legislative reform and introducing constitutional and rights discourse into the public consciousness. This is a critically important function of constitutional litigation in the Indonesian context. For much of Indonesian history, the constitution was "a rather nebulous document" (Butt 2015, 294) that was easily manipulated by Soekarno and Soeharto to justify authoritarian rule. The original 1945 Constitution contained no meaningful acknowledgement of rights. Constitutional litigation by LBH and other CSOs is therefore a crucial means to legitimise and give substance to Indonesia's democratic Constitution. In light of only minimal rights awareness among the general public (and many judges), constitutional litigation can "give practical meaning" (Meili 2009, 55) to the extensive rights protections offered by the amended constitution. It can be a strategy to demand the state live up to the promise of rights protections offered by the Constitution.

Constitutional litigation such as in the Book Banning and Blasphemy Law cases is the closest LBH has ever come to an elite/vanguard version of cause lawyering, a version of cause lawyering that has only become possible in Indonesia since the democratic transition. This is a version of cause lawyering that aims to affirm constitutional principles on behalf of unrepresented groups, and believes legal reform is capable of precipitating change in society (Hilbink 2004, 683). As stated, lawyers practicing this form of cause lawyering are typically much more removed from grassroots social movements, and tend to play a dominant role in determining strategy (Hilbink 2004, 678–79). It is significant that the Blasphemy Law challenge was submitted by a group of elite progressive religious scholars (including a former president) and prominent Jakarta-based CSOs. Asfin's comment above suggests a degree of regret over the challengers' insufficient engagement of social movements before the challenge was launched.

Some scholars have suggested that larger coalitions and public pressure can influence the court's decisions (Mietzner 2010, 400; Nardi 2018). LBH is aware of this fact, hence its tendency to submit challenges as part of broad coalitions and mobilise civil society in conjunction with its litigation efforts. As the Book Banning case indicated, even when it engages in elite/vanguard strategies, LBH is not focused solely on litigation. Current YLBHI director Muhamad Isnur (and head of advocacy under Asfin) explains the rationale behind this strategy.

> We view the courts as a place not only for the debate of legal theory, but also for the debate between the will of the people versus the will of those in power. We have a problem with judicial accountability in Indonesia. Providing opportunities for the community to pressure the courts is part of our efforts to increase independence and accountability in the court-room. Even the Law on Judicial Power states that judges must consider the 'community's sense of justice'.[64] We try to create this sense of justice by providing opportunities for the community to share their voice.[65]

LBH recognises that these cases have broader implications beyond the immediate legal outcome, and therefore attempts to deploy tactics that have more in common with grassroots models of cause lawyering alongside constitutional litigation. While these efforts have met with varying degrees of success, LBH has recognised that a purely elite/vanguard version of cause lawyering is not appropriate for contemporary Indonesia, where the rule of law remains poorly developed. In other words, even though it has engaged in these elite-focused strategies, it is not hostage to the "myth of rights" (Scheingold 2004), the idea that legal victory is an effective means of bringing about social change. Ideally, such cases are used as a catalyst for broader public education and organising activities. As Isnur said:

> The target is not only to win the case or to change policy, but also to create space for consolidation of civil society… We establish broad networks, involving many institutions and victims of the policy, and involve them in public campaigns. Every time there is a hearing, we will hold a press conference, a public discussion, a public campaign event, a demonstration. [Constitutional litigation] is a forum for us to educate the public, and judges, about human rights.[66]

The Constitutional Court has been the dominant site for this style of strategic impact litigation since the democratic transition. However, recent developments have seen even more scepticism emerge.

Erosion of the Constitutional Court's independence

It is notable that the two cases discussed here are from 2010 – a time when democratic stagnation was only beginning to be discussed among scholars of

Indonesia (Tomsa 2010). The conviction of judges Akil Mochtar (in 2014) and Patrialis Akbar (in 2017) for corruption were a major stain on the Constitutional Court. However, it has been more recent attacks on the Court's independence by the government and legislature that have led to questions about its ongoing utility as a site for strategic impact litigation. Serious concerns were raised in September 2020, when the DPR revised the 2003 Law on the Constitutional Court, extending the term limits for judges to a maximum of 15 years, up from a maximum of two five-year terms.[67] This was widely interpreted as a "gift" to judges, with the expectation that they would then rule in favour of the government and DPR when controversial legislation, such as the 2020 Omnibus Law on Job Creation, came up for review (Butt 2020). The marriage of Constitutional Court Chief Justice Anwar Usman to President Joko Widodo's sister in May 2022 also caused grumbles in civil society about potential further damage to the Court's impartiality (Lai 2022).

When the Constitutional Court was inevitably asked to review the Omnibus Law, it surprised many by ruling (in a 5–4 decision, on 25 November 2021) that the Law was "conditionally unconstitutional".[68] The challenge was submitted by three individuals and three civil society organisations (not including LBH), who argued that the drafting process of the Law was flawed, as public participation was limited and several articles in the final law differed from the text agreed by the legislature and government in the drafting process. The court ruled that the Law conflicted with the principles of [good] legislative formulation and was therefore procedurally flawed (*cacat formil*).[69] Rather than declare the Law unconstitutional and revoke it, however, the Court provided the government with two years to repeat the legislative process for the Law, allowing the Law and its implementing regulations to remain in force while this occurred.[70] If the government and legislature failed to repeat the legislative process in this time, the Law would become permanently unconstitutional.[71] The Court clearly had the government's interests in mind, referring to the Law's aim of increasing investment as justification for allowing this two-year timeframe.[72]

The Constitutional Court might have made a "compromise" decision (BBC Indonesia 2021), but the DPR was still not satisfied. In September 2022, legislators decided to "recall" one of the three judges that the DPR had nominated to the Court, Aswanto, for ruling that the Omnibus Law was conditionally unconstitutional, and replace him with another judge, Guntur Hamzah, who they thought would better represent the interests of the DPR. Indonesian Democratic Party of Struggle (*Partai Demokrasi Indonesia Perjuangan*, PDI-P) politician and Head of DPR Commission III Bambang Wuryanto said it was disappointing that the "DPR's representative" in the Court would act to invalidate laws passed by the DPR (CNN Indonesia 2022a). Despite emphatic protests from former Constitutional Court judges, including respected founding Chief Justice Jimly Asshiddiqie (Asshiddiqie 2022), President Jokowi tolerated this extraordinary attack on the Court and, two months later, the president formally appointed Guntur Hamzah to replace Aswanto (Nugraheny 2022).

Worse still, rather than deal with the troublesome task of repeating the legislative process for the Omnibus Law, on 30 December 2022, the Friday before New Year's celebrations, Jokowi circumvented the Constitutional Court's order, issuing a Government Regulation in Lieu of Law (*Peraturan Pemerintah Pengganti Undang-Undang*, Perppu). This emergency regulation simply re-enacted the law (with a few changes), while bypassing the lengthy process of repeating deliberations on the draft in the legislature.[73] This gave no opportunity for meaningful public participation, which was one of the key concerns of the Constitutional Court when it declared the Law conditionally unconstitutional.[74] A furious YLBHI director Isnur described the move as a "betrayal of the Constitution" (CNN Indonesia 2022c). He remained highly critical of attacks on the court when I spoke to him soon after.

> Government and legislative intervention into the Constitutional Court is becoming incredibly crude (*kasar*). Legislators recognise it is the only remaining institution that can limit their power. They do not respect the constitution that they themselves created! They no longer respect the Constitutional Court as an independent judicial institution.[75]

Almost immediately, the Perppu was also challenged at the Constitutional Court. Five separate challenges were submitted by dozens of labour unions, the Indonesian Labour Party, civil society organisations and individuals (although not LBH).[76] The petitioners argued that the Perppu did not allow for the "meaningful participation" that the Court had called for when it ruled in 2021 that the legislative process for the Job Creation Law should be repeated, and that there was no evidence of an "emergency" that would call for the president to issue an emergency regulation, or Perppu (Argawati 2023; Loasana 2023). But with Aswanto now out of the way, the Court ruled differently. In a 5–4 decision announced on 2 October 2023, the Court rejected all five challenges. The DPR's extraordinary intervention into the Court proved decisive. Its replacement for Aswanto, Guntur Hamzah, was one of the five judges who ruled to reject the challenges and uphold the law (Loasana 2023).

This outcome was another mark against the Court's reputation. Just two weeks later, however, the Court saw its credibility plummet to new depths, and civil society's concerns about potential conflicts of interests resulting from Chief Justice Anwar Usman's marriage to Jokowi's sister were proven to be well founded. In mid-October 2023, the Court ruled on a challenge to a provision in Law No. 7 of 2017 on Elections that stated that presidential and vice presidential candidates must be at least 40 years old. In a 5–4 decision (with Usman holding the deciding vote), the Court ruled that candidates under 40 could also run if they currently hold or have previously held elected office, including positions as regional leaders.[77] This decision gave the green light to Jokowi's son, the 36-year-old Gibran Rakabuming Raka, then mayor of Solo in Central Java, to run as vice presidential candidate alongside Prabowo

Subianto. In response to widespread public criticism, the Court's Honorary Council (*Majelis Kehormatan Mahkamah Konstitusi*, MKMK), led by the respected Jimly Asshiddiqie, investigated the matter and ruled to remove Usman as chief justice. Usman was allowed to continue to serve at the Court, but would not be permitted to rule on election disputes (Pujianti 2023).

The Omnibus Law on Job Creation discussed above is just one of several controversial laws passed by the legislature over the past few years that have been characterised by rushed deliberations, minimal opposition in the legislature, and inadequate public participation.[78] The government and legislators routinely brush off civil society concerns about deficiencies in the lawmaking process by suggesting that if civil society has any objections to these laws, they can simply challenge them at the Constitutional Court (Ihsanuddin 2020; CNN Indonesia 2022b). The recent erosion of the Court's independence has exposed the hollowness of these calls. Further, when the lawmaking process is defective, LBH now rightfully questions whether treating the problem in a legalistic manner, and taking their concerns to the Constitutional Court, only serves to legitimise problematic legal products.[79] Much like their predecessors during the Soeharto era, current LBH lawyers recognise that the government's calls to take their concerns to court is a tactic to divert potentially explosive disputes into the legal system, and discourage more radical action at the grassroots level.

As a result of government and legislature attacks on the Constitutional Court, LBH now has little faith in the institution, especially when it comes to matters of corruption, or the financial interests of oligarchs.[80] In the words of LBH lawyer Pratiwi Febry, the Constitutional Court is "no longer an ideal space to fight for rights".[81] While constitutional litigation is unlikely to be abandoned altogether, LBH is now much more circumspect about its benefits.

Administrative courts

The administrative courts were established under the New Order, through Law No. 5 of 1986, and began operating in 1991 (Bedner and Wiratraman 2019, 136), during a brief period when the New Order was making small nods toward openness (*keterbukaan*), discussed in Chapter 2. These courts introduced a new venue for citizens to challenge government decisions, defined as "decisions in writing produced by a state agency or official, based on applicable law, containing concrete, individual and final determinations, and having legal consequences for specified individuals or legal entities" (Article 1(9)).[82] However, just as cynical observers predicted, the New Order was never going to allow the judiciary to have any real control over its actions, and indeed, the number of cases seen by these courts under the New Order was small (Bedner and Wiratraman 2019, 137). There were only a few notable examples of the court ruling against the government in sensitive cases, such as when the court ruled that the government did not have the power to cancel the licence of *Tempo* news magazine (see Chapter 2). Yet even this decision was overruled on appeal to the Supreme Court, and the judge responsible for the original decision was punished with a transfer (Pompe 2005, 165).

The administrative courts have therefore only really been available as a viable forum for challenging state action since the democratic transition. Like constitutional litigation, using the administrative courts is consistent with an elite/vanguard model of cause lawyering. Under the Law on the Administrative Court, which was amended in 2004 and 2009, there are two grounds for challenging government decisions: that the decision breached "applicable law" or "principles of good governance" (Article 53(2)). The principles of good governance are elaborated in the elucidation to Article 53(2)(b) of the Law on the Administrative Court (as amended in 2004) as "legal certainty, orderly state administration, openness, proportionality, professionalism, and accountability".

LBH has occasionally used the courts to further its causes and recorded several wins. Notably, after 1998, "outright government refusals to execute injunctions and judgments became rare and government pressure on judges greatly declined" (Bedner and Wiratraman 2019, 138). In 2017, LBH Padang won a suit against the West Sumatra government, ordering the governor to cancel 26 problematic mining permits[83] and in 2019, LBH Surabaya won a case against the East Java governor to protect workers' rights to a fair wage.[84] Overall, the administrative courts have remained somewhat disappointing in democratic Indonesia (Butt and Lindsey 2018, 98–99), because of persistent corruption and limited consistency in legal interpretation (Bedner and Wiratraman 2019). The Papua Internet Shutdown case was one example, however, where the court ruled against the government and in favour of citizens' rights.

Papua Internet Shutdown case

On 15 August 2019, ethnic Papuans held demonstrations in several Indonesian cities to mark the anniversary of the 1962 New York Agreement between Indonesia and the Netherlands, which ultimately led to Indonesia's annexation of West Papua.[85] The following day (the day before Indonesian Independence Day, which falls on 17 August), nationalist protestors surrounded a Papuan student dormitory in Surabaya, East Java, claiming that Papuan students had desecrated the Indonesian flag. Protesters yelled racist insults, which were captured on video and shared widely online.[86] Police forced their way inside and arrested 43 Papuan students (Davidson 2019).

Over the following days, anti-racism and pro-Papuan independence demonstrations erupted in multiple cities across Indonesia. The protests in Papua were particularly large, and were accompanied by significant destruction of property (Davidson and Doherty 2019; Reuters 2019). In response, the Indonesian government sent more than a thousand military personnel to Papua and implemented an internet slowdown, followed by a complete shutdown, apparently to stop the sharing of "provocative" messages that could prolong unrest (Lamb and Doherty 2019; Reuters 2019). The internet was slowed on 19 August in multiple regions across Papua, then shut down completely from 21 August to 4 September in 29 districts and municipalities in Papua province and 13 districts and municipalities in West Papua province. The shutdown was then extended in four districts and municipalities in Papua and two in West Papua until 9 September.

The administrative court case emerged from civil society activism. A coalition of Indonesian CSOs, consisting of YLBHI, the Alliance of Independent Journalists (*Aliansi Jurnalis Independen*, AJI), the Southeast Asian Freedom of Expression Network (SAFEnet), KontraS, LBH Pers, Elsam and ICJR, criticised the shutdown immediately. On 23 August, the coalition issued a 'summons' (*somasi*), describing the shutdown as a violation of the right to information and right to communicate as guaranteed by Article 28F of the Constitution and calling on the Ministry of Communications and Technology and President Joko Widodo to immediately lift restrictions (Bernie 2019).

SAFEnet initiated a public campaign and Change.org petition calling on the government to #KeepItOn (or #*NyalakanLagi*). YLBHI staff also spoke regularly in the media, at one point describing the shutdown as racist, noting that such shutdowns are never used in other areas of Indonesia when demonstrations or unrest occur (Marvela 2019). In early September, YLBHI also hosted a public discussion at its office in Jakarta, exploring policy, legal and rights concerns related to the shutdown (YLBHI 2019). Papua is a highly sensitive issue in Indonesia – individuals and organisations demonstrating public support of human rights in Papua are often accused of supporting separatism. YLBHI and LBH Surabaya's support for the Papuan university students saw both their offices targeted by nationalist protesters (Faizal 2019). One afternoon when I was in the office, we were trapped inside as a group of clearly hired protestors[87] – some were just teenagers who seemed not to understand what they were protesting about – blocked the road outside the LBH Jakarta/YLBHI office. Banners were also hung on pedestrian overpasses in Jakarta, calling on YLBHI to "reveal its donors", reflecting nationalists' suspicions about international support for Papuan independence.

In any case, after the internet shutdown ended, on 21 November the CSO coalition submitted a challenge to the shutdown in the Jakarta State Administrative Court (PTUN Jakarta). The administrative courts were chosen over the Constitutional Court not only because the issue related to a government act but also because, according to LBH, these courts tend to be disciplined in terms of their administration, and decide cases quickly.[88] The challenge was submitted by AJI and SAFEnet, with the 'Team for the Defence of Press Freedom' (*Tim Pembela Kebebasan Pers*) providing legal representation. YLBHI's Muhamad Isnur was head of the legal team.

The first hearing was held on 22 January 2020. In court, the government argued that it had the authority to slow and restrict internet access under Article 40(2)(a)(b) of Law No. 11 of 2008 on Electronic Information and Transactions (as amended by Law No. 19 of 2016),[89] as well as Government Regulation in Lieu of Law (Perppu) No. 23 of 1959 on States of Disaster.

Following multiple hearings over several months, the judges decided against the government. Their decision was significant in that it made lengthy and considered reflections on human rights standards. It acknowledged that the right to information was guaranteed by the Constitution under Article 28F, as well as by Law No. 39 of 1999 on Human Rights (Article 14(1)(2)), and the ICCPR (Article 19(2)), which was ratified into domestic law in 2005. It also noted that the internet aided the fulfilment of freedom of expression, which is connected to many other rights guaranteed by the Constitution.[90] Judges stated that Article 29(2) of the Universal Declaration of Human Rights, Article 28J of the Constitution, Article 73 of the Law on Human Rights and Article 19(3) of the ICCPR allowed the restriction of rights under certain circumstances. The decision acknowledged that the government had the authority to restrict internet access under Article 40(2)(b) of Law No. 19 of 2016, but only for content that violated the law, not the entire network.[91] It also noted that although Perppu No. 23 of 1959 allows for the restriction of rights, this is only the case when a state of emergency has been declared, which had not been done. The judges ruled that the restriction of internet access in Papua and West Papua constituted an illegal act by a government agency and or official.[92] They stated that had Jokowi declared a state of emergency under Perppu No. 23 of 1959, the shutdown would not have been in violation of the law, and as such this omission was also a failure of governance.[93] It was largely a symbolic victory – the ministry and president were ordered to pay the cost of the case, which was just Rp 457,000 (US$31).

Although the case did not have any concrete outcomes in Papua – the decision was handed down on 3 June 2020, close to a year after the shutdown occurred – it had deeper significance. First, it generated significant publicity. When the general population has so little knowledge of what occurs in Papua, the case was used by YLBHI and the civil society coalition to highlight how Papuans are treated unfairly by the central government. As Isnur said: "[T]he public, and Papuans in particular, got some form of rehabilitation, a small victory. It is part of their claim to truth, that the way the government treated the community was wrong."[94]

Second, it was used to make a vigorous defence of internet freedom at a time when freedom of expression is under serious threat in the country (Freedom House 2019; Setiawan 2020). The judges made strong statements in support of internet access as a human right, highlighting its importance for the enjoyment of rights such as "the right to education and teaching, the right to obtain benefits from science and technology, arts and culture, the right to work, political rights, freedom of association and assembly, and the right to health services".[95] It was also a major public embarrassment for the government, especially Jokowi. Seemingly, the challenge has been effective. The government did not appeal the decision,[96] and at the time of writing had since made no attempts to slow or block the internet, including during mass public protests, such as those against the Omnibus Law on Job Creation.[97]

As the Gafatar and Blasphemy Law examples also demonstrated, the courts were used as a safe forum to make broader points about the rule of law, democracy and human rights. This was particularly significant in this case given the controversial nature of matters relating to Papua. Following the verdict, YLBHI continued to frame the civil society complaint in these terms of rights and the rule of law, noting that the Constitution holds that Indonesia is a rule of law state, and as such any government act in violation of the law is essentially defying the Constitution: "If the president insists [on taking action] without legal basis or breaks the law, he is declaring that he opposes the Constitution," Isnur said (Yahya 2020). YLBHI was performing its role as a check on state power. As Isnur told The Jakarta Post: "We don't want to attack the government nor disrupt the nation. We want good governance. The government should listen to us and obey the court's ruling" (Nurbaiti 2020).

The Papua internet shutdown case was a clear example of YLBHI deploying an elite/vanguard form of cause lawyering. The primary goal of litigation was to vindicate constitutional principles and embarrass the government by exposing its unlawful actions "to public and legal scrutiny" (Meili 2005, 397). Although there were some efforts to educate the public about rights abuses in Papua, the real audience for the challenge was the state and political elite (Hilbink 2004, 677). Litigation had little, if any, substantive outcomes for people in Papua, although it may affect how the government approaches similar issues in the future. This style of elite/vanguard cause lawyering was made available through the strengthened rights framework and judicial independence introduced with democratic transition. If cause lawyers were lucky enough to get a favourable outcome during the New Order period, there was every chance the government would simply ignore or circumvent the court's decision. While this risk has diminished, it certainly remains present today.

Another form of legal activism associated with elite/vanguard models of cause lawyering – and one introduced by LBH in the post-authoritarian era – is citizen lawsuits.

Citizen lawsuits

Citizen lawsuits are a new legal strategy in Indonesia. Modelled largely on the form of citizen lawsuits practiced in the United States, citizen lawsuits allow citizens to sue in the public interest if the government violates the law or fails to meet its obligations in implementing the law (Sundari 2018, 329–30; Mahkamah Agung Republik Indonesia 2009, 49). The Indonesian legal system does not regulate procedures for citizen lawsuits explicitly, although the Supreme Court has been quite supportive of them.[98] Citizens do not need to prove they have suffered direct losses as a result of government action or inaction to participate (Mahkamah Agung Republik Indonesia 2009, 49). Because citizen lawsuits are filed in the public interest and applicants do not need to show they have suffered direct losses, they do not seek monetary compensation. Instead, they typically call on the government to change its policy (Mahkamah Agung Republik Indonesia 2009, 63).

LBH pioneered the strategy in Indonesia in the years following the fall of Soeharto, beginning with the case of thousands of undocumented Indonesian migrant workers who were deported from Malaysia to the East Kalimantan district of Nunukan in 2002. The Indonesian government made few serious efforts to respond to this crisis. Many deported migrants fell seriously ill and dozens died (Ford 2006, 238–39). In response to the case, LBH and a group of other CSOs filed a citizen lawsuit on behalf of 53 citizens against nine government officials, including then President Megawati Soekarnoputri, claiming they had failed to protect and fulfil the rights of the returned migrants (Ford 2006, 240). The Central Jakarta District Court agreed that the government had failed to adequately provide protection for the returned migrants, a decision that was overturned at the Jakarta High Court (Hukum Online 2006).[99] Even though the lawsuit ultimately failed, the case and the publicity it generated, as well as ongoing legislative advocacy by a broader alliance of CSOs, resulted in the introduction of a bill on the protection of migrant workers, passed soon afterwards as Law No. 39 of 2004 on the Placement and Protection of Indonesian Workers Abroad. The Nunukan case established citizen lawsuits as a new model of public interest litigation in Indonesia. Several other prominent citizen lawsuits have since been filed in Indonesian civil courts, including on issues such as the environment, education, fuel prices and elections (Sundari 2018, 328–29). One of the more significant environmental cases in recent years has been the citizen lawsuit filed against the government over worsening air pollution in Jakarta.

Air Pollution Citizen Lawsuit

Jakarta is recognised has having some of the worst air quality in the world (Greenstone and Fan 2019, 3). Civil society organisations had conducted public advocacy on worsening air quality for a couple of years, but the government was unresponsive. On 5 December 2018, 19 citizens, in collaboration with representatives from LBH and other CSOs, announced that they would file a citizen lawsuit against the government if it did not act against air pollution in the next 60 days (Swaragita and Aqil 2018). Then in early 2019, Jakarta's air pollution levels reached worrying new heights. Media attention to the issue escalated as mobile phone apps like IQAir AirVisual placed Jakarta as ranking among the worst cities in the world for the quality of its air, and social media users began posting photos of thick smog (Lamb 2019). In April 2019, LBH Jakarta and YLBHI announced they would be going ahead with the citizen lawsuit and opened a post for one month for citizens to register to be included in the case (Prireza 2019).

The case was filed with the Central Jakarta District Court on 4 July, against the president, the minister of forestry and environment, the minister of health, the minister of home affairs, and the governors of Jakarta, West Java and Banten provinces, arguing that they had engaged in unlawful conduct and failed to fulfil the rights of citizens to a clean and healthy environment.[100] Thirty-two citizens eventually joined the citizen lawsuit. Legal action was combined with a lively public campaign, involving public demonstrations and discussions. One day when I was in the YLBHI office, the organisation held a public discussion and rooftop concert, with popular indie musicians performing, livestreamed on social media.

The citizen lawsuit dragged on for two years. A mediation process was held in late 2019, but the parties ultimately failed to reach any agreement and the case continued through to 2021. Despite the sluggish legal process, the case prompted significant policy change. Almost immediately after the case was filed, on 8 July 2019, the Jakarta government issued a revised decision on its strategic priorities to include air pollution.[101] On 1 August, Jakarta Governor Anies Baswedan also issued Governor's Instruction No. 66 of 2019 on Controlling Air Quality. And on 24 August 2020, the governor signed Jakarta Governor Regulation No. 66 of 2020 on Emissions Testing of Motorised Vehicles. In 2021, the government also published Government Regulation No. 22 of 2021 on Environmental Protection and Management.[102]

Admittedly, LBH and the other members of the CSO coalition cannot take sole credit for the changes in policy. Nonetheless, this was a clear example of LBH's adversarial style of cause lawyering holding the government to account, and strengthening citizens' awareness of air pollution issues, creating political impetus for change. As YLBHI's Muhamad Isnur explained in 2019:

> Of course, there is no guarantee we will win the case. But the courtroom is made into a space for applying pressure [on the government], for campaigning, for education. So, if, during the court case, the Jakarta government ends up publishing new regulations on air pollution, that is a good result. Were we successful in achieving our goals? We were! The government published new policies in direct response to our lawsuit. What happens later, if the case is rejected? It could occur. But our goal was achieved, there was a change. Policy responded quickly.[103]

The panel of judges postponed the sentencing hearing seven times, and it was not until LBH Jakarta reported them to the Judicial Commission for breaching ethics guidelines by delaying for so long that they finally announced their decision.[104] On 16 September 2021, the court ruled in favour of the citizen complainants, finding that the five government figures had committed unlawful acts (*perbuatan melawan hukum*) in violation of Law No. 32 of 2009 on Environmental Protection and Management.[105] Although the court found that the government defendants had neglected their duties to fulfil the right to a healthy environment, it fell short of ruling explicitly that they had violated human rights. In convoluted language, the court explained that citizen lawsuits focus on whether a defendant has contravened the law. In the court's view, it was sufficient to rule that the defendants had been negligent in upholding the right to a healthy environment – an unlawful act – without having to rule that they had violated human rights.[106] It ordered the president to enforce National Ambient Air Quality Standards (*Baku Mutu Udara Ambien Nasional*, BMUA Nasional) to protect the health of humans, ecosystems and the environment.[107] President Jokowi, the Ministry of Forestry and Environment, the Ministry of Health, and the Ministry of Home Affairs appealed the decision (Prireza 2021), but Jakarta Governor Anies Baswedan did not (Ridwan 2021). One year later, on 17 October 2022, the Jakarta High Court upheld the Central Jakarta District Court decision.[108] However, the government continued to fight, and in January 2023 Jokowi and the minister of forestry and environment submitted cassation appeals to the Supreme Court (Nababan 2023). In November 2023, the Court finally rejected their appeals (Pristiandaru 2023). Even if the cassation appeals had been successful, as Isnur's comments above indicate, LBH cause lawyers are less concerned with the outcome of the case, or even necessarily establishing a legal precedent. Their primary focus is pressuring the government to publish improved policy, and to take action to enforce its own standards. LBH lawyers may be sceptical of their chances of legal victory, yet they still view citizen lawsuits as offering certain advantages. In this case, the threat of litigation was enough to push "uncooperative foes" into making political concessions (McCann 2006, 19).

Despite their name, citizen lawsuits are largely focused upwards, toward elites, with litigation (or the threat of litigation) used to force the state to concede ground. Recall too that they do not seek financial compensation. Although LBH engaged in a vigorous public advocacy campaign designed in part to educate the public about their rights, it appeared less concerned with client empowerment than it was with speaking back to the state and attempting to establish legal or policy change (Meili 2005, 399). It is significant that rather than working through legislative or administrative channels, publishing policy papers, lobbying legislators or government officials, or drafting legislation in collaboration with the government, LBH feels one of the most viable means to deliver policy change is through the adversarial method of litigation. This is a sign of the shrinking of political opportunities over recent years, explored further in Chapter 6. Leaning on litigation (or the threat of litigation) like this to change law and policy is an inherently adversarial approach. It is true, too, that adversarial litigation may be a more natural fit for an organisation like LBH, with a long history of opposition to the state, compared to other more collaborative strategies of policy change like legislative lobbying.

Two years after the Central Jakarta District Court handed down its decision, however, the public had seen little outcome from the citizen lawsuit. The government had not taken meaningful steps to address the causes of pollution, and in July and August 2023, Jakarta was again clouded in some of the worst pollution in the world. The government largely blamed emissions from private vehicles and instituted temporary measures, like asking civil servants to work from home and artificially modifying the weather to stimulate rain (Damayanti and Maulia 2023). But the public and the media were starting to get angry, and there was increasing attention to the role of government-backed coal-fired power plants and dirty industry in contributing to pollution (Myllyvirta, Thieriot, and Hasan 2023; Kariza 2023). In the end, it appeared that this outrage on social media, and the government's embarrassment over having to host the 43rd ASEAN Summit in early September amid clouds of smog, resulted in more government action to improve air quality – albeit temporarily – than the citizen lawsuit.

Conclusions

The democratic transition in Indonesia after 1998 provided new opportunities for litigation-focused models of cause lawyering. Reforms to the Constitution to guarantee a broad array of rights, the establishment of the Constitutional Court, and increasing judicial independence made possible elite/vanguard forms of cause lawyering focused on securing progressive reforms in the courts. LBH cause lawyers have taken advantage of new avenues for mobilising the law and have used strategic impact litigation to give substance to Indonesia's new democratic constitution, to push democratic reforms further, and to challenge government inaction and overreach. And their efforts have met with some success, at least in the courtroom.

While this chapter has focused primarily on the new legal opportunities made available by democratic transition, there are also several aspects of LBH's approach to litigation that have changed little since the New Order years. In the Gafatar case, for example, LBH cause lawyers were aware that acquittal was virtually impossible, but used the courts as a safe public forum to bring illegitimate government actions to public attention, highlight injustice, popularise human rights and democratic discourse, and further the cause of religious freedom. This kind of individual case defence will never be abandoned, no matter how high the odds are stacked against LBH. In politically sensitive cases like the Gafatar case, LBH cause lawyers work within the confines of the dysfunctional Indonesian legal system to get the best outcome for their clients, while using the courts as a safe platform to advance their causes.

Despite recording several remarkable legal victories, such as in the Book Banning, Papua Internet Shutdown and Jakarta Air Pollution cases, LBH cause lawyers remain mostly sceptical of the value of litigation in bringing about social change, especially when used in isolation. The cases analysed in this chapter demonstrate the complex interplay between the quality of democracy and strategies of cause lawyering. They underscore the importance of looking beyond institutional structures considered supportive of legal mobilisation when considering the relationship between cause lawyering and democratic change.

Even though courts in post-authoritarian Indonesia have shown that they are occasionally receptive to rights claims, shortcomings in the rule of law and government accountability have meant that LBH lawyers are wary of focusing too much on litigation at the expense of social movement strategies. Their relationship with the courts is best described as ambivalent – they recognise that litigation can be beneficial to their causes but have an intimate understanding of its limitations in a legal system where even a minimalist 'thin' version of the rule of law is a work in progress (Peerenboom 2004, 25). Despite the introduction of stronger protections of human rights and new democratic structures following democratic transition, the rule of law continues to bend to the power of political and economic elites. Likewise, the Gafatar and Blasphemy Law cases showed that a focus on formal legal rights can be misleading when substantive change guaranteeing religious freedom remains lacking.

Deficiencies in Indonesian democracy have meant the elite/vanguard strategies described in this chapter have never really taken hold among LBH lawyers. This is particularly evident in the example of the Constitutional Court. While Constitutional Court judges have previously been somewhat receptive to rights claims, attacks on the Court's independence have left LBH questioning whether the Court remains a viable forum for advancing rights claims. LBH is particularly sceptical about the value of litigation when it threatens the economic interests of powerful political and business elites.

In this environment, where LBH is deeply pessimistic about the courts, the organisation finds much more value in the supplementary effects of litigation, in particular the ways in which litigation can be used to catalyse social movement activism. As such, LBH is increasingly placing considerable emphasis on its community organising and community legal empowerment work. These strategies have grown in prominence with democratic decline, and are the central concern of Chapter 5.

Notes

1 Gafatar is a portmanteau of *Gerakan Fajar Nusantara* (Archipelagic Dawn Movement). The development and history of this novel Indonesian religious movement is explained below.
2 Constitutional reforms established an independent Judicial Commission, tasked with oversight of the performance and behaviour of judges and proposing Supreme Court judge candidates to the national legislature (DPR). In 2006, a group of 31 judges challenged the provisions in Law No. 22 of 2004 on the Judicial Commission that provided for this oversight function in the Constitutional Court (Decision No. 005/PUU-IV/2006). The Constitutional Court decided that the review of judicial decisions could only be conducted by the courts, essentially limiting the Judicial Commission's role to the proposal of judge candidates to the DPR (Butt and Lindsey 2012, 96–99).
3 Since 2019, a new office has been established in Samarinda, East Kalimantan, and a 'project base' has been set up in West Kalimantan. At the time of writing, no data was available for these new sites.
4 Interview, Robertus Robet, April 2020. Robertus Robet was questioned by police in 2019 after singing a protest song critical of the military, leading him to seek temporary exile in Australia for a few months (Sudaryono 2019).
5 At the time of the arrests, homosexuality was not a criminal offence in Indonesia. The new regressive Criminal Code passed in late 2022 does not mention

same-sex acts explicitly, but a number of articles could be used to target the LGBTIQ+ community.

6 Lia Eden (also known as Lia Aminuddin) claimed to have received divine revelations and sought to establish a new religion of Eden. She was sentenced to prison for blasphemy in 2006 and 2009 (Makin 2016, 17).

7 In February 2011, Islamist vigilantes attacked the Ahmadiyah community in Cikeusik, Banten, killing three Ahmadiyah members and injuring five others. LBH lawyers represented an Ahmadiyah "security advisor" who was eventually convicted for resisting arrest and assault (Crouch 2012b, 56).

8 In December 2011, Sunni Muslims attacked the Shi'a community in Sampang, Madura, East Java. A Shi'a leader, Tajul Muluk, was eventually convicted for blaspheming Islam, while the attackers escaped prosecution (Suaedy 2016).

9 Alexander Aan made comments in an atheist Facebook group that were deemed insulting to Islam. He was sentenced to two years and six months in prison for "spreading information that could incite hatred and hostility based on an ethnic group, religion, race or group" (Hasani 2016).

10 At the same time, however, Article 28J(2) states that rights may be limited on the basis of morals, religious values, security and public order. Article 28J has been particularly contentious and is discussed further below.

11 This is described in Elucidation to Article 1 of Law No. 1/PNPS/1965. See also Lindsey and Pausacker (2016a, 3).

12 It was then converted into law in 1969, by the same law that converted the Attorney General's authority to ban books into law, Law No. 5 of 1969 (see below).

13 See point 2 of the General Elucidation to Law No. 1/PNPS/1965.

14 Article 4 of Law No. 1/PNPS/1965 and Article 156a of the KUHP.

15 Previously, in 2007, the conservative quasi-state agency the Indonesian Council of Ulama (*Majelis Ulama Indonesia*, MUI) issued fatwa declaring Qiyadah Islamiyah to be a deviant sect (Makin 2019, 89–91; Crouch 2017a, 8). Fatwa can be defined as the opinion of a suitably qualified religious scholar on a question of Islamic law. Although fatwa have no basis in Indonesian law, they can be highly influential on how state and society view minority religious groups (Lindsey and Pausacker 2016a, 32). Soon after the fatwa was issued, Mushaddeq was charged and convicted of blasphemy, and sentenced to four years in prison. On appeal to the High Court, the sentence was later reduced to two-and-a-half years. South Jakarta District Court Decision No. 227/Pid/B/2008/PN.JKT.SEL and Jakarta High Court Decision No. 135/PiD/2008/PT.DKT.

16 MUI Fatwa No. 6 of 2016 on the Gafatar Sect, dated 3 February 2016.

17 Joint Ministerial Decision No. 93 of 2016; No. KEP-043/A/JA/02/2016; No. 223–865 of 2016.

18 Decision No. 1107/Pid/Sus/2016/PN.Jkt.Tim, see, for example, statements of witnesses Muhammad Hadi Suparyono (22–25), Radix Rusdiyanto (25–26), Anton Susanto (28–29), Rio Putra (29–30) and Sujito (32–33).

19 Decision No. 1107/Pid/Sus/2016/PN.Jkt.Tim, 34–136.

20 Decision No. 1107/Pid/Sus/2016/PN.Jkt.Tim, 197.

21 A mentally ill woman charged with blasphemy for bringing her dog into a mosque was acquitted in February 2020 (ABC News 2020). According to Melissa Crouch, in 2017, all persons charged with blasphemy in Indonesia had been convicted (Topsfield and Rompies 2017).

22 Interview, Muhammad Rasyid Ridha, November 2019. Here he is referring to the fact that the Blasphemy Law is already discriminatory toward minorities, because it only recognises six official religions.

23 Interviews, Pratiwi Febry, December 2019; Asfinawati, August 2021.

24 Interview, Asfinawati, August 2021.

25 Interview, Pratiwi Febry, March 2023.
26 Interview, Asfinawati, December 2019.
27 Interview, Pratiwi Febry, March 2023.
28 Until reforms were introduced in 2020, judges could only serve for a maximum of two five-year terms (Article 22 of Law No. 24 of 2003). See discussion below.
29 Article 24C(3) of the Indonesian Constitution, Article 18 of Law No. 24 of 2003 (as amended by Law No. 7 of 2020).
30 Article 24C(1) of the Indonesian Constitution, Article 10(1) of Law No. 24 of 2003 (as amended by Law No. 7 of 2020).
31 Article 24C(1),(2) of the Indonesian Constitution, Article 10(1),(2) of Law No. 24 of 2003 (as amended by Law No. 7 of 2020).
32 Interview, Muhamad Isnur, October 2019.
33 According to former YLBHI leader Bambang Widjojanto, the Supreme Court decided just five of 26 cases submitted to it between 1992 and 1999. Of the remainder, one was struck out while 20 were never decided (Butt 2015, 30).
34 Constitutional Court Decision No. 11–17/PUU-I/2003.
35 The so-called *lèse majesté* provisions, Articles 134, 136 *bis* and 137 of the KUHP, described offences for intentionally insulting or defaming the president or vice president. These provisions were struck down by the Constitutional Court in Decision Nos. 013/PUU-IV/2006 and 22/PUU-IV/2006. The 'hate sowing' provisions, Articles 154 and 155, prohibited "public expressions of hostility, hatred or contempt towards the Government of Indonesia" and were struck down in 2007, through Decision No. 6/PUU-V/2007. See also Butt and Lindsey (2018, 199–200).
36 In a series of decisions from 2006–08, the Constitutional Court ruled that state budgets were unconstitutional for failing to meet the constitutional requirement for 20% of the state budget to be allocated to education. See Constitutional Court Decision Nos. 12/PUU-III/2005; 26/PUU-III/2005; 26/PUU-IV/2006; 24/PUU-V/2007; and 13/PUU-VI/2008.
37 Constitutional Court Decision No. 35/PUU-X/2012.
38 Constitutional Court Decision No. 22/PUU-XV/2017. In 2019, the national legislature revised the 1974 Marriage Law so that the minimum age of marriage for both men and women is now 19. See Law No. 1 of 1974 on Marriage (as revised by Law No. 16 of 2019).
39 Law No. 16 of 2004 on the Public Prosecution also allows the office to 'supervise' printed materials.
40 Constitutional Court Decision No. 6–13–20/PUU-VIII/2010, 48.
41 *Suara Gereja Bagi Umat Tertindas: Penderitaan, Tetesan Darah, dan Cucuran Air Mata Umat Tuhan di Papua Barat Harus Diakhiri* ["The Voice of the Church for the Oppressed: The Suffering, Bleeding, and Tears of God's People in West Papua Must End"] by Socratez Sofyan Yoman, *Enam Jalan Menuju Tuhan* ["Six Paths to God"] by Darmawan MM, and *Mengungkap Misteri Keragaman Agama* ["Revealing the Mystery of Religious Diversity"] by Syahrudin Ahmad.
42 *Membongkar Gurita Cikeas: Di Balik Skandal Bank Centurty* ["Unmasking the Cikeas Octopus: Behind the Century Scandal"] by George Junus Aditjondro. The book was never formally banned.
43 Three simultaneous challenges to the Law were submitted and decided together. The other two challenges were submitted by Darmawan MM (No. 6/PUU-VIII/2010), and Muhidin M. Dahlan and members of the Islamic Students Association-Assembly to Save the Organisation (*Himpunan Mahasiswa Islam-Majelis Penyelamat Organisasi*, HMI-MPO) (No. 13/PUU-VIII/2010).
44 Interview, Nurkholis Hidayat, August 2021.
45 Interview, Nurkholis Hidayat, August 2021.
46 Constitutional Court Decision No. 6–13–20/PUU-VIII/2010, 244.

47 Constitutional Court Decision No. 6–13–20/PUU-VIII/2010, 239–40.
48 Constitutional Court Decision No. 6–13–20/PUU-VIII/2010, 242.
49 Interview, Nurkholis Hidayat, August 2021.
50 Interview, Muhamad Isnur, August 2021.
51 Joint Ministerial Decision No. 3 of 2008, No. Kep-033/A/JA/6/2008 and No. 199 of 2008 on Warning and Order to Followers, Members and Administrators of Jemaat Ahmadiyah Indonesian (JAI) and to the General Public.
52 Interview, Asfinawati, December 2019.
53 Interview, Asfinawati, August 2021.
54 Constitutional Court Decision No. 140/PUU-VII/2009, 271–73.
55 Constitutional Court Decision No. 140/PUU-VII/2009, 274.
56 Constitutional Court Decision No. 140/PUU-VII/2009, 275.
57 Constitutional Court Decision No. 140/PUU-VII/2009, 275.
58 Constitutional Court Decision No. 140/PUU-VII/2009, 274.
59 Constitutional Court Decision No. 140/PUU-VII/2009, 276.
60 Constitutional Court Decision No. 140/PUU-VII/2009, 289.
61 Constitutional Court Decision No. 140/PUU-VII/2009, 288–89.
62 Constitutional Court Decision No. 140/PUU-VII/2009, 288.
63 Interview, Asfinawati, December 2019.
64 The elucidation to Article 5(1) of Law No. 48 of 2009 on Judicial Power states that judges' decisions should be consistent with the law and the sense of justice in the community.
65 Interview, Muhamad Isnur, August 2021.
66 Interview, Muhamad Isnur, August 2021.
67 Law No. 7 of 2020 revised Law No. 24 of 2003 on the Constitutional Court, removing the previous term limit for serving judges of five years with the possibility of re-appointment for a second term (by deleting Article 22 of Law No. 24 of 2003). The revised 2020 Law introduced a new minimum age for judges of 55 (Article 15(2)(d)), but also stated that judges can serve on the bench until they turn 70, unless they meet certain conditions for dismissal (Article 23). As this book went to publication, legislators were again making attempts to revise the Law.
68 Since its establishment, the court has somewhat problematically avoided striking out provisions and creating a legal vacuum, instead declaring provisions to be "conditionally unconstitutional" or "conditionally constitutional" unless interpreted in a certain way (Butt 2015, 73).
69 Constitutional Court Decision No. 91/PUU-XVIII/2020, 412.
70 Constitutional Court Decision No. 91/PUU-XVIII/2020, 416–17.
71 Constitutional Court Decision No. 91/PUU-XVIII/2020, 417.
72 Constitutional Court Decision No. 91/PUU-XVIII/2020, 413.
73 Government Regulation in Lieu of Law No. 2 of 2022 on Job Creation. It was eventually passed into law as Law No. 6 of 2023.
74 Constitutional Court Decision No. 91/PUU-XVIII/2020, 414.
75 Interview, Muhamad Isnur, February 2023.
76 Constitutional Court Decision Nos. 54/PUU-XXI/2023, 40/PUU-XXI/2023, 41/PUU-XXI/2023, 46/PUU-XXI/2023, and 50/PUU-XXI/2023.
77 Constitutional Court Decision No. 90/PUU-XXI/2023.
78 Other examples include the 2020 revisions to the Law on the Constitutional Court discussed above (Law No. 7 of 2020), Law No. 3 of 2020 on Mining, Law No. 3 of 2022 on the National Capital and Law No. 13 of 2022 on Lawmaking, which revised Law No. 12 of 2011 on Lawmaking.
79 Interview, Arif Maulana, March 2023.
80 Interviews, Asfinawati, August 2021; Erasmus Abraham Todu Napitupulu, February 2023; Muhamad Isnur, February 2023; Siti Rakhma Mary Herwati, March 2023.
81 Interview, Pratiwi Febry, March 2023.

82 Law No. 30 of 2014 on Government Administration broadened this definition to include decisions of a general nature as well as factual acts by the government (Article 87) (Bedner and Wiratraman 2019, 140).
83 Padang State Administrative Court (PTUN Padang) Decision No. 2/P/FP/2017/PTUN.PDG. The 26 permits were held by companies that had failed to obtain compulsory 'clean and clear' certificates. To get a certificate, companies need to demonstrate that they have no taxes owing, they have met their environmental commitments, and their concessions do not encroach on protected forest, or those of other companies, among other requirements (Mongabay 2017).
84 PTUN Surabaya Decision No. 44/G/2019/PTUN.SBY.
85 The 1962 New York Agreement provided for the transfer of administration of the territory from the Netherlands to Indonesia, following a seven-month period of UN administration. No Papuans participated in the negotiations for the agreement (Musgrave 2015, 217). Under the agreement, Indonesia was required to implement an act of self-determination before the end of 1969. This so-called 'Act of Free Choice', conducted in mid-1969, has been extremely contentious because only 1,022 Papuans selected by Indonesian officials were able to participate, and all voted in favour of incorporation into Indonesia (Musgrave 2015, 220).
86 Koman, Veronica (@VeronicaKoman). "Hey monyet, ke luar!" "Hey monkeys, get out!" "Hey, monkeys! Monkeys!" 19 August 2019, 23:36. Tweet.
87 Paying protestors to attend political demonstrations is common in Indonesia (Andrews 2017).
88 Interview, Muhamad Isnur, August 2021. After the administrative court case had been decided, the CSO coalition did eventually challenge Article 40(2)(b) of the Electronic Information and Transactions Law at the Constitutional Court. The Court rejected the challenge on 27 October 2021 (Decision No. 81/PUU-XVIII/2020).
89 Article 40(2)(a): "The government is obliged to prevent the dissemination and use of electronic documents and information that contain content prohibited by law". Article 40(2)(b): "In conducting prevention as described in point 2(a), the government is authorised to cut access or order operators to cut access to electronic documents and information that contain content prohibited by law".
90 Decision No. 230/G/TF/2019/PTUN-JKT, 245–46.
91 Decision No. 230/G/TF/2019/PTUN-JKT, 255.
92 Decision No. 230/G/TF/2019/PTUN-JKT, 276–77.
93 Decision No. 230/G/TF/2019/PTUN-JKT, 275–76.
94 Interview, Muhamad Isnur, August 2021.
95 Decision No. 230/G/TF/2019/PTUN-JKT, 246.
96 Notably, in the Constitutional Court's review of the Electronic Information and Transactions Law, the government's expert witness, Professor Henry Subiakto, testified that he thought the PTUN had made the correct decision, and the government did not have the authority to block the internet under Article 40(2)(b). See Summary of Court Proceedings [*Risalah Sidang*] Case No. 81/PUU-XVIII/2020, 3 August 2021, 34.
97 The government had used the strategy during previous public protests and unrest, for example following the announcement of the 2019 presidential election results in May 2019 (Putra 2019).
98 The Indonesian legal system does, however, deal with the slightly different mechanism of class action lawsuits, which are detailed in Supreme Court Regulation No. 1 of 2002 on Procedures for Class Actions. They are also mentioned in Law No. 23 of 1997 on Environmental Management (as amended by Law No. 32 of 2009 on Protection and Management of the Environment), Law No. 8 of 1999 on Consumer Protection and Law No. 41 of 1999 on Forestry.

99 Central Jakarta District Court Decision No. 28/Pdt.G/2003/PN.Jkt.Pst and Jakarta High Court Decision No. 480/PDT/2005/PT DKI.
100 Interview, Muhamad Isnur, August 2021.
101 Jakarta Governor Decision No. 1107 of 2019 on Revision of Jakarta Governor Decision No. 1042 of 2018 on Regional Strategic Activities.
102 This was issued in part as an implementing regulation for the Omnibus Law on Job Creation. But it also replaced Government Regulation No. 41 of 1999 on Control of Air Pollution, and included increased detail on the management of air pollution.
103 Interview, Muhamad Isnur, October 2019.
104 Although there have been no suggestions of bribery in this case, delaying decision making is a common tactic used by corrupt judges in Indonesia, and is thought to be a "sign" for parties to come forward and make contact with the judges (Butt and Lindsey 2018, 301).
105 Central Jakarta District Court Decision No. 374/PDT.G/LH/2019/PN JKT.PST.
106 Central Jakarta District Court Decision No. 374/PDT.G/LH/2019/PN JKT.PST, 279.
107 The decision also instructed: the minister of forestry and environment to supervise the governors of Jakarta, Banten and West Java in conducting an inventory of cross-border emissions among the three provinces; the minister of home affairs to supervise the performance of the Jakarta governor in controlling air pollution; and the minister of health to calculate the health impacts due to air pollution in Jakarta to support the Jakarta governor's preparation of an Air Pollution Control Strategy and Action Plan. The decision also included detailed instructions for the Jakarta governor on monitoring and controlling air pollution.
108 Jakarta High Court Decision No. 549/PDT/2022/PT DKI.

References

ABC News. 2020. "Indonesian Woman Who Took a Dog into a Mosque Acquitted of Blasphemy Charges on Account of Mental Health." *ABC News*, 6 February 2020. https://www.abc.net.au/news/2020-02-06/indonesian-acquitted-of-blasphemy-for-taking-dog-into-mosque/11941176.

Advocacy Coalition for Violence Against Gender and Sexual Minority Groups. 2017. "*Press Release: Condemning the Inhumane Act of Detention Against the Gay Community in Atlantis Gym & Sauna.*" 22 May 2017. https://forum-asia.org/?p=23959.

Andrews, Sally. 2017. "Packed Lunch Protesters: Outrage for Hire in Indonesia." *The Diplomat*, 3 February 2017. https://thediplomat.com/2017/02/packed-lunch-protesters-outrage-for-hire-in-indonesia/.

Argawati, Utami. 2023. "Tolak Uji Formil, Uji Materiil UU Cipta Kerja Dilanjutkan." *Mahkamah Konstitusi Republik Indonesia*, 2 October 2023. https://www.mkri.id/index.php?page=web.Berita&id=19605&menu=2.

Assegaf, Rifqi. 2018. "20 Years of Judicial Reform: Mission Not Yet Accomplished." *Indonesia at Melbourne*, 2 May 2018. https://indonesiaatmelbourne.unimelb.edu.au/20-years-of-judicial-reform-mission-not-accomplished/.

Asshiddiqie, Jimly. 2022. "The DPR Attacks the Constitutional Court – and Judicial Independence." *Indonesia at Melbourne*, 10 October 2022. https://indonesiaatmelbourne.unimelb.edu.au/the-dpr-attacks-the-constitutional-court-and-judicial-independence/.

BBC Indonesia. 2017. "Dinyatakan Menodai Agama Tokoh Eks Gafatar Divonis 3–5 Tahun." *BBC Indonesia*, 7 March 2017. https://www.bbc.com/indonesia/indonesia-39189909.

BBC Indonesia. 2021. "Presiden Jokowi 'Segera Laksanakan Putusan MK' Soal UU Cipta Kerja, MK Dinilai Cari 'Jalan Tengah.'" *BBC Indonesia*, 25 November 2021. https://www.bbc.com/indonesia/indonesia-59413391.

Bedner, Adriaan, and Herlambang P. Wiratraman. 2019. "The Administrative Courts: The Quest for Consistency." In *The Politics of Court Reform: Judicial Change and Legal Culture in Indonesia*, edited by Melissa Crouch, 133–148. Cambridge, United Kingdom: Cambridge University Press. https://doi.org/10.1017/9781108636131.006.

Bernie, Mohammad. 2019. "KontraS Hingga YLBHI Somasi Kominfo Soal Blokir Internet di Papua." *Tirto.id*, 23 August 2019. https://tirto.id/kontras-hingga-ylbhi-somasi-kominfo-soal-blokir-internet-di-papua-egTk.

Bruinessen, Martin van, ed. 2013. *Contemporary Developments in Indonesian Islam: Explaining the "Conservative Turn."* Singapore: ISEAS Publishing. https://doi.org/10.1355/9789814414579.

Butt, Simon. 2015. *The Constitutional Court and Democracy in Indonesia*. Leiden; Boston: Brill Nijhoff. https://doi.org/10.1163/9789004250598.

Butt, Simon. 2020. "The 2020 Constitutional Court Law Amendments: A 'Gift' to Judges?" *Indonesia at Melbourne*, 3 September 2020. https://indonesiaatmelbourne.unimelb.edu.au/the-2020-constitutional-court-law-amendments-a-gift-to-judges/.

Butt, Simon. 2021. "Aceh and Islamic Criminal Law in the Courts." In *Crime and Punishment in Indonesia*, edited by Tim Lindsey and Helen Pausacker, 535–558. Routledge Law in Asia. London and New York: Routledge. https://doi.org/10.4324/9780429455247-26.

Butt, Simon, and Tim Lindsey. 2018. *Indonesian Law*. Oxford; New York: Oxford University Press. https://doi.org/10.1093/oso/9780199677740.001.0001.

Butt, Simon, and Tim Lindsey. 2012. *The Constitution of Indonesia: A Contextual Analysis*. Constitutional Systems of the World. Oxford; Portland: Hart Publishing. https://doi.org/10.5040/9781509955732.

Chang, Wen-Chen. 2010. "Public Interest Litigation in Taiwan: Strategy for Law and Policy Changes in the Course of Democratisation." In *Public Interest Litigation in Asia*, edited by Po Jen Yap and Holning Lau, 136–160. London: Routledge. https://doi.org/10.4324/9780203842645.

Clarke, Ross. 2008. "The Bali Bombing, East Timor Trials and the Aceh Human Rights Court: Retrospectivity, Impunity and Constitutionalism." In *Indonesia: Law and Society*, edited by Tim Lindsey, 2nd Edition, 430–454. Sydney, Australia: The Federation Press.

CNN Indonesia. 2022a. "Aswanto Dicopot Dari Hakim Konstitusi Karena Anulir Produk DPR." *CNN Indonesia*, 30 September 2022. https://www.cnnindonesia.com/nasional/20220930164056-32-854832/aswanto-dicopot-dari-hakim-konstitusi-karena-anulir-produk-dpr.

CNN Indonesia. 2022b. "Menkumham Yasonna Respons Penolakan RKUHP: Gugat Saja ke MK." *CNN Indonesia*, 5 December 2022. https://www.cnnindonesia.com/nasional/20221205144713-12-883018/menkumham-yasonna-respons-penolakan-rkuhp-gugat-saja-ke-mk.

CNN Indonesia. 2022c. "YLBHI Kecam Perppu Ciptaker Jokowi: Ini Kudeta Konstitusi." *CNN Indonesia*, 30 December 2022. https://www.cnnindonesia.com/nasional/20221230173139-32-894254/ylbhi-kecam-perppu-ciptaker-jokowi-ini-kudeta-konstitusi.

Crouch, Melissa. 2012a. "Law and Religion in Indonesia: The Constitutional Court and the Blasphemy Law." *Asian Journal of Comparative Law* 7 (1): 1–46. https://doi.org/10.1515/1932-0205.1391.

Crouch, Melissa. 2012b. "Asia-Pacific: Criminal (in)Justice in Indonesia: The Cikeusik Trials." *Alternative Law Journal* 37 (1): 54–56. https://doi.org/10.1177/1037969x1203700115.

Crouch, Melissa. 2016a. "Constitutionalism, Islam and the Practice of Religious Deference: The Case of the Indonesian Constitutional Court." *Australian Journal of Asian Law* 16 (2): 1–15.

Crouch, Melissa. 2016b. "Legislating Inter-Religious Harmony: Attempts at Reform in Indonesia." In *Religion, Law and Intolerance in Indonesia*, edited by Tim Lindsey and Helen Pausacker, 95–112. London and New York: Routledge. https://doi.org/10.4324/9781315657356-5.

Crouch, Melissa. 2017a. "Negotiating Legal Pluralism in Court: Fatwa and the Crime of Blasphemy in Indonesia." In *Pluralism, Transnationalism and Culture in Asian Law*, edited by Gary F. Bell, 231–256. Singapore: ISEAS Publishing. https://doi.org/10.1355/9789814762724-013.

Crouch, Melissa. 2017b. "Ahok, Indonesia's 'Nemo', Sentenced to Jail." *Policy Forum*, 10 May 2017. https://www.policyforum.net/ahok-indonesias-nemo-sentenced-jail/.

Damayanti, Ismi, and Erwida Maulia. 2023. "Jakarta Orders Civil Servants to Work from Home Due to Pollution." *Nikkei Asia*, 21 August 2023. https://asia.nikkei.com/Spotlight/Environment/Jakarta-orders-civil-servants-to-work-from-home-due-to-pollution.

Davidson, Helen. 2019. "Indonesia Arrests Dozens of West Papuans Over Claim Flag Was Thrown in Sewer." *The Guardian*, 18 August 2019. http://www.theguardian.com/world/2019/aug/18/indonesia-arrests-dozens-of-west-papuans-over-claim-flag-was-thrown-in-sewer.

Davidson, Helen, and Ben Doherty. 2019. "Indonesian President Calls for Calm After Violent Protests in West Papua." *The Guardian*, 20 August 2019. https://www.theguardian.com/world/2019/aug/20/indonesian-president-calls-for-calm-after-violent-protests-in-west-papua.

Detik. 2010. "Sejarah Panjang Pelarangan Buku di Indonesia." *Detik*, 17 March 2010. https://news.detik.com/berita/d-1319432/sejarah-panjang-pelarangan-buku-di-indonesia.

Dressel, Björn, and Tomoo Inoue. 2018. "Megapolitical Cases Before the Constitutional Court of Indonesia Since 2004: An Empirical Study." *Constitutional Review* 4 (2): 157–187. https://doi.org/10.31078/consrev421.

Faizal, Achmad. 2019. "Dituding Perkeruh Kasus Mahasiswa Papua, Kantor LBH Surabaya Didemo." *Kompas.com*, 29 August 2019. https://regional.kompas.com/read/2019/08/29/20090881/dituding-perkeruh-kasus-mahasiswa-papua-kantor-lbh-surabaya-didemo.

Fenwick, Stewart. 2009. "Administrative Law and Judicial Review in Indonesia: The Search for Accountability." In *Administrative Law and Governance in Asia*, edited by Tom Ginsburg and Alan K. Chen. London and New York: Routledge. https://doi.org/10.4324/9780203888681.

Ford, Michele. 2006. "After Nunukan: The Regulation of Indonesian Migration to Malaysia." In *Mobility, Labour Migration and Border Controls in Asia*, edited by Amarjit Kaur and Ian Metcalfe, 228–247. London: Palgrave Macmillan UK. https://doi.org/10.1057/9780230503465_12.

Freedom House. 2019. "*Freedom on the Net: The Crisis of Social Media.*" Freedom House. https://freedomhouse.org/report/freedom-net/2019/crisis-social-media.

Fu, Hualing. 2011. "Challenging Authoritarianism Through Law: Potentials and Limit." *National Taiwan University Law Review* 6 (1): 339–365.

Ginsburg, Tom. 2007. "Law and the Liberal Transformation of the Northeast Asian Legal Complex in Korea and Taiwan." In *Fighting for Political Freedom: Comparative Studies of the Legal Complex and Political Liberalism*, edited by Terence C. Halliday, Lucien Karpik, and Malcolm M. Feeley, 43–63. Oxford: Hart Publishing. https://doi.org/10.5040/9781472560179.ch-002.

Goedde, Patricia. 2009. "From Dissidents to Institution-Builders: The Transformation of Public Interest Lawyers in South Korea." *East Asia Law Review* 4 (1): 63–89. https://doi.org/10.1017/9781108893947.009.

Greenstone, Michael, and Qing (Claire) Fan. 2019. "*Indonesia's Worsening Air Quality and Its Impact on Life Expectancy.*" Air Quality Life Index. Energy Policy Institute at the University of Chicago (EPIC).

Hasani, Ismail. 2016. "The Decreasing Space for Non-Religious Expression in Indonesia: The Case of Atheism." In *Religion, Law and Intolerance in Indonesia*, edited by Tim Lindsey and Helen Pausacker, 197–210. London and New York: Routledge. https://doi.org/10.4324/9781315657356-10.

Hendrianto, Stefanus. 2018. *Law and Politics of Constitutional Courts: Indonesia and the Search for Judicial Heroes.* Comparative Constitutionalism in Muslim Majority States. London and New York: Routledge. https://doi.org/10.4324/9781315100043.

Hilbink, Thomas M. 2004. "You Know the Type: Categories of Cause Lawyering." *Law & Social Inquiry* 29 (3): 657–698. https://doi.org/10.1086/430155.

Hosen, Nadirsyah. 2017. "The Indonesian Courts." In *Asia-Pacific Judiciaries: Independence, Impartiality and Integrity*, edited by HP Lee and Marilyn Pittard, 186–208. Cambridge: Cambridge University Press. https://doi.org/10.1017/9781316480946.011.

Hukum Online. 2006. "Citizen Law Suit Kasus Nunukan Kalah di Tingkat Banding." *Hukum Online*, 27 November 2006. https://www.hukumonline.com/berita/baca/hol15797/citizen-law-suit-kasus-nunukan-kalah-di-tingkat-banding/.

Hukum Online. 2010. "Silakan Gugat Wewenang Kejaksaan Melarang Peredaran Buku." *Hukum Online*, 20 January 2010. https://www.hukumonline.com/berita/baca/lt4b55f03a8ab97/silakan-gugat-wewenang-kejaksaan-melarang-peredaran-buku?page=all.

Human Rights Watch. 2013. "*In Religion's Name: Abuses against Religious Minorities in Indonesia*," 28 February 2013. https://www.hrw.org/report/2013/02/28/religions-name/abuses-against-religious-minorities-indonesia.

Human Rights Watch. 2016. "*Indonesia: Persecution of Gafatar Religious Group*," 29 March 2016. https://www.hrw.org/news/2016/03/29/indonesia-persecution-gafatar-religious-group.

Ihsanuddin. 2020. "Jokowi Persilakan Penolak UU Cipta Kerja Gugat ke MK." *Kompas.com*, 9 October 2020. https://nasional.kompas.com/read/2020/10/09/17540571/jokowi-persilakan-penolak-uu-cipta-kerja-gugat-ke-mk.

Kahfi, Kharishar. 2017. "Former MK Justice Sentenced to 8 Years in Prison." *The Jakarta Post*, 4 September 2017. https://www.thejakartapost.com/news/2017/09/04/corruption-court-to-issue-verdict-for-former-mk-justice.html.

Kariza, Divya. 2023. "Silent, Invisible Danger at Cirebon Coast." *The Jakarta Post*, 31 August 2023. https://www.thejakartapost.com/longform/2023/08/31/silent-invisible-danger-at-cirebon-coast.html.

Kompas.com. 2023. "HUT ke-30, ICEL Berikan Penghargaan kepada Organisasi dan Tokoh Hukum Lingkungan Hidup." *Kompas*, 29 August 2023. https://www.kompas.

id/baca/adv_post/hut-ke-30-icel-berikan-penghargaan-kepada-organisasi-dan-tokoh-hukum-lingkungan-hidup.

Kompas.com. 2010a. "YLBHI dan Ikapi Minta Cabut Pelarangan Buku." *Kompas.com*, 8 January 2010. https://edukasi.kompas.com/read/2010/01/08/02493611/ylbhi.dan.ikapi.minta.cabut.pelarangan.buku.

Kompas.com. 2010b. "Pelarangan Buku Diajukan ke Mahkamah Konstitusi." *Kompas.com*, 9 March 2010. https://edukasi.kompas.com/read/2010/03/09/03020961/pelarangan.buku.diajukan.ke.mahkamah.konstitusi.

Kompas.com. 2016. "Tiga Pimpinan Gafatar Ditahan, Pengacara Sebut Polisi Lakukan Kriminalisasi." *Kompas.com*, 26 May 2016. https://nasional.kompas.com/read/2016/05/26/14015311/tiga.pimpinan.gafatar.ditahan.pengacara.sebut.polisi.lakukan.kriminalisasi.

Kusuma, Edward Febriyatri. 2017. "Petinggi Gafatar Musadeq Cs Bacakan Pleidoi di PN Jaktim Siang Ini." *Detik*, 16 February 2017. https://news.detik.com/berita/d-3424074/petinggi-gafatar-musadeq-cs-bacakan-pleidoi-di-pn-jaktim-siang-ini.

Lai, Yerica. 2022. "Calls for Chief Justice to Resign Ahead of Marrying Jokowi's Sister." *The Jakarta Post*, 25 March 2022. https://www.thejakartapost.com/paper/2022/03/25/calls-for-chief-justice-to-resign-ahead-of-marrying-jokowis-sister.html.

Lamb, Kate. 2019. "Jakarta Residents to Sue Government Over Severe Air Pollution." *The Guardian*, 2 July 2019. https://www.theguardian.com/world/2019/jul/02/jakarta-residents-to-sue-government-over-severe-air-pollution.

Lamb, Kate, and Ben Doherty. 2019. "West Papua Protests: Indonesia Deploys 1,000 Soldiers to Quell Unrest, Cuts Internet." *The Guardian*, 22 August 2019. https://www.theguardian.com/world/2019/aug/22/west-papua-protests-indonesia-deploys-1000-soldiers-to-quell-unrest.

Lindsey, Tim. 2004. "Legal Infrastructure and Governance Reform in Post-Crisis Asia: The Case of Indonesia." *Asian-Pacific Economic Literature* 18 (1): 12–40. https://doi.org/10.1111/j.1467-8411.2004.00142.x.

Lindsey, Tim. 2008. "Constitutional Reform in Indonesia: Muddling Towards Democracy." In *Indonesia: Law and Society*, edited by Tim Lindsey, 2nd Edition, 23–47. Sydney, Australia: The Federation Press.

Lindsey, Tim, and Simon Butt. 2016. "State Power to Restrict Religious Freedom: An Overview of the Legal Framework." In *Religion, Law and Intolerance in Indonesia*, edited by Tim Lindsey and Helen Pausacker, 48–69. London and New York: Routledge. https://doi.org/10.4324/9781315657356-2.

Lindsey, Tim, and Melissa Crouch. 2013. "Cause Lawyers in Indonesia: A House Divided." *Wisconsin International Law Journal* 31 (3): 620–645.

Lindsey, Tim, and Helen Pausacker. 2016a. "Introduction: Religion, Law and Intolerance in Indonesia." In *Religion, Law and Intolerance in Indonesia*, 30–46. London and New York: Routledge. https://doi.org/10.4324/9781315657356-1.

Lindsey, Tim, and Helen Pausacker. eds. 2016b. *Religion, Law and Intolerance in Indonesia*. London and New York: Routledge. https://doi.org/10.4324/9781315657356.

Loasana, Nina A. 2023. "Court Upholds Jobs Law as Labor Unions Stage Rallies." *The Jakarta Post*, 2 October 2023. https://www.thejakartapost.com/indonesia/2023/10/02/court-upholds-jobs-law-as-labor-unions-stage-rallies.html.

Mahkamah Agung Republik Indonesia. 2009. *Class Action and Citizen Lawsuit*. Puslitbang Hukum dan Peradilan, Mahkamah Agung Republik Indonesia.

Makin, Al. 2016. *Challenging Islamic Orthodoxy: Accounts of Lia Eden and Other Prophets in Indonesia*. Switzerland: Springer International Publishing. https://doi.org/10.1007/978-3-319-38978-3.

Makin, Al. 2019. "Returning to the Religion of Abraham: Controversies over the Gafatar Movement in Contemporary Indonesia." *Islam and Christian–Muslim Relations* 30 (1): 87–104. https://doi.org/10.1080/09596410.2019.1570425.

Marvela. 2019. "YLBHI: Pembatasan Internet Di Papua Adalah Rasis." *Tempo.co*, 28 August 2019. https://nasional.tempo.co/read/1241286/ylbhi-pembatasan-internet-di-papua-adalah-rasis.

Maulia, Erwida. 2017. "Activists Decry Arrests of 141 Men in Alleged Jakarta Gay Sauna." *Nikkei Asia*, 22 May 2017. https://asia.nikkei.com/Politics/Activists-decry-arrests-of-141-men-in-alleged-Jakarta-gay-sauna.

McCann, Michael. 2006. "Legal Mobilization and Social Reform Movements: Notes on Theory and Its Application." In *Law and Social Movements*, edited by Michael McCann, 3–32. The International Library of Essays in Law and Society. London and New York: Routledge. https://doi.org/10.4324/9781315091983-1.

Meili, Stephen. 2005. "Cause Lawyering for Collective Justice: A Case Study of the Ampero Colectivo in Argentina." In *The Worlds Cause Lawyers Make: Structure and Agency in Legal Practice*, edited by Austin Sarat and Stuart A. Scheingold, 383–409. Stanford, California: Stanford University Press. https://doi.org/10.1515/9781503625440-015.

Meili, Stephen. 2009. "Staying Alive: Public Interest Law in Contemporary Latin America." *International Review of Constitutionalism* 9 (1): 43–71.

Menchik, Jeremy. 2014. "Productive Intolerance: Godly Nationalism in Indonesia." *Comparative Studies in Society and History* 56 (3): 591–621. https://doi.org/10.1017/s0010417514000267.

Mietzner, Marcus. 2010. "Political Conflict Resolution and Democratic Consolidation in Indonesia: The Role of the Constitutional Court." *Journal of East Asian Studies* 10 (3): 397–424. https://doi.org/10.1017/s1598240800003672.

Mongabay. 2017. "Coal Miners Owe the Indonesian Government Hundreds of Millions of Dollars." *Mongabay*, 8 May 2017. https://news.mongabay.com/2017/05/coal-miners-owe-the-indonesian-government-hundreds-of-millions-of-dollars/.

Musgrave, Thomas D. 2015. "An Analysis of the 1969 Act of Free Choice in West Papua." In *Sovereignty, Statehood and State Responsibility: Essays in Honour of James Crawford*, edited by Christine Chinkin and Freya Baetens, 209–228. Cambridge: Cambridge University Press. https://doi.org/10.1017/cbo9781107360075.017.

Myllyvirta, Lauri, Hubert Thieriot, and Katherine Hasan. 2023. "Work from Home (WFH) and Other Gimmicks Cannot Clear Jakarta's Air." *Centre for Research on Energy and Clean Air (CREA)*, 25 August 2023. https://energyandcleanair.org/work-from-home-wfh-and-other-gimmicks-cannot-clear-jakartas-air/.

Nababan, Willy Medi Christian. 2023. "Komitmen Presiden Atasi Pencemaran Udara Dipertanyakan." *Kompas*, 4 February 2023. https://www.kompas.id/baca/humaniora/2023/02/03/komitmen-presiden-atasi-pencemaran-udara-dipertanyakan.

Nardi, Dominic. 2018. "Indonesia's Constitutional Court and Public Opinion." *New Mandala*, 22 February 2018. https://www.newmandala.org/indonesias-constitutional-court-public-opinion/.

Nugraheny, Dian Erika. 2022. "Jokowi Resmi Lantik Guntur Hamzah Jadi Hakim MK Pengganti Aswanto." *Kompas.com*, 3 November 2022. https://nasional.kompas.com/read/2022/11/23/09455801/jokowi-resmi-lantik-guntur-hamzah-jadi-hakim-mk-pengganti-aswanto.

Nurbaiti, Alya. 2020. "Papuan Internet Lawsuit Intended to Push for Good Governance: Civil Groups." *The Jakarta Post*, 5 June 2020. https://www.thejakartapost.com/news/2020/06/05/papuan-internet-lawsuit-intended-to-push-for-good-governance-civil-groups.html.

Pausacker, Helen. 2021. "Homosexuality and the Law in Indonesia." In *Crime and Punishment in Indonesia*, edited by Tim Lindsey and Helen Pausacker, 430–462. Routledge Law in Asia. London and New York: Routledge. https://doi.org/10.4324/9780429455247-21.

Peerenboom, Randall. 2004. "Varieties of Rule of Law: An Introduction and Provisional Conclusion." In *Asian Discourses of Rule of Law: Theories and Implementation of Rule of Law in Twelve Asian Countries, France, and the US*, edited by Randall Peerenboom, 1–53. London and New York: RoutledgeCurzon. https://doi.org/10.4324/9780203317938_chapter_1.

Pew Research Center. 2013. "*The Global Divide on Homosexuality*." Pew Research Center. https://www.pewresearch.org/global/2013/06/04/the-global-divide-on-homosexuality/.

Pompe, Sebastiaan. 2005. *The Indonesian Supreme Court: A Study of Institutional Collapse*. Ithaca, New York: Cornell University Press. https://doi.org/10.7591/9781501718861.

Prireza, Adam. 2019. "Pencemaran Udara Jakarta, LBH Buka Posko Gugatan Ke Pemprov DKI." *Tempo.co*, 14 April 2019. https://metro.tempo.co/read/1195663/pencemaran-udara-jakarta-lbh-buka-posko-gugatan-ke-pemprov-dki.

Prireza, Adam. 2021. "Jokowi Banding Putusan Polusi Udara Jakarta, LBH Sayangkan." *Tempo.co*, 1 October 2021. https://metro.tempo.co/read/1512572/jokowi-banding-putusan-polusi-udara-jakarta-lbh-sayangkan.

Pristiandaru, Danur Lambang. 2023. "MA Tolak Kasasi, Presiden Dinyatakan Lawan Hukum Dalam Kasus Polusi Jakarta Halaman All." *Kompas.com*, 19 November 2023. https://lestari.kompas.com/read/2023/11/19/090000586/ma-tolak-kasasi-presiden-dinyatakan-lawan-hukum-dalam-kasus-polusi-jakarta.

Pujianti, Sri. 2023. "MKMK Berhentikan Anwar Usman Dari Jabatan Ketua Mahkamah Konstitusi." *Mahkamah Konstitusi Republik Indonesia*, 7 November 2023. https://www.mkri.id/index.php?page=web.Berita&id=19751&menu=2.

Putra, Eka Nugraha. 2019. "Selain Tidak Efektif Dan Ancam Demokrasi, Blokir Internet Pemerintah Tidak Punya Dasar Hukum Yang Kuat." *The Conversation*, 18 September 2019. http://theconversation.com/selain-tidak-efektif-dan-ancam-demokrasi-blokir-internet-pemerintah-tidak-punya-dasar-hukum-yang-kuat-123451.

Rahman, Abdul. 2018. "Egalitarian Agrarian Politics and the Authoritarian Challenge: The Emergence and Destruction of Indonesia's Gafatar Movement." In *Emancipatory Rural Politics Initiative (ERPI) 2018 International Conference: Authoritarian Populism and the Rural World*. The Hague, Netherlands: International Institute of Social Studies (ISS).

Rekosh, Edwin. 2008. "Constructing Public Interest Law: Transnational Collaboration and Exchange in Central and Eastern Europe." *UCLA Journal of International Law and Foreign Affairs* 13 (1): 55.

Reuters. 2016. "Indonesia Evacuates Hundreds of Members of Sect After Clashes." *Reuters*, 30 January 2016. http://blogs.reuters.com/faithworld/2016/01/30/indonesia-evacuates-hundreds-of-members-of-sect-after-clashes/.

Reuters. 2019. "Violent Protest Erupts in Capital of Indonesia's Papua." *Reuters*, 30 August 2019. https://ar.reuters.com/article/us-indonesia-papua-idUSKCN1VJ195.

Riana, Friski. 2017. "LBH Jakarta Kecam Penggerebekan Pesta Gay Di Kelapa Gading." *Tempo.co*, 22 May 2017. https://metro.tempo.co/read/877501/lbh-jakarta-kecam-penggerebekan-pesta-gay-di-kelapa-gading.
Ribka, Stefani. 2016. "Gafatar Ban Seen as a Setback." *The Jakarta Post*, 26 March 2016. https://www.thejakartapost.com/news/2016/03/26/gafatar-ban-seen-a-setback.html.
Ridwan, Taufik. 2021. "Anies: DKI Tidak Banding Terhadap Putusan Hakim Soal Polusi Udara." *Antara*, 16 September 2021. https://www.antaranews.com/berita/2397217/anies-dki-tidak-banding-terhadap-putusan-hakim-soal-polusi-udara.
Scheingold, Stuart A. 2004. *The Politics of Rights: Lawyers, Public Policy and Political Change*. Ann Arbor, Michigan: University of Michigan Press. https://doi.org/10.3998/mpub.6766.
Scheingold, Stuart A., and Austin Sarat. 2004. *Something to Believe In: Politics, Professionalism, and Cause Lawyering*. Stanford, California: Stanford University Press.
Setiawan, Ken. 2020. "A State of Surveillance? Freedom of Expression Under the Jokowi Presidency." In *Democracy in Indonesia: From Stagnation to Regression?*, edited by Thomas Power and Eve Warburton, 254–274. Singapore: ISEAS Publishing. https://doi.org/10.1355/9789814881524-018.
Siddiq, Taufiq. 2019. "Jaksa Agung Usul Razia Buku Kiri Besar-Besaran." *Tempo.co*, 23 January 2019. https://nasional.tempo.co/read/1168152/jaksa-agung-usul-razia-buku-kiri-besar-besaran.
Suaedy, Ahmad. 2016. "The Inter-Religious Harmony (KUB) Bill vs Guaranteeing Freedom of Religion and Belief in Indonesian Public Debate." In *Religion, Law and Intolerance in Indonesia*, edited by Tim Lindsey and Helen Pausacker. London and New York: Routledge. https://doi.org/10.4324/9781315657356-8.
Sudaryono, Leopold. 2019. "Military Comeback or Police Overreaction? The Arrest of Robertus Robet." *Indonesia at Melbourne*, 8 March 2019. https://indonesiaatmelbourne.unimelb.edu.au/military-comeback-or-police-overreaction-the-arrest-of-robertus-robet/.
Sundari, Elisabeth. 2018. "The Indonesian Model of the Citizen Lawsuit: Learning How to Adopt and How to Adapt." *International Trade and Business Law Review* 21: 323.
Susanto, Slamet, and Ganug Nugroho Adi. 2016. "Dozen Missing After Apparently Joining Gafatar." *The Jakarta Post*, 13 January 2016. https://www.thejakartapost.com/news/2016/01/13/dozen-missing-after-apparently-joining-gafatar.html.
Swaragita, Gisela, and Andi Muhammad Ibnu Aqil. 2018. "Citizens to Sue Govt for 'Doing Nothing' About Jakarta's Air Pollution." *The Jakarta Post*, 6 December 2018. https://www.thejakartapost.com/news/2018/12/05/citizens-to-sue-govt-for-doing-nothing-about-jakartas-air-pollution.html.
Syaifullah, Muh. 2016a. "Ada Tiga Kasus Orang Hilang Di Yogya, Ikut Gerakan Radikal?" *Tempo.co*, 6 January 2016. https://nasional.tempo.co/read/733689/ada-tiga-kasus-orang-hilang-di-yogya-ikut-gerakan-radikal.
Syaifullah, Muh. 2016b. "Orang Hilang Diduga Ikut Gerakan Radikal." *Koran Tempo*, 7 January 2016. https://koran.tempo.co/read/berita-utama-jateng/390781/orang-hilang-diduga-ikut-gerakan-radikal.
Syaifullah, Muh. 2016c. "Tiba Di Yogya, Dokter Rica Dikawal Brimob." *Tempo.co*, 11 January 2016. https://nasional.tempo.co/read/734942/tiba-di-yogya-dokter-rica-dikawal-brimob.
Taher, Andrian Pratama. 2017. "LBH Kesulitan Mengadvokasi yang Ditahan Karena Pesta Seks." *Tirto.id*, 22 May 2017. https://tirto.id/lbh-kesulitan-mengadvokasi-yang-ditahan-karena-pesta-seks-cpdF.

Tam, Waikeung. 2010. “Political Transition and the Rise of Cause Lawyering: The Case of Hong Kong.” *Law & Social Inquiry* 35 (3): 663–687. https://doi.org/10.1111/j.1747-4469.2010.01199.x.

The Jakarta Post. 2016. “Outlawed Gafatar Group May Cause Social Tension, Violence: BNPT.” *The Jakarta Post*, 13 January 2016. https://www.thejakartapost.com/news/2016/01/13/outlawed-gafatar-group-may-cause-social-tension-violence-bnpt.html.

Tomsa, Dirk. 2010. “Indonesian Politics in 2010: The Perils of Stagnation.” *Bulletin of Indonesian Economic Studies* 46 (3): 309–328. https://doi.org/10.1080/00074918.2010.522501.

Topsfield, Jewel, and Karuni Rompies. 2017. “Jakarta’s Christian Governor Ahok Jailed for Two Years for Blasphemy.” *Sydney Morning Herald*, 9 May 2017. https://www.smh.com.au/world/jakartas-christian-governor-ahok-jailed-for-two-years-for-blasphemy-20170509-gw0t4y.html.

Utama, Abraham. 2016. “Pemerintah Larang 22 Buku Pasca Reformasi.” *CNN Indonesia*, 18 May 2016. https://www.cnnindonesia.com/nasional/20160518142812-20-131643/pemerintah-larang-22-buku-pasca-reformasi.

Yahya, Achmad Nasrudin. 2020. “Jokowi Diminta Taati Putusan PTUN soal Pemblokiran Internet Papua.” *Kompas.com*, 4 June 2020. https://nasional.kompas.com/read/2020/06/04/15400531/jokowi-diminta-taati-putusan-ptun-soal-pemblokiran-internet-papua.

YLBHI. 2019. “*Pembatasan Akses Internet: Kebijakan, Batasan, dan Dampaknya.*” YLBHI. 3 September 2019. https://ylbhi.or.id/informasi/kegiatan/pembatasan-akses-internet-kebijakan-batasan-dan-dampaknya/.

YLBHI. 2020. *Reformasi Dikorupsi Oligarki: Laporan Hukum dan HAM YLBHI Tahun 2019*. Jakarta: YLBHI. https://ylbhi.or.id/bibliografi/laporan-tahunan/laporan-hukum-dan-ham-ylbhi-tahun-2019-reformasi-dikorupsi-oligarki/.

Yon, Kwan Men, and Simon Hearn. 2016. “*Laying the Foundations of Good Governance in Indonesia’s Judiciary: A Case Study as Part of an Evaluation of the Australia Indonesia Partnership for Justice.*” London: Overseas Development Institute.

5 Movement building

Community organising and legal empowerment

As LBH cause lawyers have grown increasingly sceptical of the formal legal system, they have deprioritised courtroom litigation and turned to less conventional, grassroots strategies more commonly associated with social movements. Adopting an approach typical of grassroots cause lawyers, they see their role as extending beyond a narrow focus on securing victory in the courtroom to a broad range of extra-legal strategies, such as community education and community legal empowerment, community organising, coalition-building, and direct action tactics like protests (Hilbink 2004, 686–87).

Many of these tactics, particularly community legal empowerment and community organising, have been part of the 'structural legal aid' approach since it was first articulated under the New Order, as Chapter 2 discussed. The fall of the New Order did not see a determined effort to abandon these strategies – they are too embedded in LBH's 'organisational DNA'. However, there is no doubt that the democratic transition, and the decline in some regional LBH offices created by funding challenges post-Soeharto, led to community legal empowerment and community organising approaches fading or otherwise being implemented inconsistently across the LBH network. Over the past five years or so, however, as Indonesian democracy has faltered, YLBHI has encouraged a renewed focus on community organising and community legal empowerment at the grassroots.

This chapter builds on Chapter 4, to focus on LBH's community legal empowerment and community organising strategies, which it often deploys in concert with litigation. The chapter also explores the implications of a social movement approach to cause lawyering for the lawyer-client relationship, and some of the limitations of social movement strategies, particularly in relation to gender. The chapter includes a detailed case study of the conflict between farming communities in the North Kendeng Mountains region of Central Java and cement companies seeking to establish mines and factories in the area. The Kendeng conflicts have involved lengthy and complicated legal tussles, as well as intensive investments in community legal empowerment and community organising by LBH. The case demonstrates how cause lawyers can strengthen social movements and support them to make rights claims against the government, revealing the complex interactions between

DOI: 10.4324/9781003486978-8

courtroom litigation, grassroots social movement building, public campaigning and social change.

Community legal empowerment and community organising

Under Soeharto, LBH organised communities to resist the state, recognising that collaboration among social movement groups was essential if they were to offer any real challenge to the repressive New Order. LBH's community legal empowerment and community organising work dipped in prominence following the democratic transition. This may have been partly because LBH viewed organising in opposition to the state as no longer as essential under a more open regime. But it was also because of issues not directly connected to the democratic transition, such as funding constraints and lack of oversight of regional LBH offices by YLBHI, as Chapter 3 discussed.

After Asfin assumed the YLBHI leadership in 2017, however, she and her team promoted a stronger focus on community legal empowerment and community organising as a vital part of the structural aid approach. The recent erosion in Indonesia's democratic quality has left senior YLBHI figures increasingly unconvinced about the merits of elite-focused strategies in achieving real change for the communities they represent. They believe that demands for democracy, human rights and the rule of law must come from the grassroots, and it is LBH's role to facilitate and empower communities to make these demands. Rakhma, director of knowledge management at YLBHI under Asfin and a previous director of LBH Semarang (an LBH office known for its community organising work), has been a strong voice calling for the reinvigoration of LBH's social movement building work:

> What determines whether you are successful or not at LBH is your ideology, your defence of human rights, and of poor and marginalised Indonesians. Community organising is at the core of this. If you are not doing community organising, then you are not LBH! What is the difference with private lawyers, who just receive cases from clients? They don't do community organising, community education, [or work toward] policy change. This is structural legal aid.[1]

LBH deploys community legal empowerment and community organising tactics in tandem, in a way that is both interactive and mutually reinforcing. As part of this approach to case handling, LBH cause lawyers spend time with communities, facilitating small-scale community meetings, discussions and workshops, during which they provide fundamental information on the legal system, the legislative process, human rights and democracy, as well as techniques for conducting social analysis, which involves the analysis of power structures in society. Discussions often include detailed examinations of specific legal products and their potential impacts on a community. In some cases, community members are also provided with citizen journalism training, to encourage documentation of the problems they face.[2] At their most basic level, these community legal empowerment activities aim to simply provide a forum for discussion – a space where community members can describe the problems they are facing and be heard. Yet they also have the more explicit goal of sparking greater community awareness of rights, and providing the foundations for community organising and consolidation (LBH Semarang 2016).

A goal of this legal empowerment work is to stimulate "rights consciousness" and legitimise the feelings of injustice that communities may have endured for years, building on it to encourage political mobilisation (Scheingold 2004, 131–32). These kinds of community legal empowerment activities can encourage community members to see the divergence between formal legal rights and their lived experience, which can then be used as the basis for collective action (Cummings and Eagly 2001, 468). Sarat and Scheingold summarise this well:

> The base is activated not so much by grievances, as such, which in all likelihood have been long experienced – or even by giving voice to those grievances. Instead, activation is a product of identifying grievances with the sense of legitimacy, entitlement, and the collective identity attaching to legality.
>
> (Sarat and Scheingold 2006, 11).

In other words, LBH recognises that injustice may be longstanding in the community, but hopes that community members will recognise their collective power and be motivated to take action when this injustice is connected to the sense of entitlement that is developed through rights consciousness. In doing this kind of work, LBH helps to frame problems, shape expectations, and define what forms of action are feasible. However, its role extends beyond simply advising on the risks or benefits of a certain course of legal action. As Asfin explained:

> We don't position ourselves as advocates who just inform clients about the technical aspects of the case. We position ourselves as a 'stove' – we agitate or provide enthusiasm for the clients, so that they can handle the case themselves. And so that they can see that the problems that they face are also faced by others, and this is the root of it.[3]

The intention is therefore to provide communities with the knowledge to understand the causes of the problems they face and equip them with the skills to independently advocate for the needs of their own communities into the future. At the same time as educating communities about their legal rights, LBH also encourages communities to look critically at the law. Indeed, the activities described above are often grouped under the banner of 'critical legal education' (*pendidikan hukum kritis*). Community members are urged to consider how the problems they face may be the result of an unjust political and legal system. In one LBH document on the organisation's social movement work, the law was described as a "political product". As law is created as a compromise between power holders, the document stated, law often "ends up serving the interests of these power holders" (Wijardo and Perdana 2001, 103). These kinds of statements reflect the CLS arguments that often implicitly underpin the structural legal aid approach. They also illuminate the complex relationship LBH has with the law and the Indonesian legal system. LBH is deeply critical of the law at the same time as it recognises the "liberating potential" of legal rights for marginalised groups (Levitsky 2015, 384).[4]

In many communities, LBH lawyers go beyond legal awareness discussions to train local community members as paralegals. This is a strategy that LBH has deployed since the 1980s, although it only began using the term 'paralegal' years later (Berenschot and Rinaldi 2011, 3; ILRC 2018, 3). This more intensive process involves strengthening community understandings of issues such as social justice, constitutional rights, democracy, the legal system, power relations and social analysis. Advanced paralegal training may involve aspects such as investigating facts, documentation of rights violations, drafting legal documents, and public advocacy and campaigning. Depending on the LBH office, paralegal training may last a few months, and be implemented over several phases. In LBH Bandung, for example, the process takes about three months. The goal is that these paralegals will then work in their communities to further strengthen rights awareness among their fellow community members, as well as strengthening connections between the community and LBH, thereby assisting LBH lawyers in case management (ILRC 2018, 3). Because the paralegals are drawn from the communities affected by injustice, they are considered better at understanding the needs of their communities, and better able to bridge the gap between communities and cause lawyers (Ellmann 1998, 374).

Crucially, LBH does not view the role of community-based paralegals as simply increasing access to the formal legal system for the community. This would be consistent with a proceduralist approach to cause lawyering that considers the legal system capable of delivering justice – LBH lawyers are too cynical about the Indonesian judiciary to promote such a simplistic approach. Rather, LBH-supported paralegals also perform roles common to cause lawyers who work closely with social movements, promoting community empowerment, encouraging community organising and galvanising activism (Cummings 2020, 127–28). Tellingly, LBH has been resistant to government efforts to formalise paralegals through the establishment of a formal accreditation system or stringent competency standards, recognising that they play other important roles that are not strictly legal in nature.[5]

As the discussion above indicates, LBH's community legal empowerment and community organising activities have a strong focus on power relations. Rather than viewing legal or policy reform as an end in itself, this kind of social movement work attempts to alter "the severely disproportionate allocations of power that create and reinforce the systems of oppression that produce unjust laws and policies" (Freeman and Freeman 2016, 150). Community members are supported to understand the power structures that have a role in the problems they face and develop the capacity to challenge them. YLBHI director Asfin emphasised the focus on power that underpins this aspect of LBH's approach to legal assistance.

> The notion that CSOs are intermediaries between the government and the people needs to be eliminated. LBH is an accelerator of power in the community, so that communities can be more agile. If there is no power in the community, we help to develop it. But the community must be agents themselves. They are the most important. This is actually an old perspective in YLBHI and LBH, but because of the [post-Soeharto] situation, people forgot about it.[6]

Another important aspect of this work involves efforts to strengthen community organisations and movements. LBH lawyers support and encourage local communities in forming community organisations, sometimes using their technical skills to assist community organisations seeking to formally register with the government, or develop strategic plans.[7] LBH also acts to facilitate connections with allies in civil society, such as CSOs, academics and university student groups, who can lend organisational and intellectual support and help to sustain collective action (Mukrim et al. 2015, 34–44). LBH recognises that when community organisations are consolidated and strong, they are more likely to achieve their goals, whether their tactics ultimately include legal action or not (LBH Makassar 2018).[8] As Luban observes, "powerful and well-known groups win lawsuits that enthusiastic little sects and month-old coalitions lose" (Luban 1988, 387–88). LBH lawyers may also play an important role in resolving disputes among organisations or conflicting elements of social movements, such as the often fractious labour movement, a role to which cause lawyers are suited because of their mediation skills.[9] LBH lawyers also play a key role in helping community organisations to conduct advocacy, for example by providing strategic advice about the types of action that are legally and politically possible, facilitating meetings with government officials, supporting community organisations in analysing and developing policy documents, and helping to organise street demonstrations and protests.

When Rakhma began at LBH Semarang in 2000, most of her time was spent on community organising, rather than litigation. This was during the turbulent few years immediately following the fall of Soeharto, when some LBH offices supported farmers seeking to take advantage of the weakened state to engage in radical direct action tactics of occupying and "reclaiming" disputed land (Wijardo and Perdana 2001; Lucas and Warren 2003). This experience helps to explain her ongoing affinity for community organising work. She explained how community organising work operated in a symbiotic way with the community legal education and empowerment activities described above:

> My work involved going from village to village, meeting with communities. There were many agrarian [conflict] cases at that time. We conducted critical legal education, explored how the community understood the law and the cases they were facing, things like what do we actually mean by 'law', [and how to conduct] social analysis. This was done so that communities could critically analyse their cases and recognise the role of certain policy or laws, or actors, in the cases they were facing. We read laws together critically, exploring whether particular laws could be used for advocacy or not... The goal was that communities would eventually form an organisation. Because if they fight for their rights without a strong organisational basis, it will be difficult. LBH is clear that, through structural legal aid, it aims to empower local communities, so that they develop critical awareness, so that they can change uneven power relations – with companies or the state – themselves.[10]

In some cases, LBH lawyers may even join street protests, marches, and other types of direct action alongside the communities they represent, as occurred prominently in the *Reformasi Dikorupsi* protests mentioned in Chapter 1. This is a distinctly political interpretation of the lawyer's role, in line with a grassroots approach to cause lawyering. Indeed, many LBH lawyers view themselves as political actors who are deeply embedded in social movements.[11] Yet even if many LBH lawyers see themselves as being part of social movements, they also perform essential functions for social movements.[12] When members of social movements or community organisations are arrested for participating in direct action campaigns, LBH is one of the few organisations prepared to come to their defence. This is a common function of movement lawyers (Sarat and Scheingold 2006, 9; Cummings 2020, 99), and was a key role of LBH during the *Reformasi Dikorupsi* and 'Omnibus Law' demonstrations.

Donor organisations and other CSOs have occasionally derided LBH for relying too much on disruptive direct action tactics. The typical image of LBH in the eyes of the public is that of an organisation engaged in protest out on the streets, or in front of a courtroom, with an LBH lawyer delivering a fiery oration using a megaphone. It may be a somewhat blunt advocacy tool, but LBH has used this tactic to press for accountability from the courts, to remind judges that they are being watched.[13] One indication of its effectiveness is that this approach has been replicated by LBH's opponents, such as hard-line Islamists, who often use similar tactics in blasphemy trials involving religious minorities.[14]

Social movement work and the lawyer-client relationship

Community legal empowerment and community organising is designed to empower communities to advocate for their own rights. Working in this way requires a shift in the conventional relationship between lawyers and clients, and may not come naturally to lawyers, who are likely to feel more at home in the courtroom (Sarat and Scheingold 2006, 2), or at least in a legal office. However, as discussed in Chapter 3, LBH endeavours to foster community organising and community legal empowerment skills among its lawyers through the live-in component of its Kalabahu training programme, and the lengthy 'volunteer' process required before lawyers are formally accepted as LBH staff. The result of this training is that LBH lawyers often consciously posit themselves in opposition to mainstream 'professional' lawyers in conducting this type of work. This may be expressed in a simplistic form, for example in the belief common across the network that LBH lawyers should wear simple clothing.[15] In general, however, LBH lawyers assert that they aim to create a more equal relationship between lawyers and community members. This aspect of LBH's ideology was not altered dramatically by the democratic transition – recall that as far back as the Simprug case, LBH lawyers "got their boots dirty" beside their clients (Saptono 2012, 1). Nevertheless, as community legal empowerment and community organising have grown in prominence with democratic regression, it is important to consider how LBH lawyers view their relationship with justice seekers. Consider the following comments from LBH Jakarta lawyer Muhammad Rasyid Ridha:

> New LBH lawyers are educated that LBH is not superior to justice seekers. It is not a patron-client relationship. We are partners and we struggle together. It is not simply LBH 'injecting' them with legal knowledge or social movement strategies. We learn from justice seekers as well. They provide inspiration and knowledge to us, the spirit to fight, a collective spirit… We provide legal assistance, of course, but LBH is about more than just assistance. We move together.[16]

Rasyid's comment reflects the conviction common among LBH lawyers that they not only serve social movements, but are also a part of the movement themselves. Current LBH Semarang director Eti Oktaviani also stressed the importance of partnership and collective action.

> If a community only needs assistance with litigation, and they only want to sit and be passive, they can go and look for a law firm. Because if you want to be with LBH, we will discuss the case together, evaluate what we can do together, and act together.[17]

Some might see this closeness to the community as violating standards of lawyerly professional distance. In fact, senior YLBHI lawyer Febi Yonesta (Mayong) mentioned that foreign colleagues at a regional conference once criticised LBH's approach, saying that LBH was too political, that LBH lawyers were too close to the community.[18] Yet LBH has no reservations about this proximity to the community compromising its reputation for professionalism. LBH sees itself as part of social movements, and involvement in social movements as an inherent characteristic of being an LBH lawyer.[19]

Nevertheless, it is important not to take LBH's lofty claims about partnership and client empowerment at face value. Although LBH lawyers assert that the community is in control, there is some question about the extent to which social movement constituents can ever truly drive advocacy for their communities when lawyers are involved. Even if LBH lawyers profess to playing an equal or even subordinate role in relation to the community, lawyers' professional status and more expansive knowledge of rights and the legal system can result in lawyers playing an outsized role in determining strategies and designing campaigns, with clear impacts on client autonomy (Cummings and Eagly 2001, 496). This is evident in the following statement from LBH Surabaya director Abdul Wachid Habibullah (Wachid):

> LBH lawyers must be able to do it all, they must 'pioneer' social movements, play a role in political movements in the community, everything. That is the difference between LBH lawyers and professional lawyers. Professional lawyers are pure litigators. But LBH lawyers must develop social movements. They must be prepared to get out on the streets or in communities, lead demonstrations, empower communities, negotiate with officials, or lobby powerholders. LBH lawyers therefore must have lots of skills, not just litigation. Actually, I would have to admit that our litigation skills are not that strong, because we are so rarely in the courtroom.[20]

Although Wachid emphasises the importance of LBH lawyers being on the streets and working among communities, he places LBH lawyers in a superior position in relation to community members, with lawyers initiating the development of social movements and seemingly taking the lead in advocacy activities. Another complicating factor often brought up in examinations of cause lawyering is that the cause of the community may not always align directly with the causes of lawyers – the so-called "double agent problem" (Luban 1988, 319). One example often mentioned by LBH lawyers relates to religious minorities who face conflict over construction or use of their houses of worship:

> Often we find that our goals are different to theirs. Their goal is to practice their faith. We also hope for that, but we want more. And this is sometimes where there is conflict. For example, they might be happy to pay bribes if it allows them to continue to practice their faith [for example, to ensure the construction of a disputed church can go ahead]. But we don't want to do that. We want the case to be known by the public so that they learn about the principle of freedom of religion. We have to be careful. There must be informed consent. We can't just do it because we

> have idealism, or make them into an object for LBH's goals... Consciousness raising (*penyadaran*) has to happen.[21]

Asfin's comments here reflect the tension that can sometimes exist between elite/vanguard models of cause lawyering and grassroots approaches. While she is concerned with advancing the cause of religious freedom, and shares elite/vanguard lawyers' desire to affirm fundamental human rights principles in high-level state institutions (Hilbink 2004, 683), she also retains a grassroots lawyer's commitment to client empowerment. Again, while these typologies are useful for tracking LBH's strategies over time, they are not absolute – cause lawyers can adopt multiple approaches at the same time.

Community organising and gender and minority rights

As discussed in Chapter 3, LBH is a notoriously masculine organisation. Although democratic reform and regression has not had a significant impact on how gender and minority concerns feature in LBH's community organising and community legal empowerment work, it is important to reflect on gender briefly. The typical cases and target groups handled by LBH have related to people who are marginalised economically, for example farmers who face losing their land, the urban poor facing eviction, workers and fishermen. As other scholars have observed, this focus on economic injustice can lead to these issues being emphasised at the expense of other identity issues (Cummings and Eagly 2001, 489). Further, as discussed, the presence of a 'women's and children's rights' division in some LBH offices can mean that gender matters end up sequestered in this division. Other divisions may pay scant attention to the gendered aspects of the cases they manage, viewing gender issues as already being "dealt with" in the women's and children's rights division.[22] Nevertheless, gender is not ignored by LBH altogether. Some LBH cause lawyers recognise that they face significant challenges in ensuring the participation of women in decision making when working with the labour movement.[23] Likewise, several LBH lawyers mentioned that they recognise that women often suffer the most when communities lose access to land.[24] However, much more could be done to address gender across LBH's areas of focus.

The marginalisation of gender and minority rights in community legal empowerment and community organising affects not only analysis of community problems, but also process. Community organising activities frequently fail to consider the different needs of men and women. For example, community meetings may not be organised at a time that encourages the active participation of women. In November 2019, I attended a community meeting and introduction to paralegal training conducted by LBH Surabaya with a community on Madura Island. The community faced eviction for a government 'revitalisation' project. The only community members who attended were men, and the older, senior men in the community dominated the discussion. On several occasions during the meeting, LBH facilitators expressed frustration that no women were present and said that they hoped more women would attend the next session. However, the poor turnout of women was due to the fact the community meeting was held on the day before celebrations for the prophet's birthday (*Maulid Nabi*). In Madura, this is usually marked by community members visiting their relatives' and neighbours' houses, where food is offered to guests. In a patriarchal society where women are primarily responsible for domestic tasks, this meant that women were busy cooking, and were unable to attend the community meeting. Stronger gender awareness (and probably stronger representation of women among the staff at LBH Surabaya) would have likely prevented this kind of rudimentary scheduling error from occurring.

Another concern relating to the participation of women is that simplistic understandings of gender can see them exploited in the service of movement goals. For example, several LBH staff mentioned that they involved women in community organising activities by ensuring they were on the frontline of public demonstrations. This tactic assumes that law enforcement officials will be less likely to act violently to break up demonstrations when women are on the frontline.[25] This is not an approach that sits well with all LBH staff, such as the head of the YLBHI board, Nursyahbani Katjasungkana:

> Look at the Kendeng case [see below], the advocacy was done in collaboration between the community and LBH Semarang. The majority of those leading the advocacy were women. But reports on the case never explored why the women were the ones out front. Is it only because they were women and exploited? There is a common perception that mothers will result in feelings of pity and compassion. But in my experience, this is rarely the case… Do [LBH lawyers working with communities] ever ask the extent to which the project will affect men and women differently? They never do this type of analysis.[26]

As noted, over recent years, YLBHI has put considerable effort into strengthening the capacity of LBH lawyers at the regional level to understand gender and sexuality issues and promote a more intersectional approach to their work with communities.[27] International donors have encouraged and provided financial support for these initiatives. Although LBH Semarang was criticised by Nursyahbani, in my discussions with LBH lawyers it appeared to be one of the more progressive offices in its approach to gender issues. As LBH Semarang director Zainal Arifin said:

> We have a division on minorities and vulnerable groups. But you can't just deal with these issues separately in their own division like that. You can't deal with gender equality only in relation to gender-based violence. In the farmers' movement, for example, if there are women who are not being involved in decision making, then it is LBH's responsibility to push for women and women's organisations to be involved.[28]

Some LBH offices have attempted to promote solidarity and encourage collective action among minority groups by bringing them together in community organising and legal awareness training activities. This is also an approach that has also been encouraged by YLBHI's foreign donors.[29] For example, LBH lawyers have conducted community paralegal training activities that bring together minorities like transgender women, factory workers, members of the Ahmadiyah and Shi'a minority Islamic sects, and farmers, to try to break down prejudice, build solidarity between these groups, and encourage them to conduct joint advocacy activities together. The intention is that the emerging rights consciousness fostered through legal education and legal empowerment activities will encourage community members to see themselves as sharing similar experiences of rights violations, and collaborate in a broader movement for change. This approach recently saw transgender women paralegals working with paralegals from religious minority communities in support of the Shi'a community's right to conduct Ashura celebrations in Semarang.[30] These efforts are still in their infancy, however, and LBH acknowledges it has faced challenges dismantling prejudice, even among minority groups.

The limits of social movement approaches

Community legal empowerment and community organising tactics are consistent with structural legal aid's emphasis on cases involving large numbers of victims, or where it is hoped that a win will have benefits for large numbers of people, for example by changing the law or winning a class action. This focus on communal cases is understandable to a degree. LBH has limited resources – it needs to prioritise cases somehow. The problem becomes a little murky, however, when structural legal aid is used to justify turning down a client or category of case. The example of drug cases is illustrative.

Historically, drug cases were not considered structural, and were not taken on. The older generation of LBH believed that representing individual drug cases was too close to conventional legal aid, a "charity" service from the professional class.[31] LBH lawyers with this view felt it was difficult to connect the cases and use them to drive larger, structural change. A further complicating factor is the pronounced social stigma against drug use in Indonesia. Some LBH staff consider drug use to be a moral issue. I could not find any prohibition on representing drug cases in LBH founding documents or codes of ethics, but LBH staff commonly state that they are bound by a commitment to not defend people accused of corruption, human rights violations, gender-based violence, and narcotics crime (Fajarlie 2019).[32]

It is true that if LBH offices took on every drug case that came through their doors, they would quickly become overwhelmed. To put this in perspective, about 60% of Indonesia's 200,000-strong prison population are drug convicts.[33] Without treating drugs as part of a broader strategy, one could reasonably argue that dealing with individual drug cases is not an effective use of LBH's limited resources. Nevertheless, there are ways that LBH could address the seemingly individual problem of drug crime in a structural way, for example by challenging the criminalisation of marijuana use. It might not be a politically feasible approach now, given widespread public and political hostility toward even the most minor drug crimes, but if successful it would result in a major positive change in the lives of thousands of mainly poor Indonesians, and dramatically reduce prison overcrowding, which is a serious problem in Indonesia.[34]

There does seem to be an emergent shift in how many LBH offices approach drug cases. In my interviews with LBH lawyers, I encountered broad acceptance that drug cases are indeed structural, although the reasons provided seemed to vary greatly. Some suggested drug users were victims of drug dealers, because they were addicts, or because of the government's "war on drugs",[35] while others suggested that users' right to health was being violated, or that drug cases should be considered structural because they contribute to the massive overcrowding problems in prisons.[36] Most LBH offices appear to have settled on a middle course of providing assistance for drug cases when children are involved, or where there are elements of false arrest, entrapment or torture.[37]

These examples highlight one of the key shortcomings of emphasising community organising and community legal empowerment over litigation, which is that it can divert resources and energies away from addressing the immediate legal needs of the poor. LBH lawyers recognise this but argue that structural legal aid, with its focus on the representation of groups, seeks to bring about change for larger sections of society than individual representation could ever achieve. But as Cummings and Eagly note, when lawyers are encouraged to act as community organisers it requires "less time be spent providing conventional representation to low-income clients, who are already drastically underserved" (Cummings and Eagly 2001, 491). This is a dilemma that LBH has often struggled with.

While the representation of all poor clients would be in line with LBH values, it is unreasonable to expect the organisation to see all the individual criminal cases that come through its doors. It is important to recognise that structural legal aid is not the same as organisational priorities. As Asfin explained:

> From 1,400 cases, for example, about 1,000 would be structural cases. We are not capable of handling them all. Even for labour rights cases, cases that have been associated with LBH for decades, we still decline them if they only involve one or two people. Instead, we will equip them with knowledge, or link them to a labour union. Often people think that if a case is not followed up at LBH, or it is rejected, it is because it was not a structural case. But often it is simply a matter of priorities.[38]

All the responsibility for serving poor clients should not fall on LBH. Lawyers in private firms (through pro bono work) and other legal aid organisations could be doing more to serve the community's needs for legal assistance. In an effort to promote greater uptake of pro bono cases by private lawyers, in 2018 LBH Jakarta set up a 'pro bono clearing house', and has recruited about a dozen firms to which it can refer cases (Prasetyo 2020). But these kinds of initiatives remain embryonic, and do not extend across the LBH network. In the meantime, the growing provision of government funding for legal aid (mandated by the 2011 Law on Legal Aid) may promote the expansion of legal aid, by encouraging more civil society legal aid providers to take on simple, individual cases (as discussed in Chapter 3). Nevertheless, there is such a vast need for legal services that even doubling or tripling the number of accredited providers, and the funding available to each, would still be inadequate.

The Kendeng farmers' case

The struggle of farmers in the North Kendeng Mountains of Central Java is one of the most prominent environmental and land conflicts in Indonesia over the past decade. It demonstrates how many of LBH's social movement strategies play out in practice. The case has featured multiple lawsuits, community organising and public education, collaboration (and some tension) among a broad network of CSOs, and dramatic headline-grabbing protests in Jakarta. It is an excellent illustration of LBH's approach to cause lawyering in the context of Indonesia's current democratic decline. The media might sometimes refer to 'the Kendeng case' but the main conflicts involved two district governments (Pati and Rembang), and two cement manufacturing companies (Indocement and Semen Gresik – which changed its name to Semen Indonesia).

The North Kendeng Mountains are a karst formation that extends through several districts of Central Java and East Java. The porous limestone rock of the mountains serves as an important water store, providing water for drinking and irrigation of crops for the farming communities in the area. The rocks are also a key ingredient in cement. One of the communities that calls the North Kendeng Mountains home is a traditional belief community known as the Sedulur Sikep, sometimes also called Saminists, after the founder of their faith, Surontiko Samin (Benda and Castles 1969, 210–11). The Samin movement began in the late 1800s, when Indonesia was still under Dutch colonial administration. Samin's teachings involved respect for nature and the role of farmers in society, as well as prohibitions on stealing, lying and adultery. The movement recognised no authority and aimed to abolish taxes and allow communities to have free access to forests (Korver 1976, 250). After attracting the attention of the Dutch authorities, Samin was arrested on charges of sedition and forced into exile in Sumatra, where he died in 1914 (Korver 1976, 249–50). The movement endured after his death, and followers continued to encourage defiance against the government and the refusal to pay taxes (Benda and Castles 1969, 213). The Samin community therefore has a long history of resistance to authority, one on which they have drawn in their recent struggles.

Pati district

Tension between Samin followers and cement factories began in 2006, when state-owned cement producer Semen Gresik announced plans to establish a quarry and cement factory in Pati district, Central Java. On 5 November 2008, the Pati district government issued a decision allowing Semen Gresik to mine limestone in Sukolilo subdistrct, Pati (Kompas.com 2009).[39] However, the decision was issued without the required environmental impact assessment (known in Indonesia by the term *analisis dampak lingkungan,* or *amdal*).[40] By 2008, a range of community groups and organisations had formed to oppose the mine, including the Community Network for the Kendeng Mountains (*Jaringan Masyarakat Peduli Pegunungan Kendeng*, JMPPK), a network of local farmers and Samin followers, which remains the most prominent of these community organisations today. Local and national CSOs also joined the campaign (Crosby 2009). National CSO Walhi, lawyers from LBH Semarang, and the YAPHI Legal Services Agency (*Lembaga Pengabdian Hukum Yekti Angudi Piadeging Hukum Indonesia*, LPH YAPHI) challenged the Pati district decision at the Semarang State Administrative Court (PTUN Semarang). Surprisingly, the court ruled in their favour.[41] When the decision was appealed to the Surabaya State High Administrative Court (PTTUN Surabaya), however, the community was not so fortunate.[42] But in 2010, in a cassation appeal to the Supreme Court, the civil society coalition was successful again.[43] The court ordered that the 2008 Pati district decision be revoked, and Semen Gresik was forced to look elsewhere.

However, this did not stop the efforts of cement companies in the area. Soon after the 2010 decision, the minister of energy and mineral resources issued Regulation No. 17 of 2012 on Determination of Karst Landscape Areas, which revoked a 2005 regulation on the Sukolilo Karst Landscape.[44] In May 2014, the minister then issued a new decision on the Sukolilo Karst Landscape,[45] which resulted in about 5,000 hectares of previously protected karst landscape being rezoned for mining (Komnas HAM 2016, 87–88).[46] Sahabat Mulia Sakti (SMS), a subsidiary of Indocement (whose main shareholder is the German multinational HeidelbergCement), began efforts to establish a mine in the area, by preparing an *amdal.* In September 2014, about 3,000 Pati residents came in 80 trucks to protest a public hearing on the *amdal.* They drew attention to the fact that even the *amdal* noted that two-thirds of local residents rejected the establishment of the plant (Apriando 2014). But just a few months later, in December 2014, the Pati district head provided SMS with an environmental licence to mine in the area.[47] There was much debate among the organisations that were close to the communities in Pati about whether challenging the environmental licence in court was the best course of action. Walhi, YAPHI and some other CSOs thought litigation was risky, as the licence was granted for an area now zoned for mining. If the lawsuit failed, they reasoned, it would only end up legitimising the mine.[48] LBH Semarang consulted with communities in Pati, warning them of the risks and informing them about the other organisations' reluctance to go to court. Yet the community insisted they wanted to proceed. LBH Semarang also recognised that there were other benefits of going to court.

> If they go to court then they have a reason to discuss the case every week, discuss opposition to the factory, hold environmental discussions and so on. Every week there is a hearing, but on the other six days of the week, they invite their neighbours, prepare readings, and then every Wednesday attend the hearing and hold a demonstration. This was the initiative of the community.[49]

Such an approach is in line with the grassroots vision of cause lawyering that LBH has favoured over recent years, at a time when democracy has come under threat and LBH has become more cynical about the potential impact of legal victory. Despite this scepticism of legal tactics, grassroots cause lawyers recognise that litigation can be an important catalyst, helping to invigorate collective action (Hilbink 2004, 686). Note, too, how Zainal's comments mirror Muhamad Isnur's reflections in Chapter 4 about how constitutional litigation can help consolidate social movements.

With the decision made to proceed, in 2015 a group of five Pati farmers challenged the environmental licence. A team of LBH and other civil society lawyers provided representation. Ironically, SMS was represented by the private firm of former senior YLBHI figure Abdul Hakim Garuda Nusantara.[50] To complement litigation, a broad range of nonlitigation strategies were deployed, driven largely by the JMPPK network. Outside the courtroom, Pati residents displayed the results of their harvest, gave speeches and put on theatrical performances (Apriando 2015a). Kendeng farmers and their allies visited the Gadjah Mada University (UGM) campus in Yogyakarta, where they demonstrated against academics who they believed had provided misleading expert testimony (Apriando 2015b). In the lead up to the announcement of the decision, hundreds of Pati residents embarked on a "long march" of about 100 kilometres from Pati to Semarang (Purbaya 2015), a strategy that farmers from Pati and Rembang have used multiple times throughout their struggles against the cement firms.

In November 2015, PTUN Semarang ruled in favour of the community.[51] The court found that the environmental licence conflicted with the Pati District Spatial Plan 2010–2030, Environmental Ministry Regulation No. 17 of 2012 on Involvement of the Community in the Environmental Impact Assessment and Environmental Licencing Process, and principles of good governance.[52] SMS and Indocement appealed the decision to the Surabaya State High Administrative Court. Several months later, on 1 July 2016, the Surabaya State High Administrative Court reversed the decision of PTUN Semarang, finding in favour of the company.[53] This decision was then upheld by the Supreme Court on cassation appeal on 6 March 2017.[54] Even though the community lost, the long legal process has stalled progress in Pati.[55] The community was able to delay the project, make it politically controversial, and arguably even threaten its commercial viability.[56] It was a classical cause lawyering strategy of using litigation and protest to throw "sand in the cogs" (McEvoy and Bryson 2022), an indispensable tactic of resistance in contexts where the political and legal odds are stacked against lawyers.

Rembang district

Communities in neighbouring Rembang district have also been engaged in a long, high-profile struggle against cement companies. After failing in its attempts to establish a mine in Pati in 2009, Semen Gresik, which as mentioned changed its name to Semen Indonesia in December 2012 (I will now refer to it as Semen Indonesia), began looking to Rembang. In 2010 and 2011, the Rembang district head issued three decisions in preparation for the company establishing a mine and factory in the area.[57] Semen Indonesia completed its *amdal* in April 2012, and in June of the same year, Central Java Governor Bibit Waluyo issued an environmental licence for Semen Indonesia to mine and establish a factory in the region.[58] The community subsequently claimed that they had only discovered that the *amdal* and environmental licence had been issued two years later, in June 2014.[59] Around the time when the firm commenced construction, on 16 June 2014, the community set up a tent protest at the proposed site of the mine and factory, as a form of peaceful resistance (Apriando 2016; Firdaus 2016). This tent site served as an important community hub, until it was destroyed on 10 February 2017 by a group of unknown assailants that the community, understandably, suspected were connected to Semen Indonesia (Mukti 2017).

In September 2014, Rembang community members, as well as Walhi, represented by LBH Semarang, YLBHI and other civil society cause lawyers, submitted a challenge to the environmental licence to the Semarang PTUN. As with the Pati case, Semen Indonesia was represented by the private firm of a former LBH lawyer, in this case Adnan Buyung Nasution and Partners Law Firm. Representation of a mining company would be consistent with Buyung's proceduralist tendencies in the years following the fall of the New Order. But the fact that both mining companies sought out former LBH lawyers to represent them suggests a deliberate strategy on behalf of the mining firms to counter their opponents from LBH. On 16 April 2015, PTUN Semarang rejected the Rembang community's challenge on technical grounds, stating that the community had known about the planned mine and factory since 2013, and as such the case was filed in excess of the 90-day limit for submitting challenges to the administrative courts.[60] The community appealed to the Surabaya State High Administrative Court, but in November 2015 it upheld the decision of PTUN Semarang.[61] Despite the ongoing legal tussles, the company continued with its plans to establish the factory.

Frustrated with this process, the community decided more drastic measures were needed. On 12 April 2016, a group of Kendeng women cast their feet in cement in front of the Presidential Palace, demanding to meet President Jokowi. The women dubbed themselves the 'Kartinis of Kendeng', adopting the name of Indonesian feminist symbol Kartini (Erdianto 2016). The protest attracted massive media attention and resulted in dramatic images being shared across the country, helping to garner public and civil society support for the farmers' struggle. Although the women were not successful in meeting with the president, the head of the Office of Presidential Staff at the time, Teten Masduki (yet another former YLBHI figure), met with the women and promised to arrange a meeting with Jokowi (Erdianto 2016). On 2 August, the community finally got their chance to meet with the president. Following the meeting, Jokowi reportedly ordered the Office of Presidential Staff to coordinate an independent 'Strategic Environmental Study' (*Kajian Lingkungan Hidup Strategis*, KLHS) and requested that the construction of the factory be paused until this was completed (MP 2016; Sinuko 2016b).[62] Almost immediately after this announcement, Central Java Governor Ganjar Pranowo (who replaced the previous governor in 2013, and was a losing candidate in the 2024 presidential elections) said the project could not be stopped, as it was already 95% complete (Sinuko 2016a).

Around the same time as the meeting with Jokowi, in August 2016, the community submitted a reconsideration request (*peninjauan kembali,* PK), the final avenue by which a court decision can be challenged in the Indonesian legal system. Remarkably, on 5 October 2016, the Supreme Court ruled in the community's favour.[63] In its decision, the court found that the strong community opposition to the project was evidence that community participation was insufficient and the project had not been appropriately communicated to all elements of the community.[64] The panel of judges also ruled that the *amdal* did not contain sufficient protections to ensure that mining would not negatively affect groundwater and the livelihoods of local people, which would have been consistent with the principles of good governance.[65] As such, the court found the *amdal* was procedurally flawed, and should be cancelled.[66]

The community's victory was short lived. After a lack of action from the Central Java government to put a stop to the factory's construction, on 5–8 December 2016, about 100 farmers walked the approximately 150 kilometres from Rembang to Semarang, to demand Central Java Governor Ganjar implement the court's decision (Bina Desa 2016). Instead, he simply issued a new environmental licence on 23 February 2017, allowing the company to continue operations (Nurdin 2017).[67] In March 2017, Kendeng community members travelled to Jakarta again, and again set their feet in cement (Ginanjar 2017). This time, several activists joined the protest, including YLBHI's Rakhma. Sadly, one of the demonstrating Kendeng women, Patmi, died of a heart attack on 21 March (Sapiie 2017). This tragic event forced President Jokowi to respond and the following day he met with Gunarti, another of the Kendeng women. But this time he was less receptive to the community's demands. He reportedly told the community to stop bringing their concerns to him, and that they should take up their issues up with the Central Java governor (The Jakarta Post 2017).[68]

A crucial role of LBH Semarang in the struggle of Kendeng farmers in both Pati and Rembang was in the provision of community legal education and community legal empowerment. Most Kendeng community members had no previous interaction with the formal legal system. LBH Semarang conducted village-level discussions, where it explained aspects such as the trial process, the presentation of evidence, and the accused parties in each case.[69] It also conducted environmental awareness training (which covered environmental law and the process of conducting environmental impact assessments), spatial planning processes, human rights, and even safety and security training for human rights defenders and environmental activists (LBH Semarang 2015). As discussed, these community legal empowerment activities were designed in part to encourage rights awareness and collective action. They are also key components of the structural legal aid approach. LBH Semarang director Zainal Arifin reported that when he first started at LBH in 2012, its initial attempts to offer environmental awareness training to the Kendeng farmers were not successful.[70] However, LBH's persistence eventually saw it gain the trust of the community. This trust was essential when legal challenges were subsequently launched.

Despite the Kendeng community's victory at the Supreme Court, and the president's request that construction be paused while the KLHS was completed, the Rembang mine has now begun operations. The independent KLHS study issued strong recommendations, including that the Watuputih groundwater basin and surrounding areas in Rembang should be designated a protected area. The study further recommended that while this administrative change was made, mining operations should be suspended, no new mining licences should be issued, and efforts should be made to eradicate illegal mines in the area (Kementerian Lingkungan Hidup dan Kehutanan 2017). But the government was under no obligation to follow these recommendations, and it has made no moves to do so. The struggle of the Kendeng farmers continues. They continue to resist the mine and factory in Rembang, through ongoing demonstrations, public discussions, media outreach and press conferences (CNN Indonesia 2019; JMPPK 2019; Saputra 2019; Bakara 2018). LBH remains closely connected to JMPPK, and has helped to coordinate discussions on planned revisions to the Rembang district spatial plan and facilitate meetings with the district legislature where they discussed the revised plan (Walhi Jawa Tengah 2021). The community continues to demand that the government implement the recommendations of the KLHS (JMPPK 2019).

Social movement strategies in action

The Kendeng case is not just one of the highest-profile environmental disputes of the past decade, but it also illustrates the grassroots strategies of community organising and community legal empowerment that LBH has emphasised as it became more sceptical of the state. Even so, it is important not to overstate the role of LBH in advancing these social movement strategies in Kendeng. At one point, 27 CSOs were working with communities in the region.[71] LBH occasionally performed a role typical of cause lawyers in broad coalitions, focusing on the legal realm, while other organisations and individuals played other supportive roles, such as policy-focused research to support advocacy efforts.[72] It is also essential not to minimise the agency of the Samin followers who, as discussed, have a very long history of resistance to state power. LBH did, however, invest considerable time in building a close relationship with the JMPPK network, which helped to facilitate closer connections with the community that were essential when legal action was launched. This close relationship continues today.

In the Kendeng case, litigation and social movement tactics were interactive and mutually constitutive. LBH cause lawyers did not simply seek to vindicate rights in the courts, as might have occurred through an elite/vanguard model of cause lawyering that considered litigation capable of bringing about meaningful change. But neither did they view litigation as diverting energy from grassroots activism. Rather, litigation was valued for its "indirect" effects (McCann 1994), for its capacity to galvanise broader social movement activism. Consider these comments from LBH Semarang director Zainal Arifin:

> Are we public interest lawyers, movement lawyers, or legal aid lawyers? It's an interesting discussion. There are several identities within LBH… in the Kendeng case, we are movement lawyers. We never acted without the community's blessing. We did things like providing legal education, organising the community. When we engaged in litigation, it was really a tool for community organising.[73]

Community education activities were used to encourage community mobilisation, which was in turn used to buttress formal legal action. When legal hearings were held, hundreds of community members attended. As discussed, this has long been a common method used by Indonesian cause lawyers to promote accountability in the courts. In a situation where judicial bribery is widespread (Butt and Lindsey 2018, 299–303), LBH lawyers believe "the judges have to feel that they are being watched".[74] Although he is not speaking about the Kendeng case, this comment from LBH Makassar's Ady Anugrah Pratama explains the approach:

> [T]he formal legal process and social movements are united and cannot be separated. Community organising must function at the same time as litigation, because the legal strategy in the courtroom will never be enough. We organise and encourage the community to be involved and monitor the legal process as part of this united process.[75]

Litigation, and the massive public attention the cases attracted, was also useful for recruiting allies to the cause and building a more robust social movement (Cummings 2020, 106). LBH Semarang played an important role facilitating relationships among the Kendeng community and other CSOs and cause lawyers. If it did not have the resources to provide legal representation, for example, LBH connected the community to other cause lawyers in Jakarta or Yogyakarta. Although divisions occasionally emerged among community members, LBH also played a mediating role, supporting respected community and religious leaders to travel from village to village to resolve misunderstandings or concerns ahead of legal action.[76] LBH Semarang (as well as the other organisations that accompanied the community) connected the Kendeng farmers to CSOs in Jakarta, including YLBHI. Notably, the Kendeng farmers have often featured in the '*Kamisan*' (Thursdays) human rights protests held every week in front of the Presidential Palace, as well as other human rights protests in Jakarta (Siddiq 2018; Nugroho 2019).

Consistent with broader LBH views on the lawyer-client relationship discussed above, LBH Semarang saw itself as acting in the service of movement goals, not driving the movement. Distinct from an elite/vanguard style of cause lawyering, which might see cause lawyers keeping a distance from the community, handling technical legal aspects, or advising on the most appropriate course of legal action, LBH lawyers saw themselves as embedded within the movement, and acting alongside members of the community. LBH's interaction with the community was not limited to periods when the community needed legal assistance. For example, the Kendeng women would often sleep in the YLBHI office whenever they came to Jakarta. It is telling that senior YLBHI lawyer and former LBH Semarang director Rakhma cemented her own feet in front of the Presidential Palace in the March 2017 protest. It is hard to imagine a more obvious visual metaphor for being integrated into a social movement.

There was also a strong emphasis on rights consciousness and community legal empowerment in LBH's interactions with the Kendeng farmers, with rights used "strategically and flexibly to build collective power at the grassroots level" (Cummings 2020, 121). By strengthening rights awareness in the Kendeng community, LBH aimed to promote solidarity among community members and support them in recognising their collective strength. LBH Semarang lawyers consistently stressed that the community controlled public advocacy efforts. They pointed to the fact that the movement's most prominent acts of protest – such as the establishment of the tent protest site in Rembang, the long marches, and the cementing of Kendeng women's feet in front of the Presidential Palace – were initiatives that came from the community. Likewise, in its decade of struggle against mining firms, JMPPK has since emerged as a strong, independent, active network.

Reflecting on the two legal cases, LBH's scepticism about litigation appears entirely justified. Despite the Rembang community's win at the Supreme Court, the Central Java government cynically circumvented the decision by issuing a new licence, and the mine has continued to operate (Pradana and Utomo 2022). In Pati, meanwhile, community legal empowerment and organising efforts have thus far prevented the SMS/Indocement factory from going ahead. There is no guarantee that this situation will be sustained – there is already evidence of a large number of other legal and illegal mining operations underway in the North Kendeng Mountains (Saputra 2019). Nevertheless, the effective resistance of the community in Pati does seem to vindicate the efforts that LBH and other CSOs put into strengthening the community and supporting their engagement in disruptive direct action tactics.

Conclusions

This chapter has illustrated the complex ways in which LBH is deploying community organising and community legal empowerment tactics to advance its causes in post-authoritarian Indonesia. While these political, grassroots-focused strategies were integral to LBH's structural legal aid approach during the New Order period, they were implemented inconsistently after the democratic transition began in 1998. Shifting donor priorities and funding challenges, as discussed in Chapter 3, limited the availability of funds for this resource-intensive approach to cause lawyering. Moreover, the opening up of new political and legal opportunities following the democratic transition resulted in a shift in some LBH cause lawyers' perceptions of the state and their roles as lawyers. Organising the grassroots in opposition to the state was not considered as important in a more democratic Indonesia.

The responses of LBH lawyers in this chapter have revealed declining faith in legal strategies and disillusionment with the government and its commitment to reform. Consequently, the YLBHI leadership has deprioritised litigation and made a determined effort to return to promoting community legal empowerment and community organising tactics as central components of LBH's approach to cause lawyering. While LBH has not cast aside litigation strategies completely, it has recognised the need to combine courtroom litigation with a broad range of extra-legal tactics, including community education, coalition-building, protests and demonstrations, use of the media, and political advocacy. The goal is to empower communities so that they develop an awareness of their rights, and then support them in taking direct action to challenge the underlying political or legal issues that contribute to the problems they face.

As the case of the Kendeng farmers demonstrated, litigation was an important catalyst. Legal hearings provided opportunities for strengthening rights awareness in the community and rallying community members around the cause. At the same time, non-legal strategies, like the Kendeng farmers' dramatic cement block protests and long marches, helped to attract media attention and public support, while maintaining pressure on the state. In engaging in this kind of work, LBH sees itself as both integrated into social movements, while also serving social movements, performing vital functions for them. In the Kendeng case, at least, LBH viewed the community as being in control, and responsible for determining strategies. This is consistent with a grassroots approach to cause lawyering.

LBH's commitment to a grassroots approach does not exclude elite/vanguard tactics. As Chapter 4 noted, the democratic transition provided new opportunities to mobilise the law and engage in strategic impact litigation. The following chapter looks at a different elite/vanguard approach made possible by democratic change – direct engagement with the state, for example, in policy advocacy and institutional strengthening efforts. LBH cause lawyers have struggled much more with the question of the extent to which they should deploy these more accommodative, elite-focused approaches, as the next chapter shows.

Notes

1 Interview, Siti Rakhma Mary Herwati, August 2019.
2 Interviews, Eti Oktaviani, November 2019; December 2020.
3 Interview, Asfinawati, December 2019.
4 This is an argument made by Mari Matsuda in relation to minorities in the United States (Matsuda 1987).
5 Interviews, Ajeng Wahyuni, August 2019; Febi Yonesta, September 2019.
6 Interview, Asfinawati, December 2019.
7 Interview, Asfinawati, December 2019.
8 Interview, Eti Oktaviani, November 2019.
9 Interviews, Asfinawati, December 2019; Febi Yonesta, September 2019.
10 Interview, Siti Rakhma Mary Herwati, August 2019.
11 Interviews, Abdul Fatah, Eti Oktaviani, Herlambang P. Wiratraman, Zainal Arifin, November 2019; Arif Maulana, Asfinawati, December 2019.
12 Interviews, Abdul Fatah, Emanuel Gobay, November 2019.
13 Interviews, Eti Oktaviani, Zainal Arifin, November 2019; Muhamad Isnur, August 2021.
14 For example, in 2019, I attended the trial of Suzethe Margaret, the woman charged with blasphemy for bringing her dog into a mosque (see also Chapter 4). The local mosque board mobilised dozens of members of the local community to pack the courtroom.
15 Interviews, Abdul Muit Pelu, September 2019; Poengky Indarti, October 2019; Zainal Arifin, November 2019.
16 Interview, Muhammad Rasyid Ridha, November 2019.
17 Interview, Eti Oktaviani, November 2019.
18 Interview, Febi Yonesta, September 2019.
19 Interviews, Siti Rakhma Mary Herwati, August 2019; Febi Yonesta, September 2019; Haswandy Andy Mas, November 2019; Robertus Robet, April 2020.
20 Interview, Abdul Wachid Habibullah, November 2019.
21 Interview, Asfinawati, December 2019.
22 Interview, Rezky Pratiwi, November 2019.
23 Interview, Pratiwi Febry, December 2019.
24 Interviews, Adi Anugrah Pratama, Mohamad Soleh, November 2019.
25 Interviews, Adi Anugrah Pratama, Edy Kurniawan Wahid, November 2019.
26 Interview, Nursyahbani Katjasungkana, October 2019.

27 Interviews, Renata Arianingtyas, August 2019; Asfinawati, December 2019.
28 Interview, Zainal Arifin, November 2019.
29 Interview, Renata Arianingtyas, August 2019.
30 Interview, Zainal Arifin, November 2019. Ashura is commemorated on the 10th day of Muharram, the first month of the Muslim calendar. For Shi'a Muslims, Ashura is a day of mourning, marking the martyrdom of Ali Hussein. Ashura celebrations in Indonesia have faced bans from some regional administrations and been subject to protests by some hard-line Sunni Muslim groups (Dipa 2015).
31 Interview, Luhut Pangaribuan, November 2019.
32 Interviews, Barita N. Lumbanbatu, October 2019; Mohamad Soleh, Habibus Shalihin, Abdul Wachid Habibullah, November 2019.
33 Referring to Indonesia's publicly accessible Corrections Database System (*Sistem Database Pemasyarakatan,* SDP), in October 2020, there were 238,442 total detainees and inmates, including 94,546 drug dealers and 44,369 drug users.
34 The number of inmates in Indonesian prisons swelled by more than 400% between 2003 and 2019. Harsh approaches toward drug crime have been a major contributor to this explosion in the prison population (Sudaryono 2020). As of October 2020, Indonesian prisons were at 176% overcapacity. They housed 238,442 detainees and inmates but had capacity for only 135,647.
35 Soon after he came to power in 2014, President Jokowi adopted the hard-line rhetoric of a "drugs emergency". He resumed the execution of narcotics prisoners on death row, executing 14 drug prisoners in six months, including Australians Myuran Sukumaran and Andrew Chan. While executions have since been paused, Jokowi also commented approvingly about fatal police and National Narcotics Agency (Badan Narkotika Nasional, BNN) shootings of drug suspects (McRae 2017). This "war on drugs" rhetoric is thought to be a major factor behind prison overcrowding (Wicaksana 2018; Sudaryono 2020).
36 Interviews, Abdul Muit Pelu, Willy Hanafi, September 2019; Haswandy Andy Mas, Muhammad Rasyid Ridha, November 2019; Asfinawati, December 2019.
37 Interviews, Alghiffari Aqsa, Abdul Wachid Habibullah, November 2019; Arif Maulana, Asfinawati, December 2019.
38 Interview, Asfinawati, December 2019.
39 Decision of the Head of the Pati District Integrated Permit Service Office No. 540/052/2008.
40 As required by Environment Minister Regulation No.11 of 2006 on Types of Businesses or Activities Requiring an Environmental Impact Assessment. Provisions on *amdal* have been significantly diluted with the passage of the highly controversial Omnibus Law on Job Creation, in October 2020 (Jong 2020).
41 PTUN Semarang Decision No. 04/G/2009/PTUN.SMG, 6 August 2009.
42 Surabaya State High Administrative Court Decision No. 138/B/2009/PTTUN.SBY, 30 November 2009.
43 Supreme Court Decision No. 103 K/TUN/2010, 27 May 2010.
44 Minister of Energy and Mineral Resources Decision No. 0398 K/40/MEM/2005 on the Sukolilo Karst Area.
45 Minister of Energy and Mineral Resources Decision No. 2841 K/40/MEM/2014 on the Sukolilo Karst Area.
46 Around the same time, the Pati district head issued a regulation rezoning certain areas in the district, Pati District Regulation No. 5 of 2011 on the Pati District Spatial Plan 2010–2030.
47 By issuing Pati District Head Decision No. 660.1/4767 of 2014.
48 Interview, Zainal Arifin, November 2019.
49 Interview, Zainal Arifin, November 2019.
50 Abdul Hakim Garuda Nusantara was also the head of Komnas HAM from 2002 to 2007. As discussed in Chapter 2, under the New Order, Abdul Hakim was

associated with activist, more grassroots approaches to cause lawyering. It is difficult to know whether his firm's representation of the mining company constituted a cynical abandonment of his principles for financial gain, or whether it represented a shift to a more proceduralist approach to cause lawyering, similar to Adnan Buyung Nasution. Sadly, Abdul Hakim died in early 2018, and it was not possible to interview him for this project.

51 PTUN Semarang Decision No. 015/G/2015/PTUN.SMG, 476–77.

52 PTUN Semarang Decision No. 015/G/2015/PTUN.SMG, 477–78.

53 PTTUN Surabaya No. 79/B/2016/PT.TUN-SBY.

54 Supreme Court Decision No. 4 K/TUN/2017.

55 According to Government Regulation No. 27 of 2012, if no activities have commenced three years after an environmental licence is issued then the company must request a new one.

56 In 2017, a Kendeng farmer, Gunarti, even travelled to Germany and spoke about the community's struggle at the HeidelbergCement shareholder meeting (Knight 2017).

57 These included a Mining Business Licence Area (*Wilayah Izin Usaha Pertambangan*, WIUP), through Rembang District Head Decision No. 545/68/2010, a Mining Business Exploration Licence (*Izin Usaha Penambangan (IUP) Eksplorasi*), through Rembang District Head Decision No. 545/4/2011, and a Location Licence (*Izin Lokasi*), for the construction of a cement factory, through Rembang District Head Decision No. 591/040/2011.

58 Central Java Governor Decision No. 660.1/17/2012. This was followed by the Rembang district head issuing Decision No. 545/0230/2013, on 15 February 2013, providing Semen Indonesia with a Mining Business Production Licence (*IUP Operasi Produksi*).

59 PTUN Semarang Decision No. 064/G/2014/PTUN.Smg, 26.

60 Article 55 of Law No. 5 of 1986 on the Administrative Courts (as amended by Law No. 9 of 2004 and Law No. 51 of 2009). See also PTUN Semarang Decision No. 064/G/2014/PTUN.Smg, 208–09.

61 PTTUN Surabaya Decision No. 135/B/2015/PT.TUN.SBY.

62 A KLHS is a study that should be completed by regional governments to ensure that their spatial plans, development planning, policies and programmes are consistent with sustainable development principles. See Articles 15–18 of Law No. 32 of 2009 on Environmental Protection and Management.

63 Supreme Court Decision No. 99 PK/TUN/2016, 5 October 2016.

64 Supreme Court Decision No. 99 PK/TUN/2016, 111–12.

65 Supreme Court Decision No. 99 PK/TUN/2016, 112–14.

66 Supreme Court Decision No. 99 PK/TUN/2016, 114. Semen Indonesia later submitted a reconsideration request of this reconsideration request, but it was rejected on 20 June 2017 (Supreme Court Decision No. 91 PK/TUN/2017).

67 Central Java Governor Decision No. 660.1/6/2017.

68 In May 2017, Walhi submitted a new challenge to the new environmental licence (Decree No. 660.1/6/2017), but this time the community decided not to join the lawsuit. The challenge was dismissed by PTUN Semarang. PTUN Semarang Decision No. 039/PEN-DIS/2017/PTUN.SMG, 16 June 2017, and PTUN Semarang Decision No. 039/G.PLW/2017/PTUN.SMG, 16 August 2017.

69 Interviews, Eti Oktaviani, Zainal Arifin, Herdin Pardjoangan, November 2019.

70 Interview, Zainal Arifin, November 2019.

71 Interview, Zainal Arifin, November 2019.

72 Interview, Eti Oktaviani, December 2020.

73 Interview, Zainal Arifin, November 2019.

74 Interview, Zainal Arifin, November 2019.

75 Interview, Ady Anugrah Pratama, November 2019.

76 Interview, Zainal Arifin, November 2019.

References

Apriando, Tommy. 2014. "Warga Pati Tolak Pendirian Pabrik Semen. Kenapa?" *Mongabay*, 4 September 2014. https://www.mongabay.co.id/2014/09/04/warga-pati-tolak-pendirian-pabrik-semen-kenapa/.

Apriando, Tommy. 2015a. "Warga Gugat Bupati Pati Terkait Izin Penambangan Semen. Kenapa?" *Mongabay*, 9 March 2015. https://www.mongabay.co.id/2015/03/09/warga-gugat-bupati-pati-terkait-izin-penambangan-semen-kenapa/.

Apriando, Tommy. 2015b. "Warga Rembang Dan Pati Minta Dosen UGM Jujur Selamatkan Kendeng. Ada Apa?" *Mongabay*, 22 March 2015. https://www.mongabay.co.id/2015/03/22/warga-rembang-dan-pati-minta-dosen-ugm-jujur-selamatkan-kendeng-ada-apa/.

Apriando, Tommy. 2016. "Dua Tahun Perempuan Rembang Menolak Tambang Di Tenda Perjuangan." *Mongabay*, 22 June 2016. https://www.mongabay.co.id/2016/06/22/dua-tahun-perempuan-rembang-menolak-tambang-di-tenda-perjuangan/.

Bakara, Ivan Wagner. 2018. *Kronik Kendeng Lestari: Perjuangan Rakyat Kendeng Atas Tanah Airnya*. Semarang/Jakarta: JMPPK, YLBHI, LBH Semarang, and Desantara.

Benda, Harry J., and Lance Castles. 1969. "The Samin Movement." *Bijdragen Tot de Taal-, Land- En Volkenkunde* 125 (2): 207–240.

Berenschot, Ward, and Taufik Rinaldi. 2011. *Paralegalism and Legal Aid in Indonesia: Enlarging the Shadow of the Law*. Leiden: University of Leiden Van Vollenhoven Institute.

Bina Desa. 2016. "Long March Petani Kendeng Dukung Gubernur Cabut Izin Pabrik Semen Di Rembang." *Bina Desa*, 8 December 2016. https://binadesa.org/long-march-petani-kendeng-dukung-gubernur-cabut-izin-pabrik-semen-di-rembang/.

Butt, Simon, and Tim Lindsey. 2018. *Indonesian Law*. Oxford; New York: Oxford University Press. https://doi.org/10.1093/oso/9780199677740.001.0001.

CNN Indonesia. 2019. "Warga Kendeng Kirim 'Supersemar' Untuk Ganjar." *CNN Indonesia*, 12 March 2019. https://www.cnnindonesia.com/nasional/20190311234418-20-376356/warga-kendeng-kirim-supersemar-untuk-ganjar.

Crosby, Alexandra. 2009. "Too Precious to Mine." *Inside Indonesia*, December 2009. https://www.insideindonesia.org/too-precious-to-mine.

Cummings, Scott L. 2020. "Movement Lawyering." *Indiana Journal of Global Legal Studies* 27 (1): 87–130. https://doi.org/10.2979/indjglolegstu.27.1.0087.

Cummings, Scott L., and Ingrid V. Eagly. 2001. "A Critical Reflection on Law and Organizing." *UCLA Law Review* 48 (3): 443–517.

Dipa, Arya. 2015. "Bogor Mayor Under Fire After Banning Asyura Celebration." *The Jakarta Post*, 26 October 2015. https://www.thejakartapost.com/news/2015/10/26/bogor-regent-under-fire-after-banning-asyura-celebration.html.

Ellmann, Stephen. 1998. "Cause Lawyering in the Third World." In *Cause Lawyering: Political Commitments and Professional Responsibilities*, edited by Austin Sarat and Stuart A. Scheingold, 349–430. New York; Oxford: Oxford University Press. https://doi.org/10.1093/oso/9780195113198.003.0012.

Erdianto, Kristian. 2016. "Belenggu Semen Di Kaki 'Kartini Kendeng' Dibuka Atas Permintaan Jokowi." *Kompas.com*, 14 April 2016. https://nasional.kompas.com/read/2016/04/13/21112341/Belenggu.Semen.di.Kaki.Kartini.Kendeng.Dibuka.atas.Permintaan.Jokowi.

Fajarlie, Nadia Intan. 2019. "Lebih Dari Bantuan Hukum, LBH Yogyakarta Melawan Ketidakadilan." *Warga Jogja*, 30 October 2019. http://wargajogja.

net/hukum/lebih-dari-bantuan-hukum-lbh-yogyakarta-melawan-ketidakadilan.html.

Firdaus, Febriana. 2016. "The 'Golden' Feet of Sukinah, Kendeng's Heroine." *The Jakarta Post*, 18 October 2016. https://www.thejakartapost.com/news/2016/10/18/the-golden-feet-sukinah-kendeng-s-heroine.html.

Freeman, Alexi Nunn, and Jim Freeman. 2016. "It's About Power, Not Policy: Movement Lawyering for Large-Scale Social Change." *Clinical Law Review* 23 (1): 147–166.

Ginanjar, Ging. 2017. "'Penyelundupan Hukum' Dalam Kasus Izin Pabrik Semen di Kendeng?" *BBC Indonesia*, 20 March 2017. https://www.bbc.com/indonesia/indonesia-39321180.

Hilbink, Thomas M. 2004. "You Know the Type: Categories of Cause Lawyering." *Law & Social Inquiry* 29 (3): 657–698. https://doi.org/10.1086/430155.

ILRC. 2018. *Modul Pelatihan Paralegal Tingkat Dasar: Bantuan Hukum Struktural, Bantuan Hukum Gender Struktural Dan Pendidikan Hukum Klinis*. Jakarta: Indonesian Legal Resource Centre (ILRC).

JMPPK. 2019. "*Press Release: Usulan Masyarakat Terkait Tindak Lanjut KLHS Pegunungan Kendeng Utara*." Jaringan Masyarakat Peduli Pegunungan Kendeng (JMPPK). 19 November 2019. https://lbhsemarang.id/news/pers-rilis-jaringan-masyarakat-peduli-pegunungan-kendeng-jm-ppk-usulan-masyarakat-terkait-tindak-lanjut-klhs-pegunungan-kendeng-utara72511.

Jong, Hans Nicholas. 2020. "Indonesia's Omnibus Law a 'Major Problem' for Environmental Protection." *Mongabay*, 4 November 2020. https://news.mongabay.com/2020/11/indonesia-omnibus-law-global-investor-letter/.

Kementerian Lingkungan Hidup dan Kehutanan. 2017. "*Kajian Lingkungan Hidup Strategis (KLHS): Untuk Kebijakan Pemanfaatan Dan Pengelolaan Pegunungan Kendeng Secara Berkelanjutan – Kawasan Cekungan Air Tanah (CAT) Watuputih Dan Sekitarnya, Kabupaten Rembang*." Jakarta: Kementerian Lingkungan Hidup dan Kehutanan.

Knight, Ben. 2017. "Indonesian Farmer Joins May 1 Rally to Protest German Cement." *DW*, 1 May 2017. https://www.dw.com/en/indonesian-farmer-joins-may-1-rally-to-protest-german-cement/a-38653827.

Komnas HAM. 2016. *Pelestarian Ekosistem Karst Dan Perlindungan Hak Asasi Manusia*. Jakarta: Komisi Nasional Hak Asasi Manusia.

Kompas.com. 2009. "Walhi Apresiasi Putusan Menolak Semen Gresik." *Kompas.com*, 7 August 2009. https://malang.kompas.com/read/2009/08/07/12215480/walhi.apresiasi.putusan.menolak.semen.gresik.

Korver, A. Pieter E. 1976. "The Samin Movement and Millenarism." *Bijdragen Tot de Taal-, Land- En Volkenkunde* 132 (2/3): 249–266. https://doi.org/10.1163/22134379-90002642.

LBH Makassar. 2018. "Penyuluhan Hukum; Pentingnya Organisasi Petani Dalam Kawasan Hutan." *LBH Makassar*, 11 July 2018. https://lbhmakassar.org/liputan-kegiatan/penyuluhan-hukum-pentingnya-organisasi-petani-dalam-kawasan-hutan/.

LBH Semarang. 2015. "*Catatan Akhir Tahun 2015 LBH Semarang: Membunyikan Lonceng Kematian*." LBH Semarang.

LBH Semarang. 2016. "*Merindukan Negara Hukum: Potret Kegagalan Negara Dalam Memenuhi, Melindungi Dan Menghormati Hak Asasi Manusia Terhadap 73.352 Orang Di Jawa Tengah*." Catatan Akhir Tahun LBH Semarang. LBH Semarang.

Levitsky, Sandra R. 2015. "Law and Social Movements." In *The Handbook of Law and Society*, edited by Austin Sarat and Patricia Ewick, 382–398. West Sussex: Wiley Blackwell. https://doi.org/10.1002/9781118701430.ch25.

Luban, David. 1988. *Lawyers and Justice: An Ethical Study*. Princeton: Princeton University Press. https://doi.org/10.2307/j.ctv346rrr.

Lucas, Anton, and Carol Warren. 2003. "The State, the People, and Their Mediators: The Struggle over Agrarian Law Reform in Post-New Order Indonesia." *Indonesia*, no. 76: 87–126.

Matsuda, Mari J. 1987. "Looking to the Bottom: Critical Legal Studies and Reparations." *Harvard Civil Rights-Civil Liberties Law Review* 22 (2): 323–400.

McCann, Michael. 1994. *Rights at Work: Pay Equity Reform and the Politics of Legal Mobilization*. Chicago Series in Law and Society. Chicago: University of Chicago Press.

McEvoy, Kieran, and Anna Bryson. 2022. "Boycott, Resistance and the Law: Cause Lawyering in Conflict and Authoritarianism." *The Modern Law Review* 85 (1): 69–104. https://doi.org/10.1111/1468-2230.12671.

McRae, Dave. 2017. "Indonesia's Fatal War on Drugs." *East Asia Forum*, 27 November 2017. https://www.eastasiaforum.org/2017/11/28/indonesias-fatal-war-on-drugs/.

MP, Istman. 2016. "Akhirnya, Presiden Joko Widodo Terima Petani Kendeng." *Tempo.co*, 2 August 2016. https://nasional.tempo.co/read/792613/akhirnya-presiden-joko-widodo-terima-petani-kendeng.

Mukrim, Asyari, Andi Nini Eryani, Abdul Aziz, Haswandy Andy Mas, and Nurhady Sirimorok. 2015. *Memenangkan Gerakan Rakyat: Belajar Dari Advokasi Sengketa Tanah Kassi-Kassi*. Makassar: LBH Makassar.

Mukti, Hafizd. 2017. "Tenda Perjuangan Tolak Semen Rembang Dibakar." *CNN Indonesia*, 11 February 2017. https://www.cnnindonesia.com/nasional/20170211045734-20-192790/tenda-perjuangan-tolak-semen-rembang-dibakar.

Nugroho, Rifkianto. 2019. "Photo Essay: Peringatan 26 Tahun Marsinah di Aksi Kamisan Ke 585." *Detik*, 9 May 2019. https://news.detik.com/foto-news/d-4543114/peringatan-26-tahun-marsinah-di-aksi-kamisan-ke-585.

Nurdin, Nazar. 2017. "Ganjar Pranowo Kembali Terbitkan Izin Lingkungan untuk Pabrik Semen." *Kompas.com*, 24 February 2017. https://regional.kompas.com/read/2017/02/24/06510101/ganjar.pranowo.kembali.terbitkan.izin.lingkungan.untuk.pabrik.semen.

Pradana, Aria Rusta Yuli, and Ardi Priyatno Utomo. 2022. "Disesalkan Aktivis Lingkungan, Pabrik Semen Di Rembang Tetap Beroperasi, Kok Bisa?" *Kompas.com*, 3 December 2022. https://regional.kompas.com/read/2022/12/03/142551378/disesalkan-aktivis-lingkungan-pabrik-semen-di-rembang-tetap-beroperasi-kok.

Prasetyo, Aji. 2020. "Beragam Kendala Advokat Lakukan Pro Bono." *Hukum Online*, 17 December 2020. https://hukumonline.com/berita/baca/lt5fda44185b497/beragam-kendala-advokat-lakukan-pro-bono?page=all.

Purbaya, Angling Adhitya. 2015. "Kawal Sidang Soal Pabrik Semen, Ratusan Petani Jalan Kaki Pati-Semarang." *Detik*, 17 November 2015. https://news.detik.com/berita/d-3073171/kawal-sidang-soal-pabrik-semen-ratusan-petani-jalan-kaki-pati-semarang.

Sapiie, Marguerite Afra. 2017. "Kendeng Protester's Death Triggers Calls for Activists, NGOs to Step Up Resistance." *The Jakarta Post*, 22 March 2017. https://www.thejakartapost.com/news/2017/03/22/kendeng-protesters-death-triggers-calls-for-activists-ngos-to-step-up-resistance.html.

Saptono, Irawan. 2012. "Kisah Panjang Gerakan Bantuan Hukum." In *Verboden Voor Honden En Inlanders dan Lahirlah LBH: Catatan 40 Tahun Pasang Surut Keadilan*, edited by Irawan Saptono and Tedjabayu, 1–112. Jakarta: YLBHI.

Saputra, FX Laksana Agung. 2019. "Kendeng Tagih Komitmen Presiden." *Kompas*, 19 November 2019. https://www.kompas.id/baca/utama/2019/11/19/kendeng-tagih-komitmen-presiden/.

Sarat, Austin, and Stuart A. Scheingold. 2006. "What Cause Lawyers Do For, and To, Social Movements: An Introduction." In *Cause Lawyers and Social Movements*, edited by Austin Sarat and Stuart A. Scheingold, 1–34. Stanford, California: Stanford University Press. https://doi.org/10.1515/9780804767965-003.

Scheingold, Stuart A. 2004. *The Politics of Rights: Lawyers, Public Policy and Political Change*. Ann Arbor, United States: University of Michigan Press. https://doi.org/10.3998/mpub.6766.

Siddiq, Taufiq. 2018. "25 Perempuan Desak Kasus Marsinah Diusut Lagi Setelah 25 Tahun." *Tempo.co*, 3 May 2018. https://nasional.tempo.co/read/1085430/25-perempuan-desak-kasus-marsinah-diusut-lagi-setelah-25-tahun.

Sinuko, Damar. 2016a. "Ganjar Ingin Pembangunan Pabrik Semen Kendeng Dilanjutkan." *CNN Indonesia*, 26 August 2016. https://www.cnnindonesia.com/nasional/20160825214327-20-153911/ganjar-ingin-pembangunan-pabrik-semen-kendeng-dilanjutkan.

Sinuko, Damar. 2016b. "Jokowi Utus Teten Kaji Pabrik Semen Di Pegunungan Kendeng." *CNN Indonesia*, 16 November 2016. https://www.cnnindonesia.com/nasional/20161115214141-20-172884/jokowi-utus-teten-kaji-pabrik-semen-di-pegunungan-kendeng.

Sudaryono, Leopold. 2020. "Drivers of Prison Overcrowding in Indonesia." In *Crime and Punishment in Indonesia*, edited by Tim Lindsey and Helen Pausacker, 179–206. Routledge Law in Asia. London: Routledge. https://doi.org/10.4324/9780429455247-11.

The Jakarta Post. 2017. "Jokowi's Short Meeting with Kendeng Farmers Ends in Tears." *The Jakarta Post*, 23 March 2017. https://www.thejakartapost.com/news/2017/03/23/jokowis-short-meeting-with-kendeng-farmers-ends-in-tears.html.

Walhi Jawa Tengah. 2021. "*Press Release: Dampingi JMPPK, Menagih Kabar Revisi RTRW Rembang*," 13 April 2021. https://www.walhijateng.org/2021/04/13/dampingi-jm-ppk-menagih-kabar-revisi-rtrw-rembang/.

Wicaksana, Dio Ashar. 2018. "Why Indonesia Should Stop Sending Drug Users to Prison." *The Conversation*, 23 August 2018. http://theconversation.com/why-indonesia-should-stop-sending-drug-users-to-prison-101137.

Wijardo, Boedhi, and Herlambang Perdana. 2001. *Reklaiming Dan Kedaulatan Rakyat*. Jakarta: YLBHI, RACA Institute.

6 Accommodation and opposition

Engaging with the state

This chapter is concerned with LBH lawyers' relationship with the state. Cause lawyers are often assumed to work in opposition to the state (Klug 2001, 266; Sarat 2001, 186). Cause lawyering is designed to bring about changes to "some aspect of the social, economic and political status quo" (Sarat and Scheingold 1998, 4) and, inevitably, this will involve opposing the government to a degree. Yet all cause lawyers must make decisions over the extent to which they wish to engage with the government, even those in authoritarian regimes (Fu and Cullen 2008; 2010, 25–26). The question of engagement with the state seems a particularly thorny one for cause lawyers in transitional democracies, as Chapter 3 showed. As the state becomes more open, cause lawyers must decide whether they will participate in state structures or maintain a confrontational approach (Scheingold 2001, 389). This was a dilemma that LBH initially found difficult to resolve.

Following the democratic transition, LBH discovered there was greater room to bring about reform without having to resort to legal action. There were new opportunities to engage in strategies like legislative lobbying and drafting of laws and regulations, and to collaborate directly with state institutions to improve their internal policies and practices. These strategies are consistent with an elite/vanguard form of cause lawyering. They are focused up, toward elites, and they are based on a belief in the capacity of legal and policy reform to engender broader social change.

The chapter begins by examining the impact of democratic transition and the opening up of these new legislative and bureaucratic opportunities on LBH's attitude toward engagement with the state. As the state embraced a more open stance in the years following the fall of Soeharto, cause lawyers at LBH grappled with the complex decision of how closely to engage with state institutions. LBH has, on occasion, shown a willingness to engage with authorities, and this chapter examines several examples of its attempts to collaborate with elements of the justice sector and work within state structures to further its goals. The chapter then explores in detail LBH cause lawyers' current position on engaging with the government and legislature in activities like legislative drafting and strengthening state institutions, detailing how disappointment over these past efforts has seen LBH reject engagement with the state almost entirely, and prioritise a stridently adversarial approach.

DOI: 10.4324/9781003486978-9

Cause lawyers inside the state

There are many examples of cause lawyering being transformed by democratic transition. In South Africa, for example, the transition saw cause lawyers move into the state, and into policy-making roles. After vigorously opposing apartheid-era land policy, cause lawyers helped to shape land reform under the newly democratic state (Klug 2001, 280). In South Korea, too, the democratic transition and the election of a former cause lawyer, Roh Moo-hyun, as president saw more cause lawyers recruited into senior government positions (Goedde 2009, 83; Ginsburg 2007, 54). South Korean cause lawyers in the People's Solidarity for Participatory Democracy (PSPD) organisation prioritised policy advocacy measures over litigation (Goedde 2009, 85–86), and their legislative drafting and lobbying efforts saw more than 70 pieces of legislation passed in just five years (Ginsburg 2007, 52–53). Taiwan also saw a massive movement of lawyers into the first opposition party, the Democratic Progressive Party (DPP) (Ginsburg 2007, 58), and the eventual election of prominent lawyer Chen Shui-bian as president from 2000–08.[1] The period of democratisation in the late 1980s and 1990s saw a shift in Taiwanese cause lawyers' tactics, with increasing use of constitutional litigation to bring about significant policy change, in particular for women's rights (Chang 2010, 136). However, constitutional litigation was combined with intensive legislative lobbying and policy advocacy, which were "equally, if not more important to public interest groups facing an unprecedented political liberalisation in a rapidly democratising society" (Chang 2010, 140).

Similar developments were also observed in Indonesia. Although no Indonesian cause lawyers reached the same heights as cause lawyers in South Korea or Taiwan, after 1998 they were certainly presented with new opportunities to work inside the state. In the early years following the fall of Soeharto, several former LBH staff moved into the legislature, the courts, and senior government roles. Some of the more notable examples include the current head of the YLBHI board, Nursyahbani Katjasungkana, YLBHI founder Adnan Buyung Nasution, and Abdul Rahman Saleh (known as 'Arman'), head of LBH Jakarta in the early 1980s. Nursyahbani was a member of the People's Consultative Assembly (*Majelis Permusyawaratan Rakyat*, MPR) from 1999 to 2004 and a member of the House of Representatives (DPR) from 2004 to 2009. She worked closely with other women's activists to pass the landmark Law No. 23 of 2004 on Domestic Violence. Buyung, meanwhile, was appointed to the Presidential Advisory Council (*Wantimpres*) under President Susilo Bambang Yudhoyono, from 2007–09, and was responsible for providing guidance and advice on legal affairs. Abdul Rahman Saleh went on to serve as a judge in the Supreme Court, and eventually as attorney general from 2004–07. These were important moves but, as mentioned in Chapter 3, in the first few years following the fall of Soeharto, cause lawyers external to LBH seemed much more comfortable pursuing change inside the state, where they had a role in shaping state institutions. Law-reform focused organisations like LeIP, MaPPI and PSHK put a major stamp on the reform process in the Supreme Court, helping to craft the crucial reform blueprints. Civil society activists from these organisations also joined the 'Judicial Reform Team Office', the joint Supreme Court-civil society team responsible for implementing the blueprints.

Over the past five to ten years, more former LBH lawyers have taken up strategic positions in state institutions. The most prominent is former YLBHI labour division head Teten Masduki, who at the time of writing was minister of cooperatives and small and medium enterprises, after serving as head of the Office of Presidential Staff during Jokowi's first term. Teten had previously gained a national profile after serving as director of ICW (the anti-corruption organisation founded by YLBHI in the late New Order period). The former head of YLBHI's labour division, Surya Tjandra, also served in the Jokowi administration, as deputy minister of agrarian affairs and spatial planning/deputy head of the National Land Agency (*Badan Pertanahan Nasional*, BPN), but lost his position in a cabinet reshuffle in mid-2022. Todung Mulya Lubis, YLBHI's most famous living alumnus, has held a range of short-term appointments from the state, and was briefly touted as a potential minister of law and human rights during Jokowi's first term and previously as attorney general under Yudhoyono. He was appointed to neither position, but has recently completed a term as Indonesian ambassador to Norway and Iceland. Another former YLBHI figure who has moved into the state is Benny K. Harman, now a visible and outspoken Democratic Party politician. He sits in DPR Commission III, which has responsibility for legal, human rights and security affairs. Similarly, former YLBHI lawyer Taufik 'Tobas' Basari is now a popular member of the National Democratic (Nasdem) Party, and also sits in DPR Commission III. Other former LBH and YLBHI staff have been appointed to independent quasi-state institutions, for example, Mohammad Choirul Anam, who is a commissioner at Komnas HAM, Poengky Indarti, who is a member of the National Police Commission (*Komisi Kepolisian Nasional*, Kompolnas), and Erna Ratnaningsih and Indro Sugianto, who have both served on the Prosecutors Commission (*Komisi Kejaksaan,* Komjak). These examples are by no means exhaustive, and there are many examples from the regional levels, where former LBH staff have gone on to have active roles in politics at the local level.[2]

Cause lawyers who enter the state may help shape priorities within the state, ensuring greater attention to the problems of the poor (Meili 2009, 65; NeJaime 2012, 654). They may also expand access, helping to broker relationships between social movements and government actors (NeJaime 2012, 654). Some LBH lawyers contend that their engagement with the National Law Development Agency (BPHN) has indeed been able to promote a pro-poor perspective in the institution (see below).[3] However, LBH is mostly doubtful about the ability of its alumni who have entered the state to drive change from the inside. Many LBH lawyers believe that once former LBH staff enter the state, they have been co-opted – they can no longer be faithful to social movement goals; their loyalty now lies with the government. Even if they still have "traces of LBH" on them,[4] they will be forced to follow the demands of the system. This pessimism is also partly a result of past experiences of trying to take advantage of these relationships. As Asfin commented:

> [W]hen Teten Masduki entered the Presidential Palace, we saw it as an opportunity. Although we were cautious, we used his channel, until we discovered that it was not effective. So we broke off communication.[5]

LBH appears much more comfortable maintaining a hard-line stance and operating outside state structures. It sometimes extends this oppositional approach even to its own alumni who have entered state institutions. Tobas mentioned, with a degree of disappointment, that LBH did not seem to want to take advantage of his position in the DPR.

> The people who have sought to take advantage of my position are actually from other CSOs that have chosen a strategy of working with the legislature. But for LBH, I think their antipathy to 'practical politics' is too strong to approach me. It is as though once you have entered the political realm then we must be prepared to become opponents, even as alumni of LBH. From the beginning, before I was officially installed in the legislature, I reached out to CSOs and said that I was prepared to become 'an agent on the inside'. The people who were most enthusiastic were from other organisations, not LBH. This is because of LBH's decision to take a confrontational strategy, not taking advantage of my position.[6]

One might question the wisdom of such an uncompromising approach to institutional independence. Even under the New Order, YLBHI founder Buyung sustained relationships with senior military and regime officials. Although Buyung's colleagues at LBH were highly critical of these relationships,[7] Buyung defended them, stating that for 30 years he had attempted to maintain dialogue, believing there were still people in government who would be open to proposals for reform (Aspinall 2005, 111). As Aspinall also notes, Buyung's receptiveness to engaging with senior military officials was likely influenced by his background as a member of a small elite,[8] and his experiences helping to shape ideas about the restoration of the *negara hukum* in the very early New Order period (Lev 2007, 397; Aspinall 2005, 111). Buyung was much more "politically embedded" (Michelson 2007) than contemporary LBH lawyers. The concept of political embeddedness has often been used in the Chinese context, and refers to those lawyers that have experience working in state justice sector institutions or maintain strong relationships with influential institutions or individuals (Liu and Halliday 2016, 6). Buyung had both experience inside the state as a prosecutor and ongoing ties with senior members of the New Order.

The current crop of LBH leaders is less politically embedded, even under a regime ostensibly more open than the New Order. Of course, current LBH lawyers' failure to maintain the kind of backdoor relationships kept by Buyung can be explained partly by the dramatically different social and political context. The 'elite' is much broader, and LBH lawyers do not have the same stature as the small number of lawyers did half a century ago. But, as noted, there are opportunities for maintaining direct contact with government and justice sector officials that are not being taken up by LBH. Several of my respondents commented that they felt other CSOs were far better at taking advantage of behind-the-scenes relationships with senior police, for example.[9] Besides occasionally causing LBH to miss opportunities to contribute to policy reform, this lack of engagement could adversely affect the practice of LBH lawyers. Political embeddedness has been shown to deliver advantages for Chinese lawyers in their criminal defence work (Liu and Halliday 2016, 60–64, 75–79), and it is possible that Indonesian lawyers would also see similar reductions in professional difficulties if they maintained closer relationships with senior officials. By not engaging, however, LBH can claim a degree of 'moral purity'. It can continue to stand beside its clients without any fear of being compromised, or facing questions over where its allegiances lie. The cost of not engaging is lack of influence. LBH is aware of this, too, and has occasionally endeavoured to work in closer proximity to the state, despite its unease with such approaches.

Experimentations with accommodative strategies

Under the authoritarian New Order, taking a strong, ideological position in opposition to the government was a relatively uncomplicated position to take. But as the constellation of power changed with the fall of Soeharto, patterns of relations became more complex. As discussed in Chapter 3, this forced LBH to reflect more deeply on its raison d'être and how it wanted to use the law for social change. In the early years of the democratic transition, LBH experienced a period of disorientation, where it grappled with the dilemma of maintaining a strong adversarial role that kept the government at a distance while taking a more accommodative elite-focused role involving in engaging with legislatures and bureaucracies (Scheingold 2001, 389).

It is hardly surprising that an organisation established under an authoritarian state and which spent its first three decades attempting to radically restructure that state developed a deep-rooted oppositional culture. This was particularly the case for LBH offices in the regions. During the New Order period, these offices were often on the frontline of efforts to defend communities against forced evictions, farmers from land grabbing by the state, and labour rights at a time when labour organising was severely constrained. A succinct visual representation of the emphatically oppositional culture common among regional LBH offices is that an organisational poster on the wall at LBH Bandung boldly declares 'Militancy Without Limits' in large white block letters against a black background.

Despite this entrenched oppositional culture, in the period following the fall of Soeharto, LBH did make some efforts to engage with the government. LBH undertook a considered effort to reorientate practice and get more involved in reform, consistent with an elite/vanguard model of cause lawyering. Even if it was never institutionalised as a formal organisational policy, during Patra Zen's period as leader, YLBHI spoke internally about being comfortable crossing the boundaries "between black and white".[10] The most notable example of these efforts at engagement was YLBHI's support for the development of the 2011 Legal Aid Law. During Yudhoyono's first term, when Indonesian democracy was considered to be at its healthiest (Mietzner 2021, 11–16), YLBHI, supported by donors, engaged in activities like expanding the public knowledge base on legal aid (KUBAH 2010), drafting legislation, preparing the 'academic paper' (*naskah akademik*) to accompany the draft legislation, meeting and collaborating with the national legislature and government officials, and conducting public advocacy activities.[11] YLBHI even participated as a member of a government-formed joint team tasked with drafting a government version of the legal aid bill to be submitted to the legislature (YLBHI 2009).

Since the national legal aid programme was launched in 2013, YLBHI has developed a productive working relationship with BPHN, which is responsible for implementing the legal aid scheme. YLBHI sits down regularly with BPHN to discuss methods for improving the provision of legal aid, and worked closely with the agency to prepare service standards for legal aid and a regulation recognising the role of community paralegals.[12] Regional LBH offices, including LBH Jakarta, LBH Surabaya and LBH Makassar, have also led advocacy efforts to encourage regional administrations to issue regional regulations (*peraturan daerah,* perda) on legal aid, involving the allocation of funds from subnational government budgets to legal aid. During her time as YLBHI director, Asfin was also selected as part of a joint BPHN and civil society 'Team of Seven' that evaluated legal aid providers for the 2019–21 period (BPHN 2019). This process resulted in 73 unethical legal aid providers, which had reportedly falsified documents to be accredited, being dropped (Peradi Tasikmalaya 2019).[13] LBH is invested in improving the legal aid scheme not only because it played such a prominent role in bringing the law that established it to fruition, but also because it may one day become an important source of funding for its operations. It is also undeniably in the best interests of its clients to have a strong, well-funded, functional legal aid system. YLBHI staff comment that there are people within BPHN who share its commitment to making the legal aid system work for poor people.

> There are two state institutions that are still willing to work with us: BPHN and Bappenas [National Development Planning Agency]. They

> are responsive. We can have a mature debate with them and if our arguments are strong, they will accept them. Several people in BPHN are ideologically similar to LBH, because they understand the problems. Because they work with legal aid providers, they understand that police often refuse access to lawyers... They involve us when they do not have the expertise, for example, in preparing service standards for legal aid, or the regulation on paralegals.[14]

The 2011 Legal Aid Law is the most success LBH has had with engaging in more formal policy advocacy strategies. However, the overall story of LBH's attempts at more accommodative or inside-the-state versions of cause lawyering has largely been one of disappointment. Part of the problem is that a shared understanding of accommodative, policy-focused strategies never developed across the LBH network. While some regional LBH offices, like LBH Jakarta and LBH Makassar, were at times willing to collaborate with state agencies to promote social change, other offices were resistant. This often resulted in tension and disagreements during annual meetings, as Chapter 3 discussed.[15] YLBHI's Mayong recognised that working with the government and legislature on policy reform did not come naturally for many of his colleagues:

> LBH offices [in the regions] are not used to doing 'soft advocacy'. Because these methods of advocacy have certain characteristics, such as compromise, being more restrained in delivering your message. But case management, litigation, let alone community organising, is much more confrontational. Most LBH staff are not practiced in doing soft advocacy like that.[16]

Cause lawyers have been recognised as playing important roles in institution building following democratic transition (Klug 2001; NeJaime 2012). These kinds of efforts to strengthen state institutions have been particularly controversial within LBH. As mentioned, following the democratic transition, international donors encouraged civil society to collaborate with state institutions like the Supreme Court and the police, to strengthen their internal capacity and work with them on reforming their internal procedures. In 2014 and 2015, for example, YLBHI director Asfin worked with a group of civil society representatives to collaborate with the National Police (*Kepolisian Negara Republik Indonesia*, Polri) to develop a police chief circular (*surat edaran*) on guidelines for police in the field to address hate speech.[17] This was considered important at the time, given escalating attacks on religious minorities and the violent hate speech that often preceded these attacks, typically without consequences.[18] The police elected to use an expansive definition of hate speech that included slander, defamation, blasphemy, antisocial behaviour, provocation, incitement and spreading fake news.[19] Since the circular was published, instead of focusing on hate speech against minorities, police have seemed much more interested in acting on behalf of the government and other powerful figures, who have exploited the loose definition of hate speech to pursue their critics with criminal charges (Setiawan 2020).[20]

As a move toward reform post-Soeharto, the National Police issued Police Chief Regulation (*Perkap*) No. 8 of 2009 on Implementation of Human Rights Principles and Standards in Policing Duties.[21] With encouragement from foreign donors, during 2013 and 2014, YLBHI worked with the National Police's Senior Staff and Leadership College (*Sekolah Staf dan Pimpinan Lembaga Pendidikan dan Pelatihan Polisi Republik Indonesia*, Sespimti) to strengthen police capacity to manage social conflict in line with human rights principles. It ran a series of seminars and workshops with police at the college, which is focused on training the next generation of police leaders. Similarly, LBH Jakarta worked with the National Police's counter-terrorism unit, Special Detachment 88 (*Detasemen Khusus 88*, or Densus 88), on a programme designed to mainstream human rights principles among anti-terror police, with funding from the United Kingdom.

Despite the time and effort expended in both activities, however, LBH saw little results. The police might have made these formal commitments, but examples of them failing to adhere to human rights principles are almost too many to mention. To name just one, the horrific Kanjuruhan Stadium disaster, in which 135 people died in a crowd crush following excessive police use of tear gas, focused global attention on Polri's dire human rights record (Reuters 2022). The tragedy led some observers to conclude that attempts at police reform after 1998 have been an unmitigated failure (Baker 2022). For LBH Jakarta, the negative experience of collaborating with Densus was enough to prompt it to cease working on activities designed to strengthen the capacity of state institutions. As Pratiwi Febry said: "We worked with Densus. The policy was there, we tried to increase their awareness of human rights, but there was no significant change. So we decided from that point that we will no longer work with the state."[22]

It is not hard to understand LBH's scepticism. During the *Reformasi Dikorupsi* protests in 2019, police arrested hundreds of student protesters and denied many of them access to legal counsel. When the students were eventually released, many complained of being tortured in detention (CNN Indonesia 2020). LBH now questions why it should engage with an institution that seeks to prevent it from doing its legally protected work. As a result of these disappointments, LBH has decided there is little to be gained from expending considerable funds and energy strengthening the capacity of state justice sector institutions when they do not bother adhering to the human rights principles to which they have formally agreed. LBH looks at the vast sums donors have poured into promoting reform in the Supreme Court and Polri and concludes that engaging with the state to try to promote reform 'from within' has resulted in little meaningful change.[23] Reflecting on the decisions of law reform CSOs to collaborate with, and attempt to strengthen, state institutions like the Supreme Court following the fall of Soeharto, Asfin is highly critical:

> We can't judge that decision based on the situation now. But whatever the strategy, it is crucial to involve the community. If bureaucratic reform, engaging with policymakers or reform from within, does not involve the community, it will not result in substantial change. For example, everyone knows judges are corrupt. The approach [of civil society at that time] was as if they did not understand this. They gave them training, attempted to 'inspire' them. It's not that judges do not understand – rather, it's that they are evil! So, we missed the momentum and the evil forces reconsolidated. And we can see the result 20 years later.[24]

For the current leadership of YLBHI, any dilemma about the degree to which they should engage with the state has been resolved. In the eyes of LBH, the police, the courts and the prosecution are irredeemable. Why devote their limited resources to working with elements of the state to reform the system when it is unlikely state institutions will ever genuinely implement these reforms? History agrees with this view. Twenty years after the fall of Soeharto, "there is no evidence that the entrenched system of corruption" that characterised Indonesia's justice system under the New Order has been meaningfully altered (Lindsey and Pausacker 2021, 5).

Crystallisation of an oppositional identity

LBH might have been more open to accommodative elite-focused versions of cause lawyering in years past, however it is now very wary of engaging with the state on policy reform. Distrust of the government and legislature has escalated dramatically over the past decade, as Indonesian democracy has come under attack. LBH now views the government and legislature as being dominated by oligarchs and impervious to public demands.

This attitude shift has occurred in part because of the increasing dominance of pro-government parties in the legislature. President Joko Widodo came to power in 2014 with the backing of just 37% of parties in the national legislature, or DPR. He quickly moved to strengthen his power base. The government intervened in conflicts in two opposition parties, Golkar and the United Development Party (*Partai Persatuan Pembangunan*, PPP), backing pro-Jokowi factions in both, and eventually secured their support (Mietzner 2016). These two parties backed him for re-election in 2019, along with the four parties that had supported him in 2014. After being re-elected, Jokowi strengthened his position further. By 2021, his government had secured the support of seven out of the nine parties in the DPR, equivalent to 81% of the 575 seats available (The Jakarta Post 2021). Only the Democratic Party and the Islam-based Prosperous Justice Party (*Partai Keadilan Sejahtera,* PKS) remained outside the ruling coalition. Following the 2019 election, Jokowi also brought his presidential and vice-presidential rivals, Prabowo Subianto and Sandiaga Uno, into the government as minister of defence and minister for tourism and the creative economy, respectively.

Indonesia's political parties are widely known to be corrupt and under the influence of oligarchs (Mietzner 2015). The Introduction to Part Two described how, under Jokowi, political elites in the legislature have had greater success in their efforts to dilute key reforms of the democratic era, exhibited in, for example, the revisions to the Law on the KPK, and the 2020 Omnibus Law on Job Creation. The government and legislature have also collaborated to pass a range of other statutes that harm Indonesia's democratic quality, while emboldening oligarchs and their interests. These include, but are not limited to, the 2020 Law on Mining,[25] the 2022 Law on the National Capital, the 2022 laws creating three new provinces in Papua,[26] and the revised Criminal Code (KUHP), which was eventually passed in late 2022. These laws have been characterised by hasty deliberations, minimal public participation and a lack of political opposition.

This closeness between the government and legislature has occurred at the same time as the government has presided over a dramatic reduction in civic space (Aspinall and Mietzner 2019, 314). As discussed, both the *Reformasi Dikorupsi* protests in 2019, and the 2020 protests against the Job Creation Law resulted in hundreds of demonstrating students and activists being arrested, many of whom were held without charge (Hamid and Hermawan 2020). A particularly worrying development has been the Jokowi administration's pursuit of its political opponents with criminal charges, especially with the use of defamation and hate speech provisions, although other criminal offences are also used (Setiawan 2020). By 2019, these developments were already contributing to a deep scepticism about the regime among the LBH network. Consider these comments from Asfin:

> This is the worst government in Indonesia since *reformasi*. When Megawati [Soekarnoputri] was in power, many people were also charged with insulting the president. But this government is the worst. Reports from Freedom House and so on confirm this... Maybe in 2015–16, the changes were not too dramatic. But from 2018, and especially in 2019, deterioration has been incredibly rapid. It is now almost exactly like living under the New Order.[27]

While LBH's claims of a reversion to authoritarian rule are overstated, there is no doubt that Indonesian democracy is in decline. The recent trend of the government and legislature working together to pass laws that undermine democracy and benefit oligarchic interests has been particularly difficult for LBH to stomach. Pratiwi Febri captured the LBH view:

> The government is using law to legitimise its power. The political process is no longer democratic... The legislative process is returning to what we

> saw during the authoritarian period, where the opposition parties control only a small percentage of the legislature and can only shout from the sidelines.[28]

YLBHI's head of networking and campaigns, Arip (Yogi) Yogiawan, made similar pessimistic observations about the national legislature.

> Democracy in the legislature is closed, the oligarchs have control. Look at the percentage of businesspeople in the DPR. PKS is the only real opposition party, they are too small, and their legislative agenda is vastly different to ours… We have conducted activities 'inside the room' several times but we are not heard. It is closed. So, the next option is to take action outside. If there is no space for us inside the room, we must take action outside. What does that mean? First, on the streets, and second, on social media.[29]

As these comments suggest, the narrowing of political opportunities for elite/vanguard cause lawyering has contributed to a solidification of the oppositional identity LBH forged during the New Order period, and a rejection of formal tactics in favour of more radical grassroots approaches. Given that cause lawyering is often an oppositional project, it is not entirely unexpected that democratic regression has hardened LBH's oppositional resolve. As Luban puts it, "cause lawyers are a nuisance to the state and they mean to be a nuisance" (Luban 2012, 705). Democratic regression has provided LBH with a greater degree of clarity about this aspect of its role. LBH is unapologetic about reverting to a more confrontational version of cause lawyering. Indeed, it could be said that LBH is returning to its roots after several years of experimenting with other alternatives. When I spoke to him in 2019, then LBH Jakarta director Arif Maulana was extremely cynical about the Jokowi regime, and advocated for an almost complete shutting down of engagement with the government on reform.

> I think you would be naïve if you engaged with or collaborated with the government now. LBH Jakarta's position is that it is impossible to work with the government. Because the government does not have a vision of democracy, a vision of human rights. Our values are different. How could we possibly work with people whose values, principles, and vision are different to us? It is not possible. And if you still think Jokowi offers hope, that is stupidity. With all the facts that are available. Not only naïve, but stupid.[30]

Arif's position is a principled one to take, but some might dismiss it as too uncompromising, or even lacking maturity. Every day, politicians need to sit down and discuss matters with people who believe entirely different things to them. Even defence lawyers need to be able to work with prosecutors, despite not often trusting or respecting them. Collaborating with the state on policy work will invariably involve a degree of compromise. By deciding not to engage with the government and legislature, however, LBH does not have to worry about moderating its approach – a common consequence of proximity to the state (Scheingold 2001, 390). It allows LBH to avoid the "moral complexity" of cause lawyering inside the state and the unavoidable compromises this involves (Luban 2012). LBH's affinity with oppositional approaches to cause lawyering may also be partly determined by its inherent nature as a legal, rights-focused organisation. A strict allegiance to human rights principles might be easier to maintain for reform-focused CSOs than wading into the murkier waters of political compromise (Nardi 2018, 250). LBH also frames this resolute oppositional stance as part of its commitment to the victims of rights abuses. Former YLBHI deputy Robertus Robet captured the LBH position:

> People who are working on the inside can no longer speak with one voice with victims. If they [LBH] are on the outside, they can still speak with one heart, with one voice with victims. This is a very clear, pure, position.[31]

A stridently oppositional approach allows LBH to claim the moral high ground, a certain ideological purity. While this might sometimes see LBH dismissed as too obstinate or idealistic, choosing to take this more adversarial position is a reasonable decision for LBH to make. Not everyone has to work 'inside the tent'. LBH recognises that a healthy democracy needs CSOs that can both criticise and collaborate with the state. But LBH does not need to do both itself. Instead of engaging with the government or legislature directly, LBH is more likely to leave this work up to its colleagues in civil society. It typically participates in civil society coalitions for reform but leaves it up to other members of the coalition to engage with state structures. By deciding to maintain a distance and let its colleagues in civil society take a more prominent role in negotiating with legislatures and the state bureaucracy, LBH can claim it is completely loyal to its causes, and to its clients. As Pratiwi Febry from LBH Jakarta explained:

> We know the DPR is a mess. There is significant horse-trading involving exchange of funds to pass laws. So, intervening in the legislature, lobbying political parties, we do not do it. We think it can be strategic, of course, but we do not want to interact with political parties, we leave it up to the [broader civil society] network. Why? To protect our independence. We choose to do political work on other levels, not by approaching parties.[32]

This seems to be LBH's preferred method of engaging in policy advocacy in the face of democratic regression. For example, LBH was a central member of a civil society coalition that advocated for revisions to Law No. 39 of 2004 on the Placement and Protection of Indonesian Migrant Workers Abroad. LBH contributed legal analysis and was heavily involved in the formulation of a draft civil society-prepared bill, which eventually became Law No. 18 of 2017 on the Protection of Indonesian Migrant Workers (YLBHI 2018, 106–8). Likewise, YLBHI and LBH Jakarta also joined the large civil society coalition that advocated for Law No. 12 of 2022 on the Crime of Sexual Violence, which was eventually passed in April 2022. Although women's organisations took the lead, and LBH did not participate in legislative lobbying, LBH helped to draft the civil society-prepared bill.

YLBHI and LBH Jakarta were also members of the 'National Alliance for Reform of the Criminal Code (KUHP)' (*Aliansi Nasional Reformasi KUHP*), a civil society network that was led by ICJR and Elsam. Since 2005, the alliance analysed the government and DPR prepared draft bills on the revised code, prepared policy papers, formulated lists of key concerns, and delivered its feedback to legislators. Over the past five years, YLBHI has contributed data and analysis in areas in which it has distinct experience, such as religious freedom, while IJCR took the lead in lobbying and negotiations with the legislature. As discussed, in September 2019, the DPR came very close to passing a highly problematic version of the revised code, despite ICJR's longstanding engagement with the legislature, and vociferous criticism from broader civil society. It became clear that regardless of civil society's efforts to meet regularly with the government and legislature, and to explain flaws in the draft code and the risks associated with its excessive reliance on criminal sanctions, the DPR was prepared to pass a deeply defective version. Only massive street protests eventually stopped the code being passed. Around this time, I spoke to Muhamad Isnur, then head of advocacy at YLBHI.

> Evaluating the history of civil society engagement with the government, has working from the inside been significant? That is our big question to them. To our friends who have worked with the police on reform, what do you say about the police now? Are you guaranteed to get anything from engaging with the government? We realise they are facing powerful interests. Ultimately, we understand and respect their role. As long as they don't become a part of government, as long as they can protect their critical engagement, we can still work with them. So, we share responsibilities. For example, on the issue of the KUHP, ICJR works from the inside and lobbies the DPR. But in the end, the DPR does not respect their lobbying. We understand and respect their decision, and we are still friends with them. They are still members of our larger coalition. In this time of democratic regression, we understand the importance of working together.[33]

It is worth recognising that these comments were made around the time of the 2019 *Reformasi Dikorupsi* protests, when the government and legislature were seeking to pass a range of laws that would have wound back Indonesia's democratic gains. During this heated period, it sometimes appeared as though LBH was opposed to *anyone* engaging with the state. For example, YLBHI Knowledge Management division head Rakhma had strong words for LBH's civil society allies from the Consortium for Agrarian Reform (*Konsorsium Pembaruan Agraria*, KPA) (Saptowalyono 2021). In late September 2019, at the height of the protests, KPA and other farmers' groups met with the head of the Office of Presidential Staff and former head of the Indonesian military (*Tentara Nasional Indonesia*, TNI), Moeldoko, to discuss working with the state on agrarian reform (CNN Indonesia 2019). This was at the same time as the *Reformasi Dikorupsi* protests criticised legislators' plans to pass a revised land law (*RUU Pertanahan*) and revised mining law (*RUU Minerba*) that demonstrators claimed would have weakened environmental protections and threatened community rights. Rakhma was livid. I asked her about it again in 2021.

> The agrarian reform movement cannot be separated from how we view state policy as a whole. On one hand, the state passes laws like the Omnibus Law on Job Creation and the Mining Law, but on the other, to placate civil society, they are invited to the Presidential Palace, and offered a role in agrarian reform. It is like a child being offered a lolly-pop – 'There, there, no need for you to be so angry'. Members of our network accepted this, and that's why we were so disappointed.[34]

Arguably, by working with the state on agrarian reform, KPA and other groups might secure minor wins for local communities grappling with land conflict. Of course, LBH also wants to see better outcomes for local communities too. Yet those with more uncompromising views, like Rakhma, argue that these small concessions are only tinkering around the edges, and distract social movements from demanding more transformative change. Worse still, in the minds of many LBH lawyers, these kinds of incremental reforms may end up legitimising a corrupt and unjust system, or perpetuating the status quo – the antithesis of cause lawyering. As Anna-Maria Marshall has observed, the incremental policy concessions achieved through institutional strategies seldom address the deeper structural problems that motivate social movements, and these incremental reforms are just as likely to be revoked when the state is no longer under pressure (Marshall 2006, 165). LBH would say it has its sights fixed on larger, and what it would argue are more substantive targets – the protection of Indonesia's democratic gains.

Risks of confrontational approaches

Taking a defiant adversarial position in opposition to the state may expose cause lawyers to serious risks. Other authors have described how, under authoritarian regimes, lawyers sometimes pursue less confrontational approaches, and work pragmatically within the limits of the authoritarian system to secure what small changes they can (Chua 2014; Stern 2013; O'Brien 2023). Yet there are also many examples of cause lawyers openly challenging hostile and repressive political regimes, despite the dangers this entails (Pils 2014; Liu and Halliday 2016; Cheesman and San 2013). Throughout LBH's history, it has been prepared to take a bold, political stance, in the courtroom and beyond, even when this risks provoking state repression.

While it would be inaccurate to characterise Indonesia as an authoritarian state, or suggest the threats faced by Indonesian cause lawyers are on the same scale as those faced by their counterparts in China, LBH lawyers have still faced surveillance, intimidation, harassment, arrests and even physical violence. When I was based at YLBHI in 2019, every day presumed undercover intelligence officers would work conspicuously on laptops in a food store across the road from the entrance of the office. While LBH was never entirely sure of these presumed agents' activities, the police were certainly well informed about the activities of oppositional civil society groups and occasionally approached LBH with vaguely worded threats or warnings against organising public protests.[35] Whatever their actual intelligence gathering role, however, the fact that they made little effort to mask their activities suggested that a large part of their role was to intimidate, adding to the mental stress on LBH lawyers.

Sometimes threats may be even more direct. At least once during my initial period of fieldwork in 2019, then YLBHI director Asfin's phone was hacked by unknown sources. In September of that year, Asfin was also reported to police for "spreading fake news" after making comments critical of the selection process for the new leader of the KPK.[36] The report was never followed up. Following a familiar pattern, it seemed designed simply to silence civil society criticism. Even more egregiously, from 2015 to 2021, at least 20 legal aid lawyers in the YLBHI network were arrested while doing their jobs, with three of them later charged (YLBHI 2021). Several of these lawyers were arrested when accompanying community members engaged in public protests and street demonstrations.

In an earlier incident that shocked Indonesian civil society, in September 2017 the LBH Jakarta/YLBHI office was set upon by Islamist and nationalist protestors after it hosted an academic discussion with survivors of the 1965 anti-communist violence. Police backed the protestors, forced their way into the building and tore down a backdrop banner for the discussion. Several activists were smuggled outside of the building and sought refuge at Komnas HAM until the threat passed (Hidayat 2017). While physical threats are rarer, they also occur. In September 2021, for example, a Molotov cocktail was thrown at the office of LBH Yogyakarta, the first time the office had faced such an attack in its history (Maharani 2021).[37] LBH Medan, North Sumatra, also faced a similar attack in 2019 (Dewantoro and Khairina 2019). Yet these hardships have only seemed to harden LBH's resolve. It appears to have been energised by a more repressive environment.

Independent quasi-state institutions

If LBH is so cynical about the state, what about the new independent institutions, such as Komnas HAM, the National Commission on Violence Against Women (*Komisi Nasional Anti Kekerasan Terhadap Perempuan*, Komnas Perempuan), the Indonesian Ombudsman, the Witness and Victim Protection Agency (*Lembaga Perlindungan Saksi dan Korban*, LPSK), the National Police Commission (Kompolnas), the Judicial Commission (*Komisi Yudisial*, KY), and the Prosecutors Commission (Komjak)? Aside from Komnas HAM, which was founded in 1993, these institutions are direct outcomes of the transition to democracy after 1998.[38] One might therefore expect that these new institutions would have an impact on the shape of cause lawyering practiced after the democratic transition. However, they do not feature prominently in this book simply because LBH has found them disappointing. LBH lawyers repeatedly identified Komnas HAM, Komnas Perempuan and the Ombudsman as the institutions most helpful for supporting their work but they also lamented that these institutions are constrained by their limited ability to coerce the government to follow up on their reports. YLBHI's Yogi summarised the LBH position:

> We had a lot of hope for these independent institutions. In the early 2000s, they were proactive. If there was an incident, we would report it to them, and they would respond quickly. Now Komnas HAM is sluggish – we even published a report on freedom of expression before they did... The Ombudsman is committed, it wants to work, but its reports are usually ignored... Komnas Perempuan is progressive, but it is highly centralised and can only have a real impact on issues involving women and children.[39]

LBH cause lawyers have little faith in these institutions as institutions, although there are individuals in these organisations on whom they rely as allies or sympathisers.[40] However, even the presence of former LBH staff in these institutions has not been enough for them to live up to LBH's expectations.

> Organisations like Kompolnas, the Prosecutors Commission, LPSK, they have all had former LBH members on staff, but they have not been able to achieve anything. We hoped that [the former LBH staff] would introduce the LBH perspective into these organisations, but they ended up sinking in the organisational culture.[41]

As a result, LBH does not depend on these institutions in its cause lawyering work, although it does recognise that they can help bolster advocacy activities by lending their institutional support to LBH's goals, in the form of joint statements or collaboration as part of public events. Their institutional profiles and legitimacy as state-funded independent institutions can add weight to LBH claims. More commonly, however, if these institutions are approached, it is usually as a box-checking exercise, to demonstrate that all avenues have been pursued.[42] Clearly, the fact that these new democratic institutions have not emerged as critical sites for cause lawyering is another sign – and consequence – of the democratic backsliding that has occurred over the past decade.

National Conference on Legal Aid

This chapter has described how democratic erosion in Indonesia has led to LBH developing an extremely pessimistic view of the state. It has almost completely abandoned tactics involving collaboration with government officials on legal and policy reform. To further demonstrate the practical implications of this hard-line position for the practice of cause lawyering, I wish to draw on an example from the 2019 National Conference on Legal Aid. On 20 August 2019, I attended a civil society-run national conference on legal aid on the outskirts of Jakarta. The conference was organised by YLBHI in collaboration with the Indonesian government and several CSOs,[43] with funding support from USAID (through The Asia Foundation), the Dutch government, and the International Development Law Organisation (IDLO). It was attended by more than 350 representatives from legal aid-providing organisations (in addition to LBH offices in the YLBHI network), paralegals and CSOs.

The plenary session on the first day discussed government policy on expanding access to justice and legal aid, and took the format of a talk show, with representatives from BPHN, Bappenas, the Ministry of Research and Higher Education, the National Police, the Attorney General's Office and the Supreme Court, as well as Asfin as head of YLBHI. When asked about how police are ensuring access to legal aid during the investigation process, the police representative gave a normative response, saying that when a community member faces legal troubles, police are required to provide access to legal aid in accordance with the Criminal Procedure Code (KUHAP). If they don't have a lawyer, he said, police send a letter to local legal aid providers to request representation. Likewise, the representative from the prosecutor's office claimed that if prosecutors received a file from the police where the suspect faced a sentence of five years or more, then they would immediately remind police that the suspect must be represented. For Asfin, these kinds of platitudes were insulting, and she reacted accordingly. Speaking loudly and passionately, Asfin asked the crowd:

> How many of you provide legal aid in criminal cases? [most but not all attendees raise their hands]. So, not all legal aid organisations defend criminal cases. Of those of you who raised your hands, how many of you have been prevented from accessing your client by police or prosecutors?

> [more than half of those who raised their hands raise their hands again]. I'm sure that for all of you who raised your hands, there would be others not in this room with the same experience, and many would have been rejected more than once. Equality before the law for people like me and my friends here is like a bedtime fairy tale – repeated every night but never eventuating. We have memorised it, we are familiar with it, but equality before the law is never realised.

The room erupted in rapturous applause, with many of the legal aid lawyers and paralegals jeering and laughing. Of course, Asfin was completely right. A study by ICJR reported only 2% of detainees have legal counsel present during police questioning (Domingo and Sudaryono 2016, 25).[44] And, according to the late Indonesian scholar of corrections Leo Sudaryono, in interviews with 56 inmates in prisons in Banten, Jakarta, Central Java, South Kalimantan and West Kalimantan, 41 (73%) reported being convicted without ever having access to a lawyer (Sudaryono 2020). Police and prosecutors should be called out when they claim otherwise. However, one must also question whether shaming police and prosecutors in a public forum like this is a strategic approach. These were state officials who were willing to attend a civil society-run event, suggesting they were at least open to the idea of engaging with civil society in discussions on reform. Embarrassing them in public like this would likely shut down any possibility for future collaboration, at least with the individuals on the panel. On the other hand, Asfin's response was entirely consistent with Abel's depiction of cause lawyers "speaking law to power" (Abel 1998). I am sure Asfin felt a responsibility to respond in the way she did. It was the truth, and an expression of LBH's ideology of legal aid, of holding power to account. At the same time, however, it reinforced LBH's position as an outsider.

These theatrical shows of defiance have a long history at LBH. Consider how Asfin's public excoriation of officials mirrored Buyung's outburst in court during the Dharsono case described in Chapter 2. These techniques of political theatre are also reflected in the criminal defence trials that LBH is aware that it is going to lose, such as the Gafatar blasphemy case described in Chapter 4. They are also methods common to cause lawyers operating in authoritarian contexts (McEvoy and Bryson 2022). When cause lawyers recognise the odds are stacked against them, they may use the safe space of the courtroom (or in this case, the shared stage of a conference) to make principled points in defence of human rights, democracy and the rule of law. While other organisations might be more wary of preserving their relationships and access, LBH is prepared to engage in confrontational strategies that could see it being ostracised by lawmakers and other officials.

There is another way of looking at Asfin's response. As the only civil society figure on the panel, Asfin knew she faced the additional burden of representing, or providing a voice for, the many civil society legal aid providers and community paralegals in attendance. Many junior legal aid lawyers and community paralegals would have had little or no opportunity to deliver complaints about lack of procedural fairness to senior government officials. I suspect Asfin's response was as much driven by outrage at the dishonesty of police and prosecutors as it was about building solidarity among the paralegals and legal aid lawyers, providing them with a rare public acknowledgement of the challenges they face daily in the field. It was partly about tending to and building 'the legal aid movement'. This is also consistent with an approach to cause lawyering that emphasises community organising over proceduralist approaches that seek to improve the functioning of the system, as discussed in Chapter 5.

Conclusions

There is no question that the democratic transition in Indonesia provided new political opportunities for elite/vanguard strategies of cause lawyering. After the fall of Soeharto's New Order, cause lawyers were faced with the choice of participating in state structures or maintaining a confrontational approach. As this chapter has demonstrated, LBH largely retained the oppositional approach to engaging with the state it cultivated under the New Order. This was particularly the case in regional LBH offices, which were reluctant to temper their "militancy". This meant LBH missed some opportunities to contribute to policy reform.

Although LBH has mostly maintained an oppositional stance in terms of its relations with the government, it did explore collaboration with state institutions to some extent. Indeed, YLBHI's contribution to the passage of the 2011 Legal Aid Law is one of its most important policy achievements in the democratic era, providing its network with more reliable access to government funding. But aside from this, LBH has had only minor success engaging with the legislature and executive on policy reform. In LBH's case at least, it was never as simple as democratic transition providing the conditions for cause lawyering to thrive.

The case of LBH illuminates the difficult decisions cause lawyers face over engaging with formal state structures in emerging and less stable democracies. In particular, it underscores the limits of formal democratic change for cause lawyering in an environment where political and business elites are able to continue to bend policy and governance to their benefit. These political conditions have affected LBH's approach to elite/vanguard strategies involving collaborating with officials on legislative reform. Much as Gordon observed of cause lawyers seeking to strengthen the rights of immigrant workers in sweatshops in the United States, LBH has concluded: "in a setting where every law on the books is flouted, the incentive to add one more rule to the list of those already ignored is minimal" (Gordon 2005, 27). While some avenues remain for LBH lawyers to lobby legislators or work directly with the state on drafting legislation, the substantial disconnect between the government's formal legal commitments and realities on the ground has led them to deprioritise such efforts.

This picture has become unmistakable over the past five to ten years. Democratic reform has ground to a halt and opportunities for cause lawyering through legislative and administrative channels are narrowing. While there was a period when LBH was open to exploring engaging with government agencies on reform initiatives, the past five years have seen it return to a more stridently oppositional position on engaging with the government, similar to the position it took under the New Order. LBH cause lawyers now shy away from – if not reject completely – elite-focused strategies. They are deeply sceptical of efforts to collaborate with the government and legislature on reform. LBH has concluded that more accommodative strategies are the wrong approach in an environment of democratic regression, and perhaps were never appropriate until liberal democracy was truly in place. It has no hesitation about taking this stridently oppositional approach, even if it may sometimes risk backlash from elements of the state.

Not engaging might mean LBH lacks direct influence over those in power, as its critics sometimes point out. LBH is aware of this potential but guards its independence out of a moral concern for its clients, and a belief that maintaining a distance allows it to avoid the risk of co-optation. It has decided that being able to stand on the side of justice seekers is more important than winning small concessions that are unlikely to be implemented. LBH instead contends that it is seeking broader, transformational change. This has seen it return to the kind of coalition-building role that it played under the New Order, and again emphasise the structural legal aid approach to case management. This is the focus of the next chapter.

Notes

1 Chen had risen to fame for his defence of opposition activists following the 'Kaohsiung Incident', a pro-democracy demonstration on Human Rights Day in 1979 that was met with a harsh response from police, and saw several prominent activists arrested (Ginsburg 2007, 58).
2 Interviews, Robertus Robet, April 2020; Indro Sugianto, September 2019.
3 Interviews, Febi Yonesta, September 2019; Muhamad Isnur, October 2019; Asfinawati, December 2019.
4 Interview, Siti Rakhma Mary Herwati, March 2023.
5 Interview, Asfinawati, December 2019.
6 Interview, Taufik Basari, December 2019.
7 Recall that during the 2001 leadership dispute, Bambang Widjojanto made a veiled criticism of Buyung operating in a "grey area" (see Chapter 3).
8 During Soekarno's Guided Democracy period, when Buyung was a young lawyer, the private legal profession numbered only about 250 (Lev 2007, 397).
9 Interviews, confidential, 2019.
10 The phrase '*menembus batas*' ('breaking boundaries' or 'crossing the boundary') was often used in this context. Interview, Nurkholis Hidayat, March 2023.
11 For examples of such activities see Kompas.com 2008; Hukum Online 2010b; 2011; Ratnaningsih 2011.

12 Interview, Febi Yonesta, December 2019. See also Law and Human Rights Ministry Regulations No. 3 of 2021 on Paralegals in the Provision of Legal Aid and No. 4 of 2021 on Legal Aid Service Standards.
13 A consequence of government funding becoming available is that many less ethical providers have emerged with the express purpose of chasing reimbursement for simple cases, which they may only represent half-heartedly. Interviews, Ajeng Wahyuni, August 2019; Asfinawati, Febi Yonesta, December 2019.
14 Interview, Asfinawati, August 2021.
15 Interviews, Febi Yonesta, September 2019; Nurkholis Hidayat, March 2023.
16 Interview, Febi Yonesta, September 2019.
17 Police Chief Circular No. SE/6/X/2015 on Hate Speech.
18 A famous example was former FPI leader Ahmad Sobri Lubis who was recorded publicly urging his followers to kill Ahmadiyah members, stating that it was *halal* (permitted under Islam) to spill their blood (Albanna 2021).
19 Circular No. SE/6/X/2015, 2.
20 I am only aware of one instance of police targeting a member of the majority community for "hate speech" against minorities, Islamic preacher Muhammad Yahya Waloni, who was arrested after describing the bible as "fictitious" and "fake" in a recorded sermon (Rahma 2021).
21 *Perkap* are regulations that guide police operations in the field.
22 Interview, Pratiwi Febry, December 2019.
23 Interviews, Nurkholis Hidayat, September 2019; Muhamad Isnur, October 2019; Asfinawati, Yasmin Purba, December 2019.
24 Interview, Asfinawati, December 2019.
25 In 2020, when Indonesia was grappling with the early stages of the Covid-19 pandemic, the legislature passed the revised Mining Law (Law No. 3 of 2020 amending Law No. 4 of 2009) it had failed to pass in 2019, as well as the highly contentious Omnibus Law on Job Creation, Law No. 11 of 2020. Both laws were roundly criticised by civil society for dismantling environmental protections (Mietzner 2020, 242–43).
26 Law No. 14 of 2022 on the Establishment of South Papua Province, Law No. 15 of 2022 on the Establishment of Central Papua Province, Law No. 16 of 2022 on the Establishment of Highland Papua Province.
27 Interview, Asfinawati, December 2019.
28 Interview, Pratiwi Febry, March 2023.
29 Interview, Arip Yogiawan, October 2019.
30 Interview, Arif Maulana, December 2019.
31 Interview, Robertus Robet, April 2020.
32 Interview, Pratiwi Febry, December 2019.
33 Interview, Muhamad Isnur, October 2019.
34 Interview, Siti Rakhma Mary Herwati, August 2021.
35 Interviews, confidential, 2019; 2023.
36 Then KPK spokesperson Febri Diansyah and ICW coordinator Adnan Topan Husodo were also reported to police at the same time (Ramadhan 2019).
37 At the time of writing, it remained unclear who was behind the attack. LBH Yogyakarta was defending a range of land conflict cases involving government, military and business interests (Maharani 2021).
38 Komnas Perempuan was established in 1998 in the wake of widespread sexual violence during the May 1998 riots that accompanied Soeharto's fall; the Indonesian Ombudsman was established by law in 2008; LPSK was established by law in 2006; Kompolnas was established through a presidential regulation in 2011; the Judicial Commission was established in 2001 as a result of a constitutional amendment; and the Prosecutors Commission was established through a revision to the Law on the Prosecution Service in 2004.
39 Interview, Arip Yogiawan, October 2019.

40 Interviews, Muhamad Isnur, October 2019; Pratiwi Febry, December 2019.
41 Interview, Adbul Fatah, November 2019.
42 Interviews, Nurkholis Hidayat, September 2019; Pratiwi Febry, December 2019.
43 LBH Apik, the Indonesian Legal Resource Centre (ILRC), PBHI, LBH Masyarakat, LBH Jakarta, and MaPPI.
44 Police claim they offer representation but detainees often reject it. There is reportedly a common belief among detainees that legal representation could lead to a worse experience in detention, or that police will pursue harsher charges against them (Domingo and Sudaryono 2016, 25).

References

Abel, Richard L. 1998. "Speaking Law to Power: Occasions for Cause Lawyering." In *Cause Lawyering: Political Commitments and Professional Responsibilities*, edited by Austin Sarat and Stuart A. Scheingold, 69–117. New York; Oxford: Oxford University Press. https://doi.org/10.1093/oso/9780195113198.003.0003.

Albanna, Morteza Syariati. 2021. "Sobri Lubis: Ahmadiyah Halal Darahnya untuk Ditumpahkan." *Tagar*, 1 January 2021. https://www.tagar.id/sobri-lubis-ahmadiyah-halal-darahnya-untuk-ditumpahkan.

Aspinall, Edward. 2005. *Opposing Suharto: Compromise, Resistance, and Regime Change in Indonesia*. East-West Center Series on Contemporary Issues in Asia and the Pacific. Stanford, California: Stanford University Press. https://doi.org/10.1515/9780804767316.

Aspinall, Edward, and Marcus Mietzner. 2019. "Indonesia's Democratic Paradox: Competitive Elections Amidst Rising Illiberalism." *Bulletin of Indonesian Economic Studies* 55 (3): 295–317. https://doi.org/10.1080/00074918.2019.1690412.

Baker, Jacqui. 2022. "The End of Police Reform." *Indonesia at Melbourne*, 15 November 2022. https://indonesiaatmelbourne.unimelb.edu.au/the-end-of-police-reform/.

BPHN. 2019. "Perluas Jangkauan, BPHN Loloskan 524 Organisasi Bantuan Hukum." *Badan Pembinaan Hukum Nasional (BPHN)*, 4 January 2019. https://bphn.go.id/news/2019010413001183/Perluas-Jangkauan-BPHN-Loloskan-524-Organisasi-Bantuan-Hukum.

Chang, Wen-Chen. 2010. "Public Interest Litigation in Taiwan: Strategy for Law and Policy Changes in the Course of Democratisation." In *Public Interest Litigation in Asia*, edited by Po Jen Yap and Holning Lau, 136–160. London: Routledge. https://doi.org/10.4324/9780203842645.

Cheesman, Nick, and Kyaw Min San. 2013. "Not Just Defending; Advocating for Law in Myanmar." *Wisconsin International Law Journal* 31 (3): 702–733.

Chua, Lynette J. 2014. *Mobilizing Gay Singapore: Rights and Resistance in an Authoritarian State*. Philadelphia: Temple University Press.

CNN Indonesia. 2019. "Jokowi Catat Semua Tuntutan Petani Soal Reforma Agraria." *CNN Indonesia*, 25 September 2019. https://www.cnnindonesia.com/nasional/20190924150718-20-433417/jokowi-catat-semua-tuntutan-petani-soal-reforma-agraria.

CNN Indonesia. 2020. "Kapolri Bentuk Tim Khusus Usut Dugaan Penyiksaan Lutfi." *CNN Indonesia*, 24 January 2020. https://www.cnnindonesia.com/nasional/20200124134522-12-468360/kapolri-bentuk-tim-khusus-usut-dugaan-penyiksaan-lutfi.

Dewantoro, and Khairina. 2019. "Sabtu Dini Hari, LBH Medan Dilempari Bom Molotov." *Kompas.com*, 19 October 2019. https://regional.kompas.com/read/2019/10/19/19545191/sabtu-dini-hari-lbh-medan-dilempari-bom-molotov.

Domingo, Pilar, and Leopold Sudaryono. 2016. *The Political Economy of Pre-Trial Detention: Indonesia Case Study.* London: Overseas Development Institute.
Fu, Hualing, and Richard Cullen. 2008. "Weiquan (Rights Protection) Lawyering in an Authoritarian State: Building a Culture of Public-Interest Lawyering." *The China Journal*, no. 59: 111–127. https://doi.org/10.1086/tcj.59.20066382.
Fu, Hualing, and Richard Cullen. 2010. "The Development of Public Interest Litigation in China." In *Public Interest Litigation in Asia*, edited by Po Jen Yap and Holning Lau, 9–34. London: Routledge. https://doi.org/10.4324/9780203842645.
Ginsburg, Tom. 2007. "Law and the Liberal Transformation of the Northeast Asian Legal Complex in Korea and Taiwan." In *Fighting for Political Freedom: Comparative Studies of the Legal Complex and Political Liberalism*, edited by Terence C. Halliday, Lucien Karpik, and Malcolm M. Feeley, 43–63. Oxford: Hart Publishing. https://doi.org/10.5040/9781472560179.ch-002.
Goedde, Patricia. 2009. "From Dissidents to Institution-Builders: The Transformation of Public Interest Lawyers in South Korea." *East Asia Law Review* 4 (1): 63–89. https://doi.org/10.1017/9781108893947.009.
Gordon, Jennifer. 2005. *Suburban Sweatshops: The Fight for Immigrant Rights.* Cambridge, Massachusetts and London, England: The Belknap Press of Harvard University Press. https://doi.org/10.4159/9780674037823.
Hamid, Usman, and Ary Hermawan. 2020. "Indonesia's Shrinking Civic Space for Protests and Digital Activism." *Carnegie Endowment for International Peace*, 17 November 2020. https://carnegieendowment.org/2020/11/17/indonesia-s-shrinking-civic-space-for-protests-and-digital-activism-pub-83250.
Hidayat, Nurkholis. 2017. "Democratic Emergency? Hard-Liners, Communism and the Attack on LBH." *Indonesia at Melbourne*, 18 September 2017. https://indonesiaatmelbourne.unimelb.edu.au/democratic-emergency-hard-liners-communism-and-the-attack-on-lbh/.
Hukum Online. 2010. "RUU Bantuan Hukum Jangan Kesampingkan Peran LBH." *Hukum Online*, 28 April 2010. https://www.hukumonline.com/berita/baca/lt4bd7db1e1b04a/ruu-bantuan-hukum-jangan-kesampingkan-peran-lbh/.
Hukum Online. 2011. "Sekali Lagi, Kritik Terhadap Komnas Bantuan Hukum." *Hukum Online*, 18 January 2011. https://www.hukumonline.com/berita/baca/lt4d355334935ea/sekali-lagi-kritik-terhadap-komnas-bantuan-hukum/.
Klug, Heinz. 2001. "Local Advocacy, Global Engagement: The Impact of Land Claims Advocacy on the Recognition of Property Rights in the South African Constitution." In *Cause Lawyering and the State in a Global Era*, edited by Austin Sarat and Stuart A. Scheingold, 264–286. New York: Oxford University Press. https://doi.org/10.1093/0195141172.003.0010.
Kompas.com. 2008. "Orang Miskin Butuh Bantuan Hukum." *Kompas.com*, 8 August 2008. https://nasional.kompas.com/read/2008/08/08/00292843/orang.miskin.butuh.bantuan.hukum.
KUBAH. 2010. *Bantuan Hukum Dan Pembentukan Undang-Undang Bantuan Hukum: Pertanyaan Dan Jawaban.* Jakarta: Koalisi Masyarakat Sipil Untuk Undang-Undang Bantuan Hukum (KUBAH).
Lev, Daniel. 2007. "A Tale of Two Legal Professions: Lawyers and State in Malaysia and Indonesia." In *Raising the Bar: The Emerging Legal Profession in East Asia*, edited by William P. Alford, 383–414. Cambridge, Massachusetts: Harvard University Press.

Lindsey, Tim, and Helen Pausacker. 2021. "Crime and Punishment in Indonesia." In *Crime and Punishment in Indonesia*, edited by Tim Lindsey and Helen Pausacker, 1–17. Routledge Law in Asia. London: Routledge. https://doi.org/10.4324/9780429455247-1.

Liu, Sida, and Terence C. Halliday. 2016. *Criminal Defense in China: The Politics of Lawyers at Work*. Cambridge Studies in Law and Society. Cambridge: Cambridge University Press. https://doi.org/10.1017/9781316677230.

Luban, David. 2012. "The Moral Complexity of Cause Lawyers Within the State." *Fordham Law Review* 81 (2): 705–714.

Maharani, Shinta. 2021. "Teror Bom Molotov LBH Yogya Diduga Berkaitan Advokasi Konflik Agraria." *Tempo.co*, 18 September 2021. https://nasional.tempo.co/read/1507717/teror-bom-molotov-lbh-yogya-diduga-berkaitan-advokasi-konflik-agraria.

Marshall, Anna-Maria. 2006. "Social Movement Strategies and the Participatory Potential of Litigation." In *Cause Lawyers and Social Movements*, edited by Austin Sarat and Stuart A. Scheingold, 164–181. Stanford, California: Stanford University Press. https://doi.org/10.1515/9780804767965-010.

McEvoy, Kieran, and Anna Bryson. 2022. "Boycott, Resistance and the Law: Cause Lawyering in Conflict and Authoritarianism." *The Modern Law Review* 85 (1): 69–104. https://doi.org/10.1111/1468-2230.12671.

Meili, Stephen. 2009. "Staying Alive: Public Interest Law in Contemporary Latin America." *International Review of Constitutionalism* 9 (1): 43–71.

Michelson, Ethan. 2007. "Lawyers, Political Embeddedness, and Institutional Continuity in China's Transition from Socialism." *American Journal of Sociology* 113 (2): 352–414. https://doi.org/10.1086/518907.

Mietzner, Marcus. 2015. "Dysfunction by Design: Political Finance and Corruption in Indonesia." *Critical Asian Studies* 47 (4): 587–610. https://doi.org/10.1080/14672715.2015.1079991.

Mietzner, Marcus. 2016. "Coercing Loyalty: Coalitional Presidentialism and Party Politics in Jokowi's Indonesia." *Contemporary Southeast Asia* 38 (2): 209–232. https://doi.org/10.1355/cs38-2b.

Mietzner, Marcus. 2020. "Populist Anti-Scientism, Religious Polarisation, and Institutionalised Corruption: How Indonesia's Democratic Decline Shaped Its Covid-19 Response." *Journal of Current Southeast Asian Affairs* 39 (2): 227–249. https://doi.org/10.1177/1868103420935561.

Mietzner, Marcus. 2021. *Democratic Deconsolidation in Southeast Asia*. Elements in Politics and Society in Southeast Asia. Cambridge: Cambridge University Press. https://doi.org/10.1017/9781108677080.

Nardi, Dominic. 2018. "Can NGOs Change the Constitution? Civil Society and the Indonesian Constitutional Court." *Contemporary Southeast Asia* 40 (2): 247. https://doi.org/10.1355/cs40-2d.

NeJaime, Douglas. 2012. "Cause Lawyers Inside the State." *Fordham Law Review* 81 (2): 649–704.

O'Brien, Kevin J. 2023. "Neither Withdrawal nor Resistance: Adapting to Increased Repression in China." *Modern China* 49 (1): 3–25. https://doi.org/10.1177/00977004221119082.

Peradi Tasikmalaya. 2019. "*Verifikasi dan Akreditasi Bagian Dari Mewujudkan Bantuan Hukum Sejati*." Peradi Tasikmalaya. 5 January 2019. https://peradi-tasikmalaya.or.id/verifikasi-akreditasi-bagian-dari-mewujudkan-bantuan-hukum-sejati/.

Pils, Eva. 2014. *China's Human Rights Lawyers: Advocacy and Resistance.* Routledge Research in Human Rights Law. London and New York: Routledge. https://doi.org/10.4324/9780203769061.

Rahma, Andita. 2021. "Polri Tangkap Yahya Waloni Dalam Kasus Dugaan Ujaran Kebencian." *Tempo.co*, 26 August 2021. https://nasional.tempo.co/read/1499114/polri-tangkap-yahya-waloni-dalam-kasus-dugaan-ujaran-kebencian.

Ramadhan, Ardito. 2019. "Dilaporkan Ke Polisi Bersama Jubir KPK, Ketua Umum YLBHI: Bukan Hal Baru." *Kompas.com*, 29 August 2019. https://nasional.kompas.com/read/2019/08/29/15552671/dilaporkan-ke-polisi-bersama-jubir-kpk-ketua-umum-ylbhi-bukan-hal-baru.

Ratnaningsih, Erna. 2011. "*Establishing National Legal Aid System Through Indonesia Legal Aid Act.*" YLBHI.

Reuters. 2022. "Indonesia Human Rights Body Blames Use of Tear Gas for Soccer Stampede." *Reuters*, 2 November 2022. https://www.reuters.com/world/asia-pacific/indonesia-human-rights-body-blames-use-tear-gas-soccer-stampede-2022-11-02/.

Saptowalyono, Cyprianus Anto. 2021. "Gandeng Organisasi Masyarakat Sipil, Pemerintah Percepat Penyelesaian Konflik Agraria." *Kompas*, 19 June 2021. https://www.kompas.id/baca/polhuk/2021/06/19/gandeng-organisasi-masyarakat-sipil-pemerintah-percepat-penyelesaian-konflik-agraria/.

Sarat, Austin. 2001. "State Transformation and the Struggle for Symbolic Capital: Cause Lawyers, the Organized Bar, and Capital Punishment in the United States." In *Cause Lawyering and the State in a Global Era*, edited by Austin Sarat and Stuart A. Scheingold, 186–210. New York: Oxford University Press. https://doi.org/10.1093/0195141172.003.0007.

Sarat, Austin, and Stuart A. Scheingold. 1998. "Cause Lawyering and the Reproduction of Professional Authority: An Introduction." In *Cause Lawyering: Political Commitments and Professional Responsibilities*, edited by Austin Sarat and Stuart A. Scheingold, 3–28. New York; Oxford: Oxford University Press. https://doi.org/10.1093/oso/9780195113198.003.0001.

Scheingold, Stuart A. 2001. "Cause Lawyering and Democracy in Transnational Perspective: A Postscript." In *Cause Lawyering and the State in a Global Era*, edited by Austin Sarat and Stuart A. Scheingold, 287–304. New York: Oxford University Press. https://doi.org/10.1093/0195141172.003.0015.

Setiawan, Ken. 2020. "A State of Surveillance? Freedom of Expression Under the Jokowi Presidency." In *Democracy in Indonesia: From Stagnation to Regression?*, edited by Thomas Power and Eve Warburton, 254–274. Singapore: ISEAS Publishing. https://doi.org/10.1355/9789814881524-018.

Stern, Rachel E. 2013. *Environmental Litigation in China: A Study in Political Ambivalence.* New York: Cambridge University Press. https://doi.org/10.1017/cbo9781139096614.

Sudaryono, Leopold. 2020. "*The Political Economy of Prison Overcrowding in Indonesia.*" PhD Thesis, Canberra: Australian National University.

The Jakarta Post. 2021. "Jokowi's Overweight Coalition." *The Jakarta Post*, 3 September 2021. https://www.thejakartapost.com/paper/2021/09/02/jokowis-overweight-coalition.html.

YLBHI. 2009. "*Press Release: Penyusunan RUU Bantuan Hukum*," 15 May 2009. https://ylbhi.or.id/informasi/siaran-pers/siaran-pers-tentang-penyusunan-ruu-bantuan-hukum/.

YLBHI. 2018. *Catatan Akhir Tahun Yayasan Lembaga Bantuan Hukum Indonesia Tahun 2017 (Demokrasi dalam Pergulatan)*. Jakarta: YLBHI. https://ylbhi.or.id/bibliografi/laporan-tahunan/catatan-akhir-tahun-yayasan-lembaga-bantuan-hukum-indonesia-tahun-2017-demokrasi-dalam-pergulatan/.

YLBHI. 2021. "*Arrest and Criminalization of Public Defenders Indonesian Legal Aid Foundation (YLBHI)* [*Press Release*]," 28 April 2021. https://ylbhi.or.id/informasi/siaran-pers/arrest-and-criminalization-of-public-defenders-indonesian-legal-aid-foundation-ylbhi/.

Part Three

Revival

7 The ‘revival’ of structural legal aid and YLBHI’s return as an oppositional force

Facing a compromised legal system and declining political opportunities, what other strategies do cause lawyers have at their disposal? Democratic erosion under President Joko Widodo has increasingly seen LBH deprioritise formal cause lawyering tactics and turn to oppositional activism on the streets in collaboration with social movements, recalling the kind of fervent activism that LBH became known for in the late New Order years.

This chapter uses an examination of LBH’s approach to legal aid and activism, structural legal aid, to capture this evolution in strategies, and demonstrate the close relationship between the quality of democracy and the shape that cause lawyering takes. Previous chapters have explored various cause lawyering strategies, including litigation-focused approaches (Chapter 4), nonlitigation strategies like grassroots community organising, and elite-focused tactics like policy advocacy (Chapters 5 and 6, respectively). This chapter takes a broader view, drawing these strands together and reflecting on how they resonate with structural legal aid, as well as how this concept has been interpreted and implemented over time.

An additional focus of this chapter is LBH’s efforts to organise civil society in opposition to the state, a strategy intertwined with structural legal aid. Confronted with democratic regression, LBH has returned to playing a convening and coalition-building role in civil society, and attempted to foster broader resistance against the state. This was an important role played by LBH in the late New Order period, and it is becoming important again now, as Indonesian democracy is unravelling. I examine how LBH has sought to build resistance to the serious threats to Indonesian democracy that have emerged over the past five years, focusing on the *Reformasi Dikorupsi* protests, and protests against the Omnibus Law on Job Creation in 2020. I also look at how these efforts faltered when deliberations on the revised Criminal Code heated up again in 2022.

Structural legal aid in post-authoritarian Indonesia

The decline of structural legal aid

Under the New Order, structural legal aid was largely focused on changing the ‘structure’ of the highly centralised Soeharto regime. That is, it was a

DOI: 10.4324/9781003486978-11

defiantly oppositional project aimed at breaking down authoritarian political and legal structures that repressed the poor (Nasution 1985, 36). It sought to change the law and policy, but it also had a strong focus on strengthening 'power resources' at the grassroots, and so aimed to bring about change through community organising and community legal empowerment, believing that a mobilised grassroots was essential for reform (Lev 1987, 21). Things were bound to change when the authoritarian regime was removed.

The collapse of the New Order meant structural legal aid could no longer be solely concerned with the power relationship between the people and the state, but rather it had to consider the relationships among communities, among individuals within communities, and between people and corporations. The power relationships that resulted in the legal problems of the poor were now more complex. Some within LBH questioned whether structural legal aid was even relevant anymore.[1] With the fall of the New Order, the threats to poor and marginalised Indonesians (and opportunities for civil society to contribute to reform) had changed, but structural legal aid had not. Consequently, LBH offices began to interpret structural legal aid in different ways, while some even abandoned it. As former head of LBH Jakarta Nurkholis Hidayat explained:

> One of the challenges with structural legal aid in the *reformasi* era [after 1998] is that there are too many targets to address. In the past, all our efforts were focused on taking down the regime. That is why LBH was organising farmers, labourers, academics, cultural figures and so on. The goal was to unify them to change the political situation. But after *reformasi*, there were too many systems and problems, and we were pushing for change here, there and everywhere. This resulted in structural legal aid being understood or translated in different ways.[2]

During the 1980s, senior YLBHI and LBH lawyers produced a considerable amount of work attempting to articulate the theory and methods of structural legal aid. While structural legal aid remained a broad guiding principle for the work of LBH cause lawyers in the period following the fall of Soeharto, there were limited attempts to revisit or revitalise the theory behind it for the different context. As Abdul Fatah, formerly from LBH Surabaya, said:

> The interpretation of structural legal aid should be continually perfected and contextualised. When we speak of structuralism, we can't just look at the relationship between the people and the state, because the government is not the only perpetrator of injustice. The market, for example, can also be a perpetrator. We have yet to focus much on that. Structural legal aid needs to be renewed. There are few people studying structural legal aid now.[3]

Another factor resulting in the diminished prominence of the structural legal aid approach in the post-Soeharto years was the decline in the influence of the central YLBHI umbrella body over the regional LBH offices. As discussed, financial constraints and leadership tussles within YLBHI resulted in a weaker central body and provided room for LBH offices in the regions to become more autonomous. For about a decade from the mid-2000s, there was little oversight from YLBHI over how structural legal aid was being implemented by regional offices. Many LBH offices had their own interpretations of the best way to implement structural legal aid.[4] Some offices neglected the structural legal aid approach altogether and retreated into more conventional, litigation-focused approaches to case representation. Former LBH Surabaya lawyer Herlambang P. Wiratraman reflected on this development:

> Regional [LBH] offices' work with social movements has become weaker. They don't use community organising much as a strategy for change... There are still offices committed to a movement strategy, but there are also those that do not use it at all, and all their focus is on lawyering. Community organising has been left behind.[5]

It is true that in the first few years following the democratic transition, LBH founder Buyung attempted to nudge the organisation in a more proceduralist direction, involving a greater focus on representation of individual clients. The intention was to push the legal system to meet its promise, to fortify the rule of law. This emphasis on procedural justice was a valid goal in a newly democratic state in which the courts were emerging from decades of executive interference. This might have been a view to which Buyung subscribed, at least in the latter part of his career, but it was certainly not shared by most other LBH lawyers. Rather than being a result of Buyung's encouragement, the retreat to more proceduralist approaches that occurred in some LBH offices was likely caused by a combination of the 'identity crisis' precipitated by the fall of Soeharto, a lack of oversight from YLBHI in encouraging a consistent approach to structural legal aid, and the simple fact that a litigation-focused approach comes naturally to many lawyers. Given their education, litigation may be "the line of least resistance" (Sarat and Scheingold 2006, 2). And although some LBH cause lawyers moved in a more proceduralist direction, other LBH cause lawyers experimented with elite/vanguard strategies. From 1998 through to the mid-2000s, the government was more actively engaged in liberal democratic reform, and some LBH cause lawyers had greater belief in the capacity of the law to engender social change. As current YLBHI senior lawyer and former LBH Jakarta lawyer Pratiwi Febry said, this resulted in less attention being paid to structural legal aid:

> In the early years of *reformasi*, people still had hope, so maybe structural legal aid was forgotten. But after we evaluated how we have done 20 years after *reformasi*, we realised that there has been a failure of the civil society movement to be faithful to the goals of structural legal aid. And that's why we have ended up where we are today.[6]

Perhaps, as Pratiwi suggests, democratic transition meant LBH took its eyes off the sources of power for a while. Yet recent developments have helped to sharpen LBH's focus.

The revival of structural legal aid

As discussed, recent years have seen LBH cause lawyers become increasingly disillusioned with the government and legal system, and the capacity of legal reforms to lead to actual change in the lives of the community. Mohamad Soleh, of LBH Surabaya, describes how LBH no longer considers that there is much to be gained from working with the government or legislature to draft legislation:

> In the past, LBH would prepare alternative drafts of laws. We would tell them, 'this is a better version of the regulation', and so on. But in our current political system, where the oligarchs are growing in strength, this approach is not very effective. For example, our input on the draft regional regulation (perda) on legal aid was ignored. In the end, yes, we should still do things like preparing alternative drafts, but encouraging the public to participate and act is also very important... Because the legislature and the executive do not want to listen to the aspirations of the community. Want to or not, we have to strengthen the community.[7]

To borrow the words of Freeman and Freeman, LBH has learned through harsh lessons "that even the most progressive policies are little more than words on a page to their communities if the underlying power dynamics are not altered" (Freeman and Freeman 2016, 150). Despite the occasional wins in forums like the Constitutional Court or administrative courts (described in Chapter 4), there is also considerable scepticism about the benefits of strategic impact litigation in an environment of democratic regression. The comments of LBH Jakarta director Alghiffari Aqsa demonstrate this view:

> The consolidation of elites or oligarchs in their efforts to forget about democracy and human rights has been extraordinary. Legal efforts create a stage, create a moment. They are useful as an accelerator only but should not be the end goal, because legal efforts will not resolve the problem. Structural legal aid needs to change its perspective now. In the past, if litigation efforts were effective and provisions were changed, then we would say that structural legal aid was successful. But it turns out that was not the case. The regime ignored the law.[8]

Alghiffari and Soleh articulate sentiments that are shared almost universally across the YLBHI network. They also resonate with the views of leading oligarchy scholars Jeffrey Winters (Winters 2011; 2013), Richard Robison and Vedi Hadiz (Robison and Hadiz 2004; Hadiz and Robison 2013). A key feature of these analyses of Indonesian politics is that the transition to a formal democratic structure has not been able to dismantle oligarchic power. Or, as Hadiz and Robison say, "the disintegration of authoritarian rule and the introduction of democratic and market institutions do not in themselves give rise to a broader liberal transformation of society and politics" (Hadiz and Robison 2013, 36). LBH rarely makes detailed reference to such academic works in its analysis of political events or in the design of its activities and I do not wish to suggest a sophisticated knowledge of, or close adherence to, the views of a particular thesis of oligarchy (or, for that matter, critiques of the oligarchy approach to analysis of Indonesian politics (Pepinsky 2013; Liddle 2013)). However, despite the democratic reforms of the past two decades, LBH believes democratic change has only occurred in a formal or institutional sense – the substance of democracy remains shallow. The blunt assessment of current YLBHI director Muhamad Isnur is that Indonesia is "an oligarchic country, controlled by a handful of people."[9]

Concerns over Indonesia's democratic shortcomings helped catalyse renewed attention to structural legal aid among the YLBHI leadership. When Asfin assumed the YLBHI director's position in 2017, she and her team immediately set about promoting a more consistent approach to structural legal aid across the LBH offices. YLBHI conducted an assessment to identify which regional LBH offices were weaker in terms of their application of structural legal aid, and senior YLBHI staff were even installed as temporary directors to whip these offices into shape. YLBHI's 2017 strategic planning meeting also involved efforts to revitalise the structural legal aid concept, and included reflective presentations on structural legal aid from respected figures like former Komnas HAM member MM Billah and academic Donny Danardono. A special edition of YLBHI's *Bantuan Hukum* (Legal Aid) bulletin was published on the topic, in which Asfin reflected on the theoretical and philosophical basis of the concept, and attempted to update it for the post-Soeharto period (Asfinawati 2017). This was a dense, theory-heavy document that referred to the work of sociologists including Jean-François Lyotard, Jean Baudrillard, Pierre Bourdieu and Anthony Giddens. It is doubtful that LBH lawyers in the regions would have found much of use in this, and in any case, there has been no fundamental shift in the way in which structural legal aid is being understood. In fact, senior YLBHI lawyers recognise that there is not much to be gained from these theoretical ruminations for cause lawyers on the frontline. Rather, it is more important that LBH offices across the country share a similar approach to the practice of cause lawyering.[10]

In practice, the core elements of structural legal aid have changed little since it was first formulated in the New Order period. It continues to prioritise the cases of communities over individuals, and if the cases of individuals are taken on, they must contain a structural element, that is, there must be a rights violation or an element of structural inequality that resulted in the victim's suffering. One of the most important shifts has been a renewed emphasis on community legal empowerment and community organising as fundamental aspects of structural legal aid. When a case is managed using the structural legal aid approach, there are attempts to ensure individuals or communities emerge stronger from the process, and develop the skills to independently advocate for the needs of their communities. Consider this comment from former LBH Jakarta director Arif Maulana:

> The thing that differentiates structural legal aid from conventional legal aid is that LBH promotes community legal empowerment, it encourages broader society to take action to realise social justice. To change policy, to create spaces for the full participation of the community in a democratic country, want to or not, public interest lawyers cannot act alone. They need public support. Structural legal aid is not just an activity but a movement. And as a movement, it must be able to push for the involvement of the broader community.[11]

Yet the 'revival' of structural legal aid has not only involved a greater emphasis on community legal empowerment and community organising. One of the most striking aspects of the renewed attention to structural legal aid is that it is again being interpreted as a resistance movement against the state. Former LBH Jakarta director Nurkholis Hidayat summarises the evolution of the concept since the fall of Soeharto:

> Over the past 20 years, some have said that structural legal aid means changing policy, but others said it means empowering the community so they develop a political perspective and can take the lead themselves. Now that the political situation is returning to what we faced under the New Order, structural legal aid will naturally return to its previous function. With the Presidential Palace as the source of all the problems we are facing, structural legal aid will become more sharply focused... In the past, all energy was focused on getting rid of the New Order regime, and [donor] funding support backed these efforts. After the New Order fell, people had their own definitions of structural legal aid and [the relative importance of] litigation versus nonlitigation approaches. There were many changes, wins in the courts. I won a few times at the Constitutional Court, for example in the Book Banning case. But it is now clear *reformasi* has been derailed, or even failed. We were never able to drastically change the political structure. This gave the old powers the opportunity to reconsolidate. There was a time when they were afraid but now they have created a fortress of impunity. We have no choice but to try to destroy this structure again.[12]

Nurkholis's comments clearly demonstrate a shift from an elite/vanguard model of cause lawyering, in which there is faith in the power of the law to change society, to an overtly political, grassroots approach. As stated, lawyers in the grassroots category typically view the legal system as "corrupt, unjust or unfair" and view legal action as "only one weapon in a widespread assault on injustice" (Hilbink 2004, 681). LBH lawyers repeatedly told me that they recognised litigation was insufficient on its own, and that concurrent efforts also needed to be made to "mobilise political strength".[13] As former YLBHI leader Asfinawati said:

> [I]f you use litigation alone it is very 'dry', and that is not where the real battle is anyway. We are returning to the old way of doing things, with the courts used just as an arena, not a place to resolve problems. We have reverted to a form of structural legal aid similar to the model implemented under the New Order.[14]

These comments from academic Robertus Robet, formerly of YLBHI, also help to summarise this shift to a grassroots approach to cause lawyering:

> When the legal infrastructure of the state is no longer the most appropriate arena, when it can no longer be used fairly by civil society in the struggle for justice, then, want to or not, almost instinctively, organising the community and involving social and marginalised groups, becomes more important.[15]

It is important not to paint the developments described above too simplistically. Typologies of cause lawyering are not sharply circumscribed and, of course, cause lawyers can share characteristics or approaches from multiple categories at the same time. For example, LBH cause lawyers continue to conduct strategic impact litigation, for example in relation to the Papua Internet Shutdown or the Jakarta Air Pollution cases discussed in Chapter 4, even if they have little faith in the ability of the courts to constrain state power. They also continue to frame their work in terms of constitutional and human rights principles, consistent with an elite/vanguard approach (Hilbink 2004, 676). At the same time, however, LBH views real transformation as coming from the grassroots, not just by affirmation of human rights or constitutional principles in the courts, as would be typical with an elite/vanguard approach (Hilbink 2004, 683).

Despite these qualifications, the recent revival of structural legal aid points to the central argument of this book – that there is a clear relationship between the reform orientation and openness of the regime and the style of cause lawyering practiced. Now that the government is becoming more repressive, and the challenges faced are more like those faced under the New Order, LBH is in a sense reverting to type. This tradition of grassroots resistance is so deeply rooted in LBH's organisational lore that it is not surprising that it is again becoming a central feature of its approach to cause lawyering. The fact that there were few efforts to update structural legal aid after the democratic transition has ended up being inconsequential. As democracy has begun to erode, LBH has been able to revive the approach, and found it quite well suited to the current moment.

One unanticipated element of this reversion to type is that YLBHI has begun playing a convening, coalition-building role in civil society as it did in years past. During the final years of the Soeharto regime, YLBHI became the de facto centre of Indonesia's human rights and pro-democracy movement. It would be wrong to suggest that YLBHI has assumed *the* dominant role in civil society, but it is certainly playing a more confident convening role, making a conscious effort to facilitate connections among social movement groups and build coalitions of pro-democracy organisations. As current YLBHI director Muhamad Isnur stated:

> LBH lawyers are educated not just to be lawyers, they must be the 'axis' of social movements. LBH is a social movement, it will always be a part of social movements. LBH lawyers are always encouraged to accelerate the achievement of social movements' aspirations. This is a competency all LBH lawyers must have. It is a challenge. At times, some LBH offices have been too focused on litigation, and we have criticised them for this.[16]

It might seem odd for a legal aid organisation to be so critical of litigation. But Isnur's comment speaks to the political understanding of LBH's role that is again dominant among the YLBHI leadership. These attempts to become the "axis" of social movements have grown since 2019, as attacks on Indonesian democracy have intensified.

Escalating threats to democracy and return to a 'locomotive of democracy' role?

Reformasi Dikorupsi

In September 2019, threats to Indonesian democracy reached a crisis point, culminating in the student and civil society demonstrations that captured global headlines and featured in the opening of this book. Several major challenges occurred in quick succession, and civil society activists described feeling like they were being assaulted from all directions. One of the first major events occurred in August, when thousands of ethnic Papuans participated in massive anti-racism and pro-independence protests in major cities in the Papuan provinces, following racist insults directed at Papuan students in Surabaya (Reuters 2019). Around the same time, devastating forest and peatland fires were burning across Sumatra and Kalimantan, putting human health at risk and eventually leading to more than 3.1 million hectares of land being burned (Jong 2021). Then, to the dismay of activists, in mid-September the national legislature selected Firli Bahuri, a police general who had faced accusations of "gross ethical violations" while serving in the KPK, as the anti-graft institution's next leader (Ghaliya 2019). Just days later, following a hasty and opaque deliberation process, legislators approved revisions to the 2002 Law on the KPK that seriously weakened its powers (Butt 2019).

Following this "legislative assault" on the KPK (The Jakarta Post 2019), lawmakers then said they intended to pass a revised Criminal Code (RKUHP or RUU KUHP) before their terms expired on 30 September. This regressive draft code represented a serious attack on personal freedoms, with provisions outlawing insulting the president and state institutions, as well as prohibitions on co-habitation of unmarried couples and all extramarital sex (Ristianto 2019). Lawmakers also proposed a new land law, as well as revisions to laws on mining, corrections and labour, while claiming that they did not have time to pass long discussed bills on the prevention of sexual violence, the rights of Indigenous Peoples and the protection of domestic workers (Cahya 2019; Detik 2019).

Civil society opposition began heating up following the passage of the revised KPK Law. On 17 September, Jakarta-based activists held a mock funeral for the institution, with then YLBHI director Asfin giving an emotional speech outside the KPK building in front of hundreds of onlookers (CNN Indonesia 2019). Larger student and civil society demonstrations then broke out in front of the national legislative complex in Jakarta on 19 September, and continued over several days, with the largest protests on 23 and 24 September, when thousands of students protested (Dongoran 2019). Similar protests erupted in multiple cities across the country (Haryanto 2019).

During this tense period, the YLBHI and LBH Jakarta offices were a hive of activity. LBH made a determined effort to encourage collective action and promote broader mobilisation against the proposed legal changes. YLBHI and LBH Jakarta hosted public discussions and press conferences almost daily, usually in collaboration with other pro-democracy CSOs, gathering academic and civil society experts to discuss issues such as the controversial revised criminal code, attacks on the KPK, and the Jokowi administration's prioritisation of business interests over human rights. Senior staff provided regular updates for journalists and other civil society actors, and Asfin appeared regularly on television, squaring off against government ministers and legislators ("Ujian Reformasi" 2019). The location of the YLBHI office in Central Jakarta meant it was an easily accessible gathering point for organisations from across the city. Even if YLBHI or LBH Jakarta did not always take the lead in press conferences or discussions, the office was a safe space. LBH's name and legacy, and the fact that the building is occupied by lawyers, offered a degree of protection. Historically, this has largely, but not entirely, prevented it from attacks.[17]

As the student protests gathered momentum, YLBHI and LBH staff spoke at university campuses, providing students with information on problematic legislation and informing them about recent developments.[18] YLBHI already had relationships with some Student Executive Bodies (*Badan Eksektutif Mahasiswa*, BEM), such as those from the University of Indonesia and Trisakti University, but its connections with student leaders were initially somewhat limited. To rectify this, LBH sent staff to meet with student leaders, to better coordinate efforts among student groups and between students and civil society.[19] LBH also encouraged other social movement groups like labour groups to join the students after they did not take part in the early demonstrations.[20] As mentioned, LBH even engaged in direct action tactics itself, joining the students on the streets. I was in the YLBHI office on the morning of one of the largest protests, as YLBHI staff coordinated their clothing and packed their bags with toothpaste, which they smeared under their eyes as a remedy for tear gas. This was the day when, as discussed in the opening to this book, Asfin delivered a rousing speech to demonstrating students, railing against legislators who were willing to ignore citizens' demands. LBH again assumed the role of a gathering point, as it had nearly three decades before. As YLBHI senior campaigner Arip Yogiawan (Yogi) said:

> We should not be leading the movement, but we should be a hub. In the movement there are labour groups, the urban poor, farmers, LGBTIQ+ groups, women, environmental groups. We must become a hub for them. Ultimately, the leaders of the protest movement are the students, but we need to create opportunities for these groups to meet, to connect ideas.[21]

Although the protests in September 2019 were largely driven by students, LBH and broader civil society helped to shape the narrative and sharpen focus. For example, the rallying call *Reformasi Dikorupsi* was coined at a late night meeting of dozens of civil society representatives in the YLBHI office in Jakarta. The influence of LBH and other CSOs was also seen in the list of seven demands circulated by protestors.[22] There were therefore concerted efforts to foster the students' growing rights consciousness and channel their anger into clear demands that could be delivered to lawmakers. Activists said they chose the term *Reformasi Dikorupsi* because they wanted a term that conveyed frustration over the gains of *reformasi* crumbling away, while highlighting the attacks on the KPK.[23] The use of the word *dikorupsi* had additional advantages. Activists recognised that corruption was one of the few issues that had motivated large numbers of people to get out onto the streets in protest in the past. 'Corruption' was shorthand for government abuse of power. As one lawyer said, "it is like mentioning Satan".[24] The echoing of the 1998 pro-democracy movement was also evident in regional protests. Massive demonstrations in Yogyakarta were initially dubbed 'Gejayan Calling' (*Gejayan Memanggil*), explicitly referencing the site of a major anti-Soeharto protest in May 1998, at which a student was killed.

While LBH assumed a coalition-building role, it is crucial not to forget about the lawyering it does in this context. A key role of cause lawyers in broad coalitions is providing defensive litigation, especially in those campaigns that use protest as a core strategy (Cummings 2020, 118). The police response to the *Reformasi Dikorupsi* demonstrations was incredibly repressive. Five students died and at least 232 required medical treatment (Anjar

2019; Sahara 2021). For several tense days following the largest protests on 23 and 24 September, at least 93 students had been reported 'missing', detained by police with no access to their families or legal counsel (Saputri 2019). LBH responded by forming the 'Advocacy Team for Democracy' along with other CSOs, calling on police to follow correct procedures and demanding that they release the detained students (Komnas HAM 2019). One of those arrested in the wake of the protests was former journalist and musician Ananda Badudu, who was active on Twitter at the time of the demonstrations and had raised money to support demonstrators on the crowdfunding site kitabisa.com (BBC Indonesia 2019). Ananda was arrested in the early hours of 27 September. Several hours earlier, police had also arrested activist filmmaker Dandhy Laksono for tweeting about tensions in the Papuan provinces (Galih 2019). Police and legislators were clearly shocked by the scale of the protests and, recognising that many students had been mobilised online, hoped to reduce the size of protests planned for 30 September. However, Ananda and Dandhy were close to many activists who rushed to their defence, including lawyers from LBH and YLBHI (Asih 2019). Ananda was released the next morning without charge after an enormous backlash on social media and, tellingly, the charges against Dandhy were never pursued after the protests died down in October.

LBH's prominent role in defending these political cases not only prevented them from being processed further by authorities but also positioned LBH at the centre of the pro-democracy movement. As one of LBH's colleagues in civil society said: "We have a saying in our office, 'The last CSO standing should be LBH'… We know that if we are criminalised, if we are arrested by police, we will go to LBH."[25]

I do not want to overstate the role of LBH in the *Reformasi Dikorupsi* protests. Indeed, LBH cause lawyers themselves are wary of making bloated claims about their influence.[26] Many students were mobilised through social media and had little contact with pro-democracy CSOs. But LBH was effective in fostering alliances and trust among CSOs, and strengthening relations between CSOs and students. There is no doubt that CSOs provided an intellectual contribution to the protests – creating the rallying cry *Reformasi Dikorupsi* and formulating the list of seven demands. It is also vital to recognise that even if the protest movement represented a strong student and civil society defence of Indonesia's democratic gains, it was unsuccessful. The protests were (temporarily) effective in preventing the regressive revised Criminal Code from being passed, but that was the limit of their success. At the height of the protests, Jokowi attempted to placate demonstrators, saying that it was possible he would use his presidential authority to cancel the amendments to the Law on the KPK. However, as soon as the heat dissipated, he quietly acknowledged that the amendments would stay (Gorbiano 2019). Similarly, legislators had little difficulty securing their controversial revisions to the 2009 Mining Law in May 2020, when Indonesia was grappling with the early stages of the Covid-19 pandemic (Jong 2020a).

Omnibus Law on Job Creation

Major street demonstrations returned in 2020, against Law No. 11 of 2020 on Job Creation. President Jokowi first announced plans for an omnibus "Law on Job Creation" (*Undang-Undang Cipta Lapangan Kerja*) during a swearing-in ceremony to mark his second term in office in October 2019. When the government handed over the draft bill to the legislature in February, it became clear that despite being named 'Job Creation', it would regulate a vast range of issues, from employment to investment, environmental protection, mining, public participation, and even the authority of regional governments. The Job Creation Law revised 79 different statutes at the same time, and affected more than 1,200 provisions (Thea 2020). LBH and civil society groups complained that the new Law seemed designed to benefit the interests of oligarchs, referring to its dilution of workers' rights and weakening of environmental standards for business activities (Ungku, Suroyo, and Christina 2020). In particular, it restricted public participation in the environmental impact assessment process, and did away with environmental licenses, instead integrating them with business licences. Activists worried that communities would have fewer legal avenues to challenge environmentally destructive projects, such as the cement mine and factories discussed in Chapter 5 (Jong 2020b). Meanwhile, labour activists were furious about the bill's exclusion of unions from the process for calculating the minimum wage, provisions allowing increased use of outsourced labour and continuous contracting of workers, and reductions in severance pay, among other issues (Fajrian 2023). Civil society also raised procedural concerns over the opaque and rushed nature of the legislature's deliberations on the bill.

With growing confidence over the role it played during the *Reformasi Dikorupsi* protests, LBH positioned itself as a "facilitator of a people's movement",[27] and immediately set about coordinating civil society activism against the bill. In the first few months of 2020, before the Covid-19 pandemic struck, YLBHI and LBH held a series of public discussions and seminars with academic and civil society experts, in Jakarta and major regional centres like Bandung, Semarang, Yogyakarta, Surabaya, Lampung, Banda Aceh, Makassar and Manado.[28] Labour activists immediately abbreviated the bill's name to 'Cilaka' (***CiptaLapanganKerja***, a pun on the Indonesian word for disaster, *celaka*), annoying the government so much that it eventually renamed the bill *Cipta Kerja* (Putri 2020). YLBHI realised it was important that other groups aside from the labour movement were involved in opposition to the bill and helped to form a large coalition of pro-democracy CSOs called the 'Indonesian People's Faction' (*Fraksi Rakyat Indonesia*) (Fraksi Rakyat Indonesia 2020). This name referenced political party factions in the DPR, and was a clear sign that civil society believed political opportunities in the legislature had completely evaporated. It also recalled one of the functions of LBH under the New Order, where in a highly constrained environment it provided an outlet for political action. LBH also helped to coordinate scholarly opposition to the bill,[29] supporting academics to deliver a joint statement to the DPR (Nugraheny 2020).

The first major protests against the Law on Job Creation were held in March 2020, just before Indonesia implemented Covid-19 social distancing measures (Nurhadi et al. 2020). Labour groups planned May Day protests for 30 April, but they were called off when the government and DPR said that they would delay discussion of labour-related provisions (Erwanti 2020). The pandemic offered some unexpected advantages for YLBHI. As work moved online, senior YLBHI staff reached out to student, labour and other civil society groups in the regions. Asfin, Isnur and Yogi shared responsibilities, meaning it was possible to speak at up to a dozen different events across the country every day,[30] helping to amplify and focus opposition to the bill. Protests were again held in August, but the largest demonstrations occurred from 6–20 October, after the Law was passed on 5 October. Thousands of workers, students, academics and civil society activists took to the streets in more than 18 provinces across the country (BBC Indonesia 2020). This time, protests were held under the banner of *#MosiTidakPercaya* (Motion of No Confidence), a sign of the contempt in which protestors held the legislature (although the *#ReformasiDikorupsi* hashtag was also used again).[31]

In the wake of the *Reformasi Dikorupsi* protests, LBH had developed much broader and stronger connections with student groups. YLBHI worked with other CSOs to train "street paralegals", covering issues like documentation of rights abuses, the rights of the accused, how to accompany people detained by police, personal safety and more.[32] The police response to the Job Creation Law protests was also brutal. Days before the Law was passed, the police chief issued an internal telegram calling on police to "prevent, subdue and divert" protests, ostensibly under the guise of controlling Covid-19.[33] The telegram even asked police to offer a "counter-narrative" in support of the Law, which to YLBHI was a depressing indication of the extent to which the force had been politicised under Jokowi (YLBHI 2020). Police were captured on video beating protestors, and Amnesty International Indonesia recorded that at least 402 people were injured (Amnesty International Indonesia 2020). Police themselves acknowledged they detained nearly 6,000 protestors by 12 October, about four times the number detained during the *Reformasi Dikorupsi* protests (Bustomi 2020). The 'Advocacy Team for Democracy' sprang into action again (Kamil 2020). It received thousands of requests for legal assistance, and reports of violent and demeaning treatment of demonstrators in police detention.[34]

At the same time as LBH was helping to organise mass protests, it also deployed legal strategies. In April 2020, the Advocacy Team for Democracy, led by YLBHI, submitted a challenge to the Jakarta State Administrative Court (PTUN Jakarta) over the Presidential Letter (*Surat Presiden,* Surpres) that appointed ministers to represent the government in deliberations on the omnibus bill in the DPR.[35] LBH claimed that the government had breached the principles of good governance, as well as its obligations under the 2011

Law on Lawmaking, by appointing ministers to begin deliberations before there had been sufficient public participation (Kustiasih 2020). The court heard the case, and even examined witnesses, but ultimately ruled that the presidential letter was not within its authority.[36] The decision was made public on 19 October 2020, two weeks after the Law on Job Creation was passed. The timing of the decision, and the fact that the PTUN had bothered to hear witness testimony but eventually ruled that it did not have the authority to hear the matter at hand, suggested to LBH that the court lacked the courage to threaten the interests of oligarchs.[37] After the Law's passage, LBH decided against taking it to the compromised Constitutional Court.[38]

LBH and broader civil society emerged from the Job Creation Law saga bruised and exhausted. Not only did protestors face an incredibly repressive police response, but LBH and its allies were also reminded how much the legal system was stacked against them. Even when constitutional litigation was partially successful, and the Constitutional Court declared the Law "conditionally unconstitutional", Jokowi issued an emergency regulation that bypassed the Court's decision, rendering it basically meaningless. This had implications when discussions on the revised Criminal Code intensified again in 2021 and 2022.

The 'locomotive of democracy' stalls: the revised Criminal Code (KUHP)

After the mass protests of 2019 forced the government to shelve the draft revised Criminal Code (RKUHP), when discussions on the code were revived in 2021, the government said it would focus on revising just 14 issues that had been the focus of civil society concerns (CNN Indonesia 2022b). In late 2020, Jokowi appointed Gadjah Mada University (UGM) criminal law scholar Edward Omar Sharif Hiariej as deputy minister of law and human rights, and tasked him with representing the government in discussions on the RKUHP. Throughout 2021, the government held a limited number of public events, supposedly to gain public and civil society feedback as it prepared its revisions to the code (Kementerian Hukum dan HAM 2021). Discussions picked up pace in 2022, and the government claimed it planned to pass the new code by July of that year (CNN Indonesia 2022c). But by late June, an updated draft had still not been made public (Purnamasari 2022), and it was only following significant pressure from student and civil society groups[39] that a draft was finally released in the first week of July (BBC Indonesia 2022; CNN Indonesia 2022a).

Two further drafts were made public in November, and the revised Criminal Code was eventually passed on 6 December 2022.[40] Human Rights Watch immediately described it as "disastrous for rights" (Human Rights Watch 2022). Some of the most problematic articles of the new code include provisions making it an offence to insult the president and vice president (Article 218) and state institutions (Article 240) (similar articles had previously been struck down by the Constitutional Court in the mid-2000s), provisions criminalising sex outside marriage (Article 411)[41] and cohabitation of unmarried couples (Article 412), a provision criminalising conducting public protests without first notifying authorities (Article 256), and vague and broad provisions on 'fake news' (Articles 263–64). At the last minute, and without the knowledge of civil society, legislators also inserted a new provision outlawing the spread or development of ideas that contradict the national ideology, Pancasila (Article 188). Despite these major disappointments, it is true that civil society advocacy was able to mitigate some of the Code's worst aspects.

Civil society was more divided than in 2019. Some of LBH's contemporaries felt that the government was determined to pass the revised code no matter what, and noted that the new deputy minister was more open to civil society input than the government and legislature had been in 2019.[42] They therefore believed civil society should try to influence discussions as much as possible. ICJR, several other civil society groups, and academics engaged intensively with the government to try to moderate some of the worst articles. For example, civil society activism ensured that provisions on insulting the president, vice president and state institutions are now 'complaint offences', meaning that charges can only be brought if the president, vice president or head of the relevant state institution files a police report.[43] Similarly, charges for adultery or cohabitation can only be brought if a complaint is made by the spouse, parent or children of the offender – the 2019 draft would have allowed village or neighbourhood heads to file reports on cohabitation. In the wake of significant civil society and international criticism, the government has also said that it does not intend to apply these provisions to same sex couples (Saptohutomo 2022), although there are serious concerns about how law enforcement officials on the ground will interpret these morality articles. While LBH largely maintained an oppositional stance,[44] it was prepared to sit down with the deputy minister on a couple of occasions. Input from LBH and others resulted in the word "blasphemy" (*penodaan agama*) being dropped from the provisions on religion (Article 300).[45] This civil society advocacy also ensured that the language used in Article 300 was more consistent with provisions addressing religious discrimination, hostility and violence in the International Covenant on Civil and Political Rights (Bagir 2023).[46]

Even though the new KUHP constitutes a serious attack on rights, mass protests failed to materialise in 2022. LBH Jakarta led a protest of students and CSOs in front of the national legislature on 5 December, the day before the Code was passed, but demonstrations were more muted than in previous years (Pardede 2022). The active participation of some CSOs in consultations with the government on the KUHP in 2022 dampened civil society resistance, as some organisations sought to protect their relationships with state agencies. Even so, the lack of large demonstrations in 2022 should not be seen as a 'failure' of LBH's community organising and convening power. Other factors were at play. After the harrowing protests of 2019 and 2020, students were emotionally and physically scarred. When LBH reached out to its networks, "students said they were traumatised by the violence of 2020, in terms of both the physical injuries they faced and the stress of being detained".[47] Another factor was timing. The Code was passed in late 2022, following three years of Covid-19 social restrictions in which university students were mostly studying online. There were fewer opportunities for students to consolidate, share knowledge and train the next generation of student activists.[48] The passage of the KUHP also coincided with end of semester exams and the period of leadership transition in BEMs and other student bodies. Students were not ready to launch another major round of protests.[49] Passing the KUHP at this time was likely a tactical move by legislators, similar to the way Jokowi issued his emergency regulation on the Job Creation Law just before the New Year's holiday.

LBH and civil society still have opportunities to influence the implementing regulations for the Code before it comes into force in early 2026. Some of the most harmful articles will no doubt be challenged at the Constitutional Court. It is also worth acknowledging that a new Criminal Code was desperately needed, and the new KUHP contains several progressive changes, such as a mandatory ten-year probation period for the death sentence. Yet it is hard not to conclude that the revised Criminal Code is one of the most serious setbacks for Indonesian democracy in the post-authoritarian era.

What enabled YLBHI to play a convening role in 2019 and 2020?

Even though civil society failed to prevent the KUHP from being passed in 2022, it is worth reflecting on the revival of LBH as an oppositional force in 2019 and 2020. It was striking the degree to which LBH was able to encourage collective action and bring diverse civil society groups together during the *Reformasi Dikorupsi* and Job Creation Law protests. While LBH's generalist orientation has sometimes been seen as a weakness in the post-Soeharto period (see Chapter 3), it has been an asset as LBH has sought to play a convening role. As LBH Jakarta director Arif Maulana said:

> LBH provides a space that can be accepted by all parties, by all elements of civil society. We are considered 'neutral', not just focused on labourers, women, or religious minorities. We advocate for them all. The role of LBH Jakarta in *Reformasi Dikorupsi* is therefore to unify these elements of civil society.[50]

Former YLBHI lawyer Robertus Robet also recognised that LBH's broad networks were a key factor in the organisation being able to facilitate connections:

> LBH pays attention to its networks… [This] meant that during *Reformasi Dikorupsi*, LBH-YLBHI could still play the role of – I don't want to say a

> 'locomotive of democracy', that was Buyung's term – but it was a hub, a gathering point. Because it has strong networks and has maintained its ethos as a social movement, a movement for human rights, using the law.[51]

Cause lawyers are well placed to build coalitions on the basis of rights – the universal nature of human rights, and cause lawyers' skill and knowledge of human rights, mean they can be a tool to cut across narrower interests and help to form diverse coalitions (Cummings 2008, 1018). Human rights were certainly a central element of the *Reformasi Dikorupsi* and Job Creation Law protests, but it was largely LBH's democratic advocacy that was able to unify diverse groups. Importantly, despite experiencing about a decade of decline, LBH has retained significant legitimacy among broader civil society because of its reputation as a 'locomotive of democracy' under the New Order. As former YLBHI lawyer and Nasdem legislator Taufik Basari said, "LBH's greatest strength is its legacy, its reputation. With its big name it is still capable of unifying issues and movements."[52] As has been observed of other human rights lawyers (Baik 2010, 130), through LBH's historical opposition to Soeharto and its defence of political detainees, it also developed a degree of moral legitimacy. LBH lawyers are considered representative of the public interest. Subsequently, in times of crisis, when there are serious threats to human rights and democracy, civil society and the media (and, to a smaller extent, the public) look to LBH. Undeniably, LBH's oppositional nature, its reluctance to compromise with the government, has meant it can be more easily accepted by a broad range of civil society groups as a site of resistance. It is an obvious 'home' for opposition to the state.

Another crucial reason LBH has maintained legitimacy among broader civil society is because of its role in giving birth to other prominent CSOs, such as ICW and KontraS. Many former LBH staff have left the organisation and gone on to establish or lead other influential CSOs engaged in oppositional activism. This includes those organisations founded in the late New Order and early *reformasi* period,[53] as well as more recent examples, such as LBH Masyarakat, LBH Pers or ILRC. Consequently, YLBHI is considered a "mother" to the democratic civil society movement.[54] It is often said that one of the normative roles of civil society is to recruit and train future political leaders (Diamond 1994, 9). LBH has performed this role, that is true. More important, however, has been its contribution to the development of oppositional civil society. While Buyung's claims of LBH being a 'locomotive of democracy' were inflated, LBH can certainly claim to be an 'incubator of civil society'.

LBH also has the advantage of being considered politically neutral. A feature of Indonesia's democratic backsliding has been growing polarisation along pluralist and Islamist lines. Although this ideological divide has deep roots, it has become particularly pronounced since 2014, exacerbated by the fiercely contested 2014 presidential election, the 2017 Jakarta gubernatorial election and the 2019 presidential election (Warburton 2020, 25). A result of these contests is that President Jokowi has come to be seen as representing the pluralist camp. As discussed, Islamist mobilisations in late 2016 and early 2017 rattled the Jokowi regime, and it responded in part by targeting Islamist leaders and revising the Law on Mass Organisations to allow it to disband CSOs without having to go through the courts. The government soon used this power to disband Islamist groups HTI and FPI. When this occurred, it became clear that polarisation had also extended to civil society. Some religious freedom-focused CSOs stayed silent about this serious threat to freedom of association, and some individuals even praised the government's efforts, demonstrating that these organisations and individuals were willing to prioritise defence of religious pluralism over defence of democratic principles (Mietzner 2021, 167). By contrast, LBH, despite being a fierce advocate of religious freedom, was highly critical of the government's anti-democratic actions, even speaking out against the ban on FPI, an organisation that had attacked LBH in the past (Iswinarno and Isdiansyah 2020). This has meant that LBH has largely avoided accusations of partisanship, allowing it to play this convening role in civil society.[55]

Another undeniable factor that allowed YLBHI to return to playing a convening and coalition-building role in civil society was Asfin and the strong leadership team around her. In fact, the former director of LBH Masyarakat, Ricky Gunawan, described Asfin's leadership as pivotal:

> Asfin's leadership can be accepted by all other organisations in civil society. It would be a different story if the leader were a sexist male, for example. He would not be accepted. [Asfin's leadership] has allowed YLBHI to play that role, to focus and protect the *Reformasi Dikorupsi* movement… Sometimes there are egos and competition between CSOs, that is difficult to avoid. Asfin's leadership means ego problems can be overcome. On one hand, she can rally everyone but, on the other, there is a huge burden on her. If we rely too much on YLBHI, the movement could be threatened. But if it is not YLBHI, which other organisation has that kind of legitimacy? Who else can lead?[56]

Fortunately, Asfin's successor, Muhamad Isnur, has thus far maintained widespread support among civil society since assuming the leadership in early 2022. Yet Ricky's final comments here remain important. YLBHI's coalition-building role can also be a significant burden for the organisation when other organisations in civil society expect it to take the lead. It also drags LBH further into political action and away from strategies that draw on the law.

Convening and coalition-building was one of YLBHI's most vital functions under the New Order. LBH sought to promote collaboration among pro-democracy groups, believing a consolidated movement was essential in the drive for democratic change. This role is typical of a grassroots style of cause lawyering and is re-emerging as an important role of LBH as Indonesian democracy has begun to falter. Like other grassroots lawyers, LBH sees itself as deeply embedded within social movements (Hilbink 2004, 689). LBH cause lawyers firmly believe that if marginalised groups and CSOs are ever going to exert any influence over oligarchic power, they must be empowered and organised, in broad coalitions. Muhamad Isnur's comments underscore this point:

> In the future it will be vital to consider how we can encourage the community to organise, to act, and exert pressure. We shouldn't be too egocentric; these movements should not be led by CSOs. But we should encourage students, farmers, workers, and so on, to access information, to increase their awareness, strength, and independence to take action. We should encourage them to become a new force and apply pressure, and they may even be able to balance the power of the oligarchs.[57]

In fostering pro-democratic opposition to the state in this way, LBH starts to move beyond grassroots cause lawyering and into pure political activism. Although LBH usually deploys the language of law, there is often very little lawyering involved. The line between cause lawyering and political activism becomes much harder to distinguish. Comments from Arif Maulana, the former head of LBH Jakarta, capture this shift toward strategies focused on fostering resistance to those in power.

> Whoever has control of politics has control of the law. LBH is deeply aware of this. From the beginning, LBH has never fully believed in the law, because law is only a tool. Therefore, when we do litigation, we never forget about community legal empowerment, community organising, public campaigns. We do community organising because given the current state of affairs, that is how we can have the most impact. Do we want to play the regime's game, where it uses the law as a tool of power, a tool to maintain the status quo? We will lose. We try to compete in other arenas, where it is still possible to provide balance.[58]

This was one of the key functions of LBH under authoritarianism. When political parties were not representative, it was LBH that helped organise groups in society and channel political action (see Chapter 2). In the present day, in the near-complete absence of any meaningful political opposition in the legislature, oppositional civil society – and LBH as one of its most prominent actors – offers one of the few voices of resistance against further democratic backsliding.

While it is important to acknowledge that civil society protests have had limited success in arresting further democratic decline, it is clear that, as the discussion on structural legal aid above showed, LBH has been reinvigorated by democratic regression. Although LBH might be reluctant to call itself a locomotive of democracy now, it certainly comes to life when its main cause is democracy and the rule of law. Or in other words, as democracy has begun to evaporate, LBH has found its relevance again in its defence. Pro-democratic groups in civil society and the media have responded positively. Senior LBH staff are now sought regularly for comment in the media on rule of law and democracy issues, just as they were in the past (The Jakarta Post 2020; Wee 2022). YLBHI directors Asfin and Isnur have become respected national figures. At times, the return of LBH as a strong civil society voice against state repression has seemed like muscle memory. It is as though LBH instinctively knows what to do under a more repressive regime.

Conclusions

This chapter has demonstrated that it is possible to trace a distinct arc to LBH's approach to structural legal aid, showing a clear relationship between the form that cause lawyering takes and the quality of democracy. Under Soeharto, structural legal aid was a highly political concept. LBH aimed to dismantle the authoritarian structure of the New Order state and believed an empowered grassroots was essential in pushing for change. After the fall of the New Order, during the period when the government was more open and engaged in reform, some LBH offices shifted away from structural legal aid, adopting more conventional, litigation-focused roles. Others experimented with direct engagement with the state, working on policy reform or capacity building within state structures.

As the quality of Indonesia's democracy has deteriorated, however, LBH has developed a deep pessimism about the capacity of legal reforms to lead to real change in the lives of its constituents. Although democratic transition brought structural changes supportive of cause lawyering, LBH believes change only occurred in a formal, institutional sense – the substance of democracy remains poorly developed. Stated in cause lawyering terms, LBH has realised that in an oligarchic political system, elite-focused strategies designed to change the law, either through strategic litigation or collaborating with the state through legislative or administrative channels, will never be effective on their own. This has seen structural legal aid and, in particular, its grassroots community organising and community legal empowerment aspects, grow in prominence again. LBH still demands that the state uphold the rule of law, but believes change will come from grassroots pressure – it will not come from working within state structures. In a way, LBH's limited effort to revise the structural legal aid concept in the years after the fall of Soeharto has almost been a blessing. As the government has become more authoritarian, the strategy of empowering the community at the grassroots has been revived. LBH has been able to simply pick up the old concept and use it again for the current moment.

LBH now sees that it has little alternative but opposition. Democratic regression has led to LBH reverting to the kind of coalition-building, convening role it played during the late New Order years, illustrated during mass demonstrations in 2019 and 2020. The 2019 *Reformasi Dikorupsi* protests were significant for helping to reinvigorate civil society as an oppositional force, and encouraging civil society groups to overcome the ideological polarisation that had emerged following bitter elections in 2014 and 2019. LBH's legacy as a site of resistance to the Soeharto regime, as well as its perceived neutrality, allowed it to be accepted by broader civil society and again play a role in mobilising resistance against the government. LBH sees no conflict between adopting this overtly political, even radical position and its function as a cause lawyering organisation, even if law and legal strategies do not always feature prominently.

What sets the LBH case apart from other accounts of cause lawyering is that the relationship between cause lawyering and democratic change does not closely align with expectations. The introduction of structures supportive of cause lawyering saw LBH struggle, while the narrowing of legal and political opportunities has seen the organisation rise in prominence, and again play a central role in organising collective opposition against the state. Paradoxically, the organisation is seemingly thriving in a more hostile environment. Whether LBH will be able to sustain this role in encouraging collective action against the state remains an open question.

Notes

1 Interview, Febi Yonesta, September 2019.
2 Interview, Nurkholis Hidayat, September 2019.
3 Interview, Abdul Fatah, November 2019.
4 Interviews, Muhamad Isnur, October 2019; Prawiti Febry, December 2019; Siti Rakhma Mary Herwati, August 2019; Yunita, September 2019.
5 Interview, Herlambang P. Wiratraman, November 2019.
6 Interview, Pratiwi Febry, December 2019.
7 Interview, Mohamad Soleh, November 2019.
8 Interview, Alghiffari Aqsa, November 2019.
9 Interview, Muhamad Isnur, October 2019.
10 Interview, Febi Yonesta, September 2019.
11 Interview, Arif Maulana, December 2019.
12 Interview, Nurkholis Hidayat, September 2019.
13 Interview, Alghiffari Aqsa, November 2019.
14 Interview, Asfinawati, March 2023.
15 Interview, Robertus Robet, April 2020.
16 Interview, Muhamad Isnur, October 2019.
17 Recall the 2017 attack on the LBH office, discussed in Chapter 6.
18 Interview, Arip Yogiawan, October 2019.
19 Interview, Asfinawati, March 2023; Muhamad Isnur, February 2023.
20 Interview, Alghiffari Aqsa, November 2019.
21 Interview, Arip Yogiawan, October 2019.
22 Interview, Arif Maulana, December 2019. The seven demands, widely circulated on social media, were: 1) Reject the planned revisions to the Criminal Code (RKUHP), the revised Mining Law, the revised Land Law, the revised Corrections Law and the revised Labour Law; cancel the [revisions to] the Law on the KPK and the Natural Resources Law; and pass the anti-sexual violence bill and bill on the protection of domestic workers; 2) Cancel the appointment of the problematic

KPK chief; 3) Reject the appointment of military (TNI) and police figures in civilian posts; 4) Stop militarism in Papua and other areas, and release all Papuan political prisoners immediately; 5) Stop criminalising activists; 6) Stop forest fires in Kalimantan and Sumatra, prosecute responsible corporations and revoke their licences; 7) Resolve human rights violations, try rights violators, including those in positions of power, and restore victims' rights immediately (Akbar 2019).

23 Interview, Arif Maulana, December 2019.
24 Interview, Alghiffari Aqsa, November 2019.
25 Interview, Erasmus AT Napitupulu, February 2023.
26 Interviews, Muhamad Isnur, Arip Yogiawan, October 2019.
27 Interview, Muhamad Isnur, February 2023.
28 Interviews, Muhamad Isnur, February 2023; Arif Maulana, March 2023.
29 Interview, Arif Maulana, March 2023.
30 Interviews, Muhamad Isnur, February 2023; Asfinawati, March 2023.
31 The rallying cry 'Motion of No Confidence' was borrowed from the title of a song by indie band Efek Rumah Kaca that had become the unofficial anthem of the *Reformasi Dikorupsi* protests a year earlier (Taufiqurrahman 2019).
32 Interview, Asfinawati, March 2023.
33 Police Chief Telegram No. STR/645/X/PAM.3.2./2020, dated 2 October 2020.
34 Interviews, Muhamad Isnur, February 2023; Arif Maulana, March 2023.
35 Jakarta State Administrative Court Decision No. 97/G/2020/PTUN-JKT.
36 Jakarta State Administrative Court Decision No. 97/G/2020/PTUN-JKT, 157–59.
37 Interview, Arif Maulana, March 2023.
38 The Constitutional Court challenge to the Law mentioned in Chapter 4 was submitted by labour groups, not by LBH.
39 Including the National Alliance for Reform of the Criminal Code, of which YLBHI was a member.
40 Passed in December 2022, it was added to the State Gazette as Law No. 1 of 2023 on the Criminal Code.
41 Adultery was already an offence under the existing Criminal Code, but this offence only applied to married couples, and charges could only be brought at the request of their spouse.
42 Interview, Erasmus AT Napitupulu, February 2023.
43 Another important change compared to previous drafts was that 'state institutions' in the provision on insulting state institutions is now considered to refer only to the People's Consultative Assembly (MPR), People's Representative Council (DPR), Regional Representative Council (DPD), the Supreme Court and the Constitutional Court (elucidation to Article 240(1)).
44 Interviews, Muhamad Isnur, February 2023; Pratiwi Febry, March 2023.
45 Interviews, Muhamad Isnur, Erasmus AT Napitupulu, February 2023; Asfinawati, March 2023.
46 Even so, Article 300 states that it is an offence to commit an act of hostility, or incite hostility, violence, or discrimination, against a religion or belief. That is, religions and beliefs are still included as objects of the offence of hostility, violence or discrimination. This means individuals may still be arrested for what is, in effect, a crime of 'blasphemy'.
47 Interview, Muhamad Isnur, February 2023.
48 Interviews, Muhamad Isnur, Erasmus AT Napitupulu, February 2023; Asfinawati, March 2023.
49 Interviews, Muhamad Isnur, February 2023; Arif Maulana, March 2023.
50 Interview, Arif Maulana, December 2019.
51 Interview, Robertus Robet, April 2020.
52 Interview, Taufik Basari, December 2019.
53 Such as PBHI, KIPP, Elsam and LBH Apik.

54 Interview, Ricky Gunawan, November 2019.
55 An important caveat here is that former YLBHI director Bambang Widjojanto served as lead lawyer for Jokowi's opponent in the 2019 presidential election, Prabowo Subianto, as he challenged his election loss at the Constitutional Court. Nevertheless, this development did not seem to affect YLBHI as an institution, or lead to accusations of partisanship.
56 Interview, Ricky Gunawan, November 2019.
57 Interview, Muhamad Isnur, October 2019.
58 Interview, Arif Maulana, March 2023.

References

Akbar, Riyan Rahmat. 2019. "Demonstrasi Mahasiswa, Protes Revisi UU KPK Dan Tujuh Desakan." *Tempo.co*, 2 October 2019. https://grafis.tempo.co/read/1834/demonstrasi-mahasiswa-protes-revisi-uu-kpk-dan-tujuh-desakan.

Amnesty International Indonesia. 2020. "*Press Release: Usut Bukti-Bukti Kekerasan Polisi Sepanjang Demo Tolak Omnibus Law*," 2 December 2020. https://www.amnesty.id/usut-bukti-bukti-kekerasan-polisi-sepanjang-demo-tolak-omnibus-law/.

Anjar, Angelina. 2019. "CekFakta #30 Demo Mahasiswa, Apa Yang Sebenarnya Terjadi?" *Tempo.co*, 15 December 2019. https://newsletter.tempo.co/read/1283938/cekfakta-30-demo-mahasiswa-apa-yang-sebenarnya-terjadi.

Asfinawati. 2017. "Bantuan Hukum Struktural: Sejarah, Teori, dan Pembaruan." *Buletin Bantuan Hukum* 2: 69–99.

Asih, Ratnaning. 2019. "LBH Pers Hingga KontraS Bersatu Dampingi Ananda Badudu di Polda." *Liputan6.com*, 27 September 2019. https://www.liputan6.com/showbiz/read/4072848/lbh-pers-hingga-kontras-bersatu-dampingi-ananda-badudu-di-polda.

Bagir, Zainal Abidin. 2023. "Half-Hearted Progress: Religious Freedom After the New Criminal Code." *Indonesia at Melbourne*, 17 January 2023. https://indonesiaatmelbourne.unimelb.edu.au/half-hearted-progress-religious-freedom-after-the-new-criminal-code/.

Baik, Tae-Ung. 2010. "Public Interest Litigation in South Korea." In *Public Interest Litigation in Asia*, edited by Po Jen Yap and Holning Lau, 115–135. London: Routledge. https://doi.org/10.4324/9780203842645.

BBC Indonesia. 2019. "Ananda Badudu: Saat Penahanan Mengklaim Melihat Banyak Mahasiswa 'Diproses Secara Tidak Etis.'" *BBC Indonesia*, 27 September 2019. https://www.bbc.com/indonesia/indonesia-49848208.

BBC Indonesia. 2020. "Demo Tolak Omnibus Law di 18 Provinsi Diwarnai Kekerasan, YLBHI: 'Polisi Melakukan Pelanggaran.'" *BBC Indonesia*, 9 October 2020. https://www.bbc.com/indonesia/indonesia-54469444.

BBC Indonesia. 2022. "Keengganan Pemerintah Buka Draf Terbaru RKUHP Dinilai Tunjukkan 'Gejala Otoriter.'" *BBC Indonesia*, 13 June 2022. https://www.bbc.com/indonesia/indonesia-61774257.

Bustomi, Muhammad Isa. 2020. "Totalnya, Polisi Tangkap 5.918 Orang Dalam Demo Tolak Omnibus Law Di Indonesia." *Kompas.com*, 12 October 2020. https://megapolitan.kompas.com/read/2020/10/12/18553771/totalnya-polisi-tangkap-5918-orang-dalam-demo-tolak-omnibus-law-di.

Butt, Simon. 2019. "Amendments Spell Disaster for the KPK." *Indonesia at Melbourne*, 18 September 2019. https://indonesiaatmelbourne.unimelb.edu.au/amendments-spell-disaster-for-the-kpk/.

Cahya, Gemma Holliani. 2019. "Knocking on the House's Door: Victims, Activists Send Flowers for the Sexual Violence Bill." *The Jakarta Post*, 7 September 2019. https://www.thejakartapost.com/news/2019/09/06/knocking-on-the-houses-door-victims-activists-send-flowers-for-the-sexual-violence-bill.html.

CNN Indonesia. 2019. "Tembakan Laser Dan Bendera Kuning Warnai 'Pemakaman KPK.'" *CNN Indonesia*, 17 September 2019. https://www.cnnindonesia.com/nasional/20190917204911-20-431391/tembakan-laser-dan-bendera-kuning-warnai-pemakaman-kpk.

CNN Indonesia. 2022a. "Pemerintah Tidak Transparan, BEM UI Desak Buka Akses RKUHP ke Publik." *CNN Indonesia*, 14 June 2022. https://www.cnnindonesia.com/nasional/20220519134251-12-798526/pemerintah-tidak-transparan-bem-ui-desak-buka-akses-rkuhp-ke-publik.

CNN Indonesia. 2022b. "Pasal-Pasal Krusial RKUHP Yang Jadi Sorotan Masyarakat Sipil." *CNN Indonesia*, 23 June 2022. https://www.cnnindonesia.com/nasional/20220622142741-12-812221/pasal-pasal-krusial-rkuhp-yang-jadi-sorotan-masyarakat-sipil.

CNN Indonesia. 2022c. "Pemerintah Ragu RUU KUHP Disahkan Bulan Juli." *CNN Indonesia*, 24 June 2022. https://www.cnnindonesia.com/nasional/20220624084543-32-812931/pemerintah-ragu-ruu-kuhp-disahkan-bulan-juli.

Cummings, Scott L. 2008. "The Internationalization of Public Interest Law." *Duke Law Journal* 57 (4): 891–1036.

Cummings, Scott L. 2020. "Movement Lawyering." *Indiana Journal of Global Legal Studies* 27 (1): 87–130. https://doi.org/10.2979/indjglolegstu.27.1.0087.

Detik. 2019. "Daftar RUU Disahkan, Ditangguhkan Dan Yang Masih Jadi Tuntutan Mahasiswa." *Detik*, 25 September 2019. https://news.detik.com/berita/d-4720722/daftar-ruu-disahkan-ditangguhkan-dan-yang-masih-jadi-tuntutan-mahasiswa.

Diamond, Larry. 1994. "Rethinking Civil Society: Toward Democratic Consolidation." *Journal of Democracy* 5 (3): 4–17. https://doi.org/10.1353/jod.1994.0041.

Dongoran, Hussein Abri. 2019. "Mahasiswa Bergerak." *Tempo Magazine*, 28 September 2019. https://majalah.tempo.co/read/laporan-utama/158486/mahasiswa-bergerak.

Erwanti, Marlinda Oktavia. 2020. "Jokowi: Pemerintah-DPR Tunda Bahas RUU Cipta Kerja Klaster Ketenagakerjaan." *Detik*, 24 April 2020. https://news.detik.com/berita/d-4989972/jokowi-pemerintah-dpr-tunda-bahas-ruu-cipta-kerja-klaster-ketenagakerjaan.

Fajrian, Happy. 2023. "Buruh Gugat UU Ciptaker Ke Mahkamah Konstitusi Untuk Diuji." *Katadata*, 9 April 2023. https://katadata.co.id/happyfajrian/berita/6432772112330/buruh-gugat-uu-ciptaker-ke-mahkamah-konstitusi-untuk-diuji.

Fraksi Rakyat Indonesia. 2020. "*Press Release: Pernyataan Sikap Atas Draft Omnibus Law RUU Cilaka Fraksi Rakyat Indonesia (FRI)*." https://bantuanhukum.or.id/demokrasi-dihabisi-omnibus-law-mematikan-demokrasi/.

Freeman, Alexi Nunn, and Jim Freeman. 2016. "It's About Power, Not Policy: Movement Lawyering for Large-Scale Social Change." *Clinical Law Review* 23 (1): 147–166.

Galih, Bayu. 2019. "Dandhy Dwi Laksono Ditangkap Polisi atas Tuduhan Menebarkan Kebencian." *Kompas.com*, 27 September 2019. https://nasional.kompas.com/read/2019/09/27/00462591/dandhy-dwi-laksono-ditangkap-polisi-atas-tuduhan-menebarkan-kebencian.

Ghaliya, Ghina. 2019. "House Elects Controversial Police General to Lead KPK." *The Jakarta Post*, 13 September 2019. https://www.thejakartapost.com/news/2019/09/13/house-elects-controversial-police-general-to-lead-kpk.html.

Gorbiano, Marchio Irfan. 2019. "'Trust Me': Jokowi Cancels Plan to Revoke KPK Law Amendment." *The Jakarta Post*, 1 November 2019. https://www.thejakartapost.com/news/2019/11/01/trust-me-jokowi-cancels-plan-to-revoke-kpk-law-amendment.html.

Hadiz, Vedi R., and Richard Robison. 2013. "The Political Economy of Oligarchy and the Reorganization of Power in Indonesia." *Indonesia*, no. 96: 35–57. https://doi.org/10.5728/indonesia.96.0033.

Haryanto, Alexander. 2019. "Gelombang Demo Mahasiswa: Dari Palembang, Semarang, Solo dan Medan." *Tirto.id*, 24 September 2019. https://tirto.id/gelombang-demo-mahasiswa-dari-palembang-semarang-solo-dan-medan-eiDY.

Hilbink, Thomas M. 2004. "You Know the Type: Categories of Cause Lawyering." *Law & Social Inquiry* 29 (3): 657–698. https://doi.org/10.1086/430155.

Human Rights Watch. 2022. "Indonesia: New Criminal Code Disastrous for Rights." *Human Rights Watch*, 8 December 2022. https://www.hrw.org/news/2022/12/08/indonesia-new-criminal-code-disastrous-rights.

Iswinarno, Chandra, and Bagaskara Isdiansyah. 2020. "Kantornya Pernah Dikepung Laskar, YLBHI Malah Lantang Tolak Pembubaran FPI." *Suara.com*, 31 December 2020. https://www.suara.com/news/2020/12/31/185806/kantornya-pernah-dikepung-laskar-ylbhi-malah-lantang-tolak-pembubaran-fpi.

Jong, Hans Nicholas. 2020a. "With New Law, Indonesia Gives Miners More Power and Fewer Obligations." *Mongabay*, 13 May 2020. https://news.mongabay.com/2020/05/indonesia-mining-law-minerba-environment-pollution-coal/.

Jong, Hans Nicholas. 2020b. "Indonesia's Omnibus Law a 'Major Problem' for Environmental Protection." *Mongabay*, 4 November 2020. https://news.mongabay.com/2020/11/indonesia-omnibus-law-global-investor-letter/.

Jong, Hans Nicholas. 2021. "2019 Fires in Indonesia Were Twice as Bad as the Government Claimed, Study Shows." *Mongabay*, 16 December 2021. https://news.mongabay.com/2021/12/2019-fires-in-indonesia-were-twice-as-bad-as-the-government-claimed-study-shows/.

Kamil, Irfan. 2020. "Tim Advokasi untuk Demokrasi Siapkan Bantuan Hukum bagi Demonstran Tolak UU Cipta Kerja." *Kompas.com*, 8 October 2020. https://nasional.kompas.com/read/2020/10/08/18113371/tim-advokasi-untuk-demokrasi-siapkan-bantuan-hukum-bagi-demonstran-tolak-uu.

Kementerian Hukum dan HAM. 2021. "*Sosialisasi Dan Diskusi Publik Rancangan Undang-Undang Hukum Pidana*." Kementerian Hukum dan Hak Asasi Manusia.

Komnas HAM. 2019. "Tim Advokasi Untuk Demokrasi Datangi Komnas HAM." *Komisi Nasional Hak Asasi Manusia*, 3 October 2019. https://www.komnasham.go.id/index.php/news/2019/10/3/1195/tim-advokasi-untuk-demokrasi-datangi-komnas-ham.html.

Kustiasih, Rini. 2020. "Surat Presiden tentang Omnibus Law Digugat ke PTUN." *Kompas*, 3 May 2020. https://www.kompas.id/baca/polhuk/2020/05/03/surpres-omnibus-law-digugat-ke-ptun/.

Lev, Daniel. 1987. *Legal Aid in Indonesia*. Working Papers (Monash University. Centre of Southeast Asian Studies) 44. Clayton, Victoria: Monash University.

Liddle, R. William. 2013. "Improving the Quality of Democracy in Indonesia: Toward a Theory of Action." *Indonesia*, no. 96 (October): 59–80. https://doi.org/10.5728/indonesia.96.0057.

Mietzner, Marcus. 2021. "Sources of Resistance to Democratic Decline: Indonesian Civil Society and Its Trials." *Democratization* 28 (1): 161–178. https://doi.org/10.4324/9781003346395-9.

Nasution, Adnan Buyung. 1985. "The Legal Aid Movement in Indonesia: Towards the Implementation of the Structural Legal Aid Concept." In *Access to Justice: Human Rights Struggles in South East Asia*, edited by Laurie S. Wiseberg and Harry M. Scoble, 31–39. London: Zed Books.

Nugraheny, Dian Erika. 2020. "Puluhan Akademisi Tolak Pengesahan UU Cipta Kerja." *Kompas.com*, 5 October 2020. https://nasional.kompas.com/read/2020/10/05/19075301/puluhan-akademisi-tolak-pengesahan-uu-cipta-kerja.

Nurhadi, Jamal A. Nashr, Shinta Maharani, Friski Riana, and Robby Irfany. 2020. "Demonstrasi Penolakan RUU Cipta Kerja Meluas." *Koran Tempo*, 12 March 2020. https://koran.tempo.co/read/nasional/450937/demonstrasi-penolakan-ruu-cipta-kerja-meluas.

Pardede, Raynard Kristian Bonanio. 2022. "Jelang Pengesahan RKUHP Besok, Massa Lintas Elemen Unjuk Rasa." *Kompas*, 5 December 2022. https://www.kompas.id/baca/polhuk/2022/12/05/rkuhp-dinilai-minim-partisipasi-publik-dan-mempersempit-ruang-demokrasi.

Pepinsky, Thomas B. 2013. "Pluralism and Political Conflict in Indonesia." *Indonesia*, no. 96 (October): 81–100, 220. https://doi.org/10.1353/ind.2013.0014.

Purnamasari, Dian Dewi. 2022. "Tak Hanya Publik, DPR Juga Belum Terima Draf RKUHP Terbaru." *Kompas*, 27 June 2022. https://www.kompas.id/baca/polhuk/2022/06/27/komisi-iii-tetap-targetkan-rkuhp-disahkan-juli.

Putri, Budiarti Utami. 2020. "Pemerintah Ubah Nama Omnibus Law Cilaka Jadi Cipta Kerja." *Tempo.co*, 12 February 2020. https://nasional.tempo.co/read/1306717/pemerintah-ubah-nama-omnibus-law-cilaka-jadi-cipta-kerja.

Reuters. 2019. "Violent Protest Erupts in Capital of Indonesia's Papua." *Reuters*, 30 August 2019. https://ar.reuters.com/article/us-indonesia-papua-idUSKCN1VJ195.

Ristianto, Christoforus. 2019. "RKUHP Soal Penghinaan Presiden, Kumpul Kebo, Hingga Unggas, Ini Penjelasan Menkumham." *Kompas.com*, 21 September 2019. https://nasional.kompas.com/read/2019/09/21/08562241/rkuhp-soal-penghinaan-presiden-kumpul-kebo-hingga-unggas-ini-penjelasan.

Robison, Richard, and Vedi R. Hadiz. 2004. *Reorganising Power in Indonesia: The Politics of Oligarchy in an Age of Markets*. London and New York: RoutledgeCurzon. https://doi.org/10.4324/9780203401453.

Sahara, Wahyuni. 2021. "Mengenang Mereka Yang Meninggal Dalam Aksi #Reformasi-Dikorupsi." *Kompas.com*, 20 September 2021. https://nasional.kompas.com/read/2021/09/20/13081761/mengenang-mereka-yang-meninggal-dalam-aksi-reformasidikorupsi.

Saptohutomo, Aryo Putranto. 2022. "Pemerintah Tegaskan Tak Ada Pasal Pidana LGBT Di KUHP Baru." *Kompas.com*, 16 December 2022. https://nasional.kompas.com/read/2022/12/16/06284131/pemerintah-tegaskan-tak-ada-pasal-pidana-lgbt-di-kuhp-baru.

Saputri, Maya. 2019. "Sampai Kapan Polisi Tutup Akses Soal Mahasiswa Ditahan dan Hilang?" *Tirto.id*, 28 September 2019. https://tirto.id/sampai-kapan-polisi-tutup-akses-soal-mahasiswa-ditahan-hilang-eiSs.

Sarat, Austin, and Stuart A. Scheingold. 2006. "What Cause Lawyers Do For, and To, Social Movements: An Introduction." In *Cause Lawyers and Social Movements*, edited by Austin Sarat and Stuart A. Scheingold, 1–34. Stanford, California: Stanford University Press. https://doi.org/10.1515/9780804767965-003.

Taufiqurrahman, M. 2019. "#MosiTidakPercaya: How an Obscure Indie Song Became the Anthem of a Generation." *The Jakarta Post*, 30 September 2019. https://www.thejakartapost.com/academia/2019/09/29/mositidakpercaya-how-an-obscure-indie-song-became-the-anthem-of-a-generation.html.

The Jakarta Post. 2019. "Legislative Assault on KPK - Editorial." *The Jakarta Post*, 18 September 2019. https://www.thejakartapost.com/academia/2019/09/18/legislative-assault-on-kpk.html.

The Jakarta Post. 2020. "Legal Experts Scoff at Indonesia's Improved Rule of Law Ranking." *The Jakarta Post*, 20 March 2020. https://www.thejakartapost.com/news/2020/03/20/legal-experts-scoff-at-indonesias-improved-rule-of-law-ranking.html.

Thea, Ady. 2020. "RUU Cipta Kerja Dinilai Lemahkan Perlindungan Lingkungan Hidup." *Hukum Online*, 18 February 2020. https://hukumonline.com/berita/a/ruu-cipta-kerja-dinilai-lemahkan-perlindungan-lingkungan-hidup-lt5e4b949650958/.

"Ujian Reformasi". 2019. *Mata Najwa*. Narasi.tv and Trans7. www.narasi.tv/mata-najwa/perlawanan-mahasiswa.

Ungku, Fathin, Gayatri Suroyo, and Bernadette Christina. 2020. "Explainer: Indonesia's Jobs Law Endangers Environment, Say Activists, Investors." *Reuters*, 9 October 2020. https://www.reuters.com/article/us-indonesia-economy-protests-explainer-idUSKBN26U0MB.

Warburton, Eve. 2020. "Deepening Polarization and Democratic Decline in Indonesia." In *Political Polarization in South and Southeast Asia: Old Divisions, New Dangers*, edited by Thomas Carothers and Andrew O'Donohue, 25–39. Carnegie Endowment for International Peace. https://carnegieendowment.org/2020/08/18/deepening-polarization-and-democratic-decline-in-indonesia-pub-82435.

Wee, Sui-Lee. 2022. "In Sweeping Legal Overhaul, Indonesia Outlaws Sex Outside Marriage." *The New York Times*, 6 December 2022. https://www.nytimes.com/2022/12/06/world/asia/indonesia-sex-gay-rights.html.

Winters, Jeffrey A. 2011. *Oligarchy*. Cambridge: Cambridge University Press. https://doi.org/10.1017/cbo9780511793806.

Winters, Jeffrey A. 2013. "Oligarchy and Democracy in Indonesia." *Indonesia*, no. 96 (October): 11–33. https://doi.org/10.5728/indonesia.96.0099.

YLBHI. 2020. "*Press Release: Polri Dan Pemerintah Harus Hormati Undang-Undang Dasar Negara RI Tahun 1945 Yang Menjamin Hak Menyampaikan Pendapat Di Muka Umum*," 5 October 2020. https://ylbhi.or.id/informasi/siaran-pers/polri-pemerintah-harus-hormati-undang-undang-dasar-negara-ri-tahun-1945-yang-menjamin-hak-menyampaikan-pendapat-di-muka-umum/.

8 Conclusion

In September 2021, LBH released a slick, 39-minute video documentary to commemorate the 50th anniversary of its establishment in 1970, titled "*LBH – Meniti Jalan Terjal Demokrasi*" ('LBH – Climbing the Steep Path to Democracy'). The documentary was made by Watch Doc, the production company of activist filmmaker Dandhy Laksono who, as discussed in the previous chapter, was defended by LBH figures in 2019 when he was briefly arrested for tweeting about unrest in the Papuan provinces. The film focused primarily on the major cases represented by LBH under the New Order, and LBH's emergence as a centre of civil society resistance to the Soeharto regime. There was a brief discussion of democratic change and the efforts of the organisations established by YLBHI, like KRHN,[1] to make an impression on the early post-authoritarian reforms. The film then jumped forward to the presidency of Joko "Jokowi" Widodo, noting how since he came to power in 2014, Jokowi has presided over the erosion of Indonesian democracy, despite the hopes of civil society. Against a stirring soundtrack, the film presented footage of the 2019 *Reformasi Dikorupsi* student and civil society protests, including Asfin's fiery speech to demonstrators quoted at the beginning of this book, as well as the 2020 protests against the Omnibus Law on Job Creation. It was almost as though the intervening 15 years or so between the fall of the New Order and the election of Jokowi never happened. LBH was only depicted as existing in relation to authoritarianism. This is an overly simplistic narrative that ignores several of the important reforms secured by LBH in the post-authoritarian era. But there is hardly a better representation of how LBH views its return to relevance under a more repressive regime. LBH has been energised by the cause of democracy defence.

This book tells a more complex version of this story, through investigation of the impact of democratic change on cause lawyers and the practice of cause lawyering in Indonesia. Cause lawyering exists in a variety of forms and practice contexts in Indonesia, but LBH has been the most prominent and important cause lawyering organisation in Indonesia for more than half a century. Examining the trajectory of LBH and its practice of cause lawyering from the authoritarian New Order era through to the current moment, this book argues that there is a clear relationship between the quality of

DOI: 10.4324/9781003486978-12

democracy and the form of cause lawyering practiced. In particular, democratic backsliding has prompted LBH to reject direct engagement with the state and prioritise an aggressive, grassroots version of cause lawyering, similar to the style of cause lawyering it practiced under Soeharto.

This chapter summarises the main findings of this book, beginning by reviewing LBH's establishment under the New Order and the development of the grassroots, structural legal aid approach to cause lawyering that became its hallmark. It then continues to review how the democratic transition that began in 1998, and more recent democratic decline, has affected cause lawyers at LBH and their strategies of cause lawyering. Finally, I reflect more broadly on what LBH's experience can illustrate about cause lawyering in countries experiencing democratic decline, like Indonesia, and look at some of the challenges that lie ahead.

LBH: An Indonesian cause lawyering organisation

LBH was founded in 1970 under Soeharto's authoritarian New Order to provide legal aid to poor and marginalised Indonesians. In establishing LBH, Adnan Buyung Nasution was also motivated by a struggle to limit state power through instatement (or reinstatement) of the *negara hukum*. Yet this somewhat proceduralist emphasis on the *negara hukum* quickly gave way to a more expansive, political orientation. Mirroring the CLS arguments that were becoming popular around the same time (Hunt 1986; Kelman 1987), LBH believed that providing conventional legal aid under the New Order's grossly unjust system would only end up legitimising the status quo (Lubis 1985, 136). It maintained that legal aid should be directed at addressing the structural causes of inequality that were at the root of the legal problems of the poor (Aspinall 2005, 104). These ideas served as the foundation for the development of 'structural legal aid', LBH's ideology of legal aid, which continues to guide its practice today.

Structural legal aid was an overtly activist, political, grassroots approach to cause lawyering. As the New Order state was considered the source of many of the problems faced by Indonesia's poor, it was also infused with a deeply oppositional ethos. It was focused on dismantling the authoritarian New Order system and replacing it with a more democratic framework. As the New Order period progressed, LBH began actively advocating for democratic reform, styling itself as a "locomotive of democracy". It developed into a focal point for opposition to the regime, a gathering point for activists, journalists, students and other pro-democracy forces (Nasution 2011, 12). The New Order eventually collapsed in May 1998. Even if civil society (and LBH as one of its most prominent pro-democratic elements) could not claim to have toppled Soeharto, it was able to chip away at the ideological pillars of authoritarian rule (Aspinall 2004, 82). This history of political struggle is key to understanding how LBH adapted to the democratic transition that followed, and how it sees the role of cause lawyering in Indonesia's current flawed democracy.

After the New Order collapsed in 1998, Indonesia restructured its institutions to provide the foundations for more democratic governance (Lindsey 2004, 19). This process involved many structural changes considered supportive of legal mobilisation. The constitution was amended to formally grant citizens a broad range of rights, a Constitutional Court was established, judicial independence was strengthened, and restrictions on civil society were dismantled. Yet the experience of LBH after 1998 demonstrates that the introduction of democratic institutions does not necessarily provide a more conducive environment for cause lawyering.

As Chapter 3 detailed, there are several reasons why LBH struggled after Soeharto fell, despite institutional reforms supportive of legal mobilisation. One of the most significant was that following the fall of the New Order, LBH suffered from an identity crisis. So much of its identity, and its approach to legal aid and activism, structural legal aid, was tied up in opposing authoritarianism. With Soeharto out of the picture, LBH had to reassess its reason for being. It is little wonder LBH faced an identity crisis. The world was eager to treat Indonesia as a democratic country, while LBH viewed the country's institutions as still being dominated by New Order-era elites. It had a lot of difficulty reinventing itself as an organisation open to more accommodative, elite-focused forms of cause lawyering in this environment.

A related factor was funding. Following the democratic transition, there was less donor support for adversarial forms of cause lawyering and a greater emphasis on governance programming designed to strengthen the institutions of the newly democratic state (Munarman et al. 2004, 56; Aspinall 2010, 6). After Soeharto fell, a very large number of new human rights and law reform-focused organisations were established. Some of them were more comfortable about taking up accommodative forms of cause lawyering and building the legitimacy of the state, and they attracted donor attention. Donor funds were therefore available for a certain version of cause lawyering; there was a preference for the state-centred elite/vanguard version of cause lawyering over the grassroots version historically favoured by LBH.

For a period of about 10–15 years the central body, YLBHI, also experienced multiple debilitating leadership crises and related weaknesses in organisational management. These leadership crises often reflected disputes over the most appropriate form of cause lawyering for the political moment, but they also had roots in the organisational challenges that occurred following democratic transition. Once the organisation lost donor funding and began questioning its organisational identity, tensions arose.

However, the story of LBH in the post-authoritarian era is more than a story of funding shortfalls, management weaknesses and internal conflict. As Chapter 4 discussed, since 1998, LBH has remained an essential provider of legal assistance to Indonesia's marginalised, disadvantaged and unpopular communities. Notably, in the years after Soeharto fell, when the government was more engaged in reform, there was broader civil society enthusiasm for the ability of legal reforms to lead to social change. Cause lawyers, including some from LBH, hoped to reform law and policy to improve the position of marginalised groups. LBH was always somewhat ambivalent about this elite-centred version of cause lawyering, although it did attempt to engage in some activities consistent with an elite/vanguard approach.

LBH, along with other civil society organisations, attempted to take advantage of the new opportunities for constitutional litigation offered by the establishment of the Constitutional Court in 2003. LBH lawyers launched a successful challenge to the Attorney General's legal authority to ban books in 2010, and were central players in the high-profile, but ultimately unsuccessful, challenge to the so-called Blasphemy Law in 2009 and 2010. Chapter 4 also showed how LBH has drawn on strategies like citizen lawsuits and litigation at the administrative courts to challenge government acts. This kind of strategic litigation has been a key means by which LBH attempts to give life to the new democratic constitution – and demand the state meet its rights commitments.

As the state became more open following the democratic transition, LBH was also presented with new political opportunities to engage directly with the legislature and the state bureaucracy on policy reform, and strengthen the capacity of state institutions. Chapter 6 detailed how cause lawyers at LBH experimented with these elite-oriented strategies, and had some success in securing progressive reforms, most notably the 2011 Legal Aid Law. YLBHI has since developed a productive working relationship with the national legal aid system's implementing agency, the National Law Development Agency (BPHN), and they continue to collaborate to improve the implementation of the scheme. On the whole, however, LBH has found efforts to collaborate with the government and legislature on drafting laws and regulations, or strengthening the capacity of state institutions, to have been deeply disappointing. For example, YLBHI and LBH Jakarta made genuine attempts to improve attention to human rights among police. But police continue to regularly violate the rights of citizens, as the mass demonstrations of 2019 and 2020, and the horrific Kanjuruhan Stadium tragedy, brought into stark relief. YLBHI and LBH now look at the huge sums donors have spent over the past two decades attempting to promote reform in state institutions like Polri, the Attorney General's Office, and the Supreme Court, and have concluded that these efforts have resulted in very little meaningful change.

LBH's scepticism about collaborating with the state has only increased as the quality of Indonesian democracy has declined. Democratic backsliding has helped LBH put aside questions about the degree to which it should engage with state structures. It has settled on a highly oppositional, political form of cause lawyering, in which it maintains a distance from the state. LBH is less interested in the incremental changes that can be achieved through collaboration with the government. LBH lawyers see little incentive in securing piecemeal legal reforms when they believe the state will just ignore reforms when it wishes. As the responses of LBH lawyers in Chapter 6 demonstrated, however, the organisation acknowledges that engaging with the state can be a valid approach – it is just not a strategy LBH wishes to pursue itself. The blossoming of civil society following democratic transition has allowed other organisations to take up more accommodative strategies, leaving LBH free to oppose the state without having to worry about maintaining relationships with state officials or institutions. LBH's reluctance to compromise has also provided it with a certain moral heft – there is no question that LBH remains on the side of victims.

Democratic regression has prompted LBH to return to a strong focus on structural legal aid as its guiding approach to legal and political action. Along with their deep distrust of the state, in an environment where the rule of law remains underdeveloped, LBH lawyers have also become disillusioned about the capacity of legal victory to lead to substantive social change, and they have increasingly emphasised extra-legal grassroots strategies, like community legal empowerment and community organising.

This strong grassroots orientation does not mean that LBH has jettisoned litigation altogether, but rather that litigation is valued for its supplementary effects. Even if LBH has deep reservations about the utility of litigation strategies, it views the courts as a political forum (Hilbink 2004, 687), a safe space to raise and debate sensitive issues, to pressure state institutions to take action, or to promote mobilisation and consolidation of marginalised communities. Legal action is seen as beneficial for its ability to spark public education, community organising, public protests and media campaigning, as Chapter 5 showed. In pursuing grassroots strategies, LBH continues to demand that the state uphold the rule of law but believes change will only occur with sustained pressure from the grassroots. It identifies with and considers itself to be part of social movements. It believes elite-focused reform strategies will never be successful unless there are concurrent efforts to empower local communities and shift unequal power relations.

LBH is now blurring the lines between cause lawyering and more straightforward political activism, playing a convening, coalition-building role in civil society, similar to the kind of movement building role it played under the New Order. Partly because of its legacy as a focal point of resistance to the Soeharto regime, LBH has retained legitimacy among broader civil society. Now that democracy is under threat, civil society is again looking to LBH as a pro-democratic force. Further, by largely keeping a distance from the state over the past two decades, LBH has mostly avoided compromising with power holders. This has meant it has retained broad acceptance among wider civil society.

Cause lawyering and democratic change

This book offers several unique insights relevant to the broader literature on cause lawyering. A main contention of the book is that the experience of LBH in the post-authoritarian period demonstrates a clear relationship between the type of cause lawyering practiced and the quality of democracy.

It is not surprising that cause lawyering responds to shifts in the political context, as other scholars have previously identified (Sarat and Scheingold 2001, 13; Nesossi 2015; Fu 2018; Tam 2013). But in the case of LBH, the relationship between cause lawyering and democratic change is not what one might expect. Cause lawyering and legal mobilisation literature suggest that institutional developments associated with democratisation should provide more fertile terrain for cause lawyering (Munger, Cummings, and Trubek 2013; Cummings and Trubek 2008; Scheingold and Sarat 2004, 131; Sarat and Scheingold 1998, 5; Yap and Lau 2010, 2). This book instead found that democratisation involved significant challenges for LBH, while democratic regression has reinvigorated it.

Developments associated with democratisation described as being supportive of cause lawyering include the introduction of constitutional and other legal guarantees of rights, the presence of constitutional courts, judicial independence, political openness, the development of a strong civil society sector and free media, among other political and institutional factors (see, for example, Baik 2010; Chang 2010; Cummings and Trubek 2008; Ginsburg 2007; Goedde 2009; Meili 2009; Munger, Cummings, and Trubek 2013; Vieira 2008). These structural factors are important, but they are not the whole story.

The trajectory of LBH after 1998 demonstrates the need to be realistic about the limits of formal institutional change in post-authoritarian states where the rule of law is a work in progress. Cause lawyering and legal mobilisation literature might emphasise the importance of formal structures for cause lawyering but in a country that is transitioning to democracy like Indonesia, the presence of appropriate institutional structures for legal mobilisation is not sufficient for cause lawyering to thrive. It is also critical to pay attention to the quality of democracy when considering the relationship between cause lawyering and democratisation. One of the central problems in the Indonesian case is that political and economic elites have easily adapted to, and dominated, new democratic structures (Winters 2011; Robison and Hadiz 2004). The power relationships that result in many of the legal problems of the poor remain relatively unchanged in democratic Indonesia. Government and business interests routinely override the rights of local communities, and the justice sector almost invariably comes down on the side of the powerful (Winters 2013, 19). The legal system has been – to borrow Winters' pessimistic phrasing – "relentlessly trampled by oligarchs" (Winters 2011, 210). Oligarchs also have control over political parties and dominate the legislature (Winters 2011, 179–82; Mietzner 2015), and as the discussion in the previous chapter showed, LBH now views the policymaking process as being completely captured by oligarchic interests. In this kind of oligarchic system, the formal legal opportunities for cause lawyering made available by the democratic transition are less appealing for cause lawyers. LBH believes elite-focused strategies involving collaboration with the state to achieve incremental reforms will not have a significant impact as long as these oligarchic power structures remain in place. In the face of deficiencies in the structural factors said to be supportive of cause lawyering, LBH has turned to grassroots strategies.

The findings of this study also enrich understandings of how cause lawyers respond to creeping authoritarianism. As noted, there is growing academic attention to cause lawyering in authoritarian environments (McEvoy and Bryson 2022; Mustafina 2022; Moustafa 2007a). Advocating for rights against authoritarian regimes can provoke serious state repression, and this may discourage legal mobilisation, or lead cause lawyers to deploy less confrontational strategies (Gobe and Salaymeh 2016; Stern and O'Brien 2012, 183–85; Fu 2018, 558). In some cases, the constrained political opportunities under authoritarian regimes have been shown to lead cause lawyers to retreat to litigation, and use the relatively safe space of the courtroom as a site of protest and resistance (McEvoy and Bryson 2022, 89–94; Fu 2011, 353; Munger, Cummings, and Trubek 2013, 378; Moustafa 2007b, 40). In other contexts, activists and lawyers avoid overt confrontation with the ruling regime to ensure survival, and focus on securing incremental and cumulative victories when they can (Chua 2014; Stern 2013, 157; O'Brien 2023). The Indonesian example is particularly interesting in this context because when confronted with shrinking democratic space, LBH cause lawyers have instead become more political and more confrontational. Recent years in Indonesia have seen activists and protestors subject to criminal charges, cyberattacks, intimidation and even physical attacks (Amnesty International Indonesia 2022). Yet LBH has turned to risky acts of dissent as the state has become more repressive, emphasising grassroots mobilisation and direct action tactics, including street protests, which put them in direct opposition to the state.

It is true that this kind of response to repression is not entirely unique to Indonesia. Increasing repression can sometimes lead to increased grassroots mobilisation by lawyers (Nesossi 2015). In China, for example, scholars have documented how lawyers frustrated with the constraints of operating within an authoritarian legal system have undertaken collective action beyond the courtroom (Pils 2014, 105; Fu 2018, 555; Liu and Halliday 2016, 144–70; Nesossi 2015, 971), and have forged alliances with and organised civil society groups (Fu 2018, 564; Fu and Cullen 2010, 9–34). Compared to China, however, civil society in Indonesia faces fewer restrictions. This openness has enabled Indonesian cause lawyers to engage in more openly defiant acts of resistance, like the *Reformasi Dikorupsi* and Job Creation Law protests, in sharp contrast to the narrower scope of collective action tolerated in the Chinese context.

As this book has demonstrated, there are potent historical reasons as to why LBH has returned to confrontational grassroots strategies in response to repression. LBH's legacy as a hub of resistance to the New Order has remained deeply embedded in its organisational identity. Indeed, this oppositional aspect of its identity proved difficult to shake when conditions were more amenable to accommodative forms of cause lawyering. In the end, however, this historical legacy has put the organisation in good stead as democracy has unravelled. As Indonesia's democratic gains have come under attack, LBH was there, ready to spring back into action. In contrast to what one might expect, LBH is thriving precisely because of a deterioration in the institutional factors supposedly supportive of cause lawyering.

An important contribution of this book is that it draws attention to the profound ways in which internal organisational factors can affect cause lawyering. As this study identified, LBH's failure to flourish in the years after the democratic transition was largely due to internal organisational and management challenges. Charles Epp (1998) described rights advocacy organisations as a pivotal component of the support structure for legal mobilisation, in addition to rights advocacy lawyers and adequate funding. Epp recognised that the strength and institutionalisation of rights advocacy groups can affect the contribution that they can make to supporting legal mobilisation (Epp 1998, 48, 107, 150, 190). Yet these organisational capacity issues were not dealt with in depth. The LBH case underscores the importance of examining factors like organisational vision, leadership, organisational governance and management, funding, recruitment and training, and research capacity when examining cause lawyering. I am reluctant to simply conclude that more managerialism would benefit cause lawyering organisations. Development scholars have for some time recognised that the international development sector's excessive focus on managerialism can distract organisations from pursuing transformative change (Girei 2016; Banks, Hulme, and Edwards 2015). Nevertheless, the experience of LBH suggests that more comprehensive interrogation of the organisational sites for cause lawyering can provide rich insights into how cause lawyering is practiced.

There are also important lessons to be learned from the funding challenges LBH faced following the democratic transition. The Indonesian case suggests foreign donors should be careful about promoting strategies to meet form when function is not yet there. Donors encouraged LBH to work on elite/vanguard strategies of policy advocacy, legislative drafting and institutional strengthening as if they were operating in a functional liberal democracy. But LBH believed the substance of democracy had yet to catch up with formalistic change (and public rhetoric). LBH recognised that elements of Indonesia's authoritarian regime were able to survive and continue to dominate state institutions following the transition. It is not hard to understand how donors would want to encourage progressive civil society organisations to engage with the state and assert a degree of influence over the democratic reform process. Even so, the case of LBH demonstrates that in new and transitional democracies, donors should be wary about placing too much emphasis on strengthening government institutions and prematurely leaving behind support for critical, pro-democracy CSOs focused on holding the state to account.

Emerging challenges

LBH's charismatic founder, Adnan Buyung Nasution, died in 2015. In the days before Buyung died, Todung Mulya Lubis, the other towering LBH figure of the New Order years, visited him in hospital. Buyung was unable to speak, but between tears, he scrawled a dramatic and moving final message to Lubis on a piece of paper, which read: "Keep watch over LBH/YLBHI. Keep thinking and struggling for the poor and oppressed" (Kuwado 2015).[2] It is clear now that Buyung need not have worried about LBH ever fading from relevance – its ability to bounce back after years of decline has proved as much. However, the future will offer significant challenges for LBH.

As discussed, Muhamad Isnur was selected to replace Asfin as YLBHI director in early 2022. He is a different style of leader to Asfin, but he also enjoys broad acceptance among pro-democratic civil society groups and has strived to continue the internal organisational reform project she initiated. Fortunately, most of the hard work of organisational repair has been done. YLBHI has established stronger internal policies and procedures, and the organisation is now in good financial health. Yet there is some unfinished business. A particular challenge for YLBHI in the post-Soeharto years has related to research and knowledge management. Isnur has retained Asfin, Siti Rakhma Mary Herwati and Arip Yogiawan (Yogi) in a 'Knowledge Management Team' that is acting like a 'shadow board'. They have been tasked with conducting research on past cases and approaches, identifying successful practices that can be shared across the YLBHI network, rewriting and strengthening training curricula, and paying greater attention to the regeneration of staff and leadership. This is wise. If LBH is to continue to play a convening and coalition-building role in civil society, it will be required to manage and disseminate knowledge among social movement groups – the important work of framing grievances and repressive government actions in terms of rights, and ensuring students, communities and other civil society organisations are armed with accurate and up-to-date information. The Knowledge Management Team is also attempting to build stronger connections between pro-democracy civil society and the academic community, which has provided inconsistent support for democratic reform in the post-authoritarian period. The retention of experienced figures like Asfin, Rakhma and Yogi is also an acknowledgement of the fact that LBH has not always maintained strong relationships with its alumni after they leave the organisation. These accomplished team members have provided continuity and minimised disruption during the transition to the new leadership team – a serious weak spot for LBH in previous years. The fact that Isnur has continued to engage Asfin in his leadership team is a break from the past and undoubtedly a sign of maturity, but it is also an implicit acknowledgement of her ongoing influence among LBH offices and in broader civil society.

A particularly formidable challenge faced by LBH is that the trend of democratic decline that has emerged over the past decade now seems entrenched. When President Jokowi leaves office in October 2024, he will be replaced by former general Prabowo Subianto, the former son-in-law of Soeharto. While it is difficult to predict with certainty what kind of president Prabowo will be, he has given few indications that he will take steps to reverse the democratic backsliding of the past decade. In fact, it is much more likely that with Prabowo in the top job, Indonesia will see an intensification of efforts to erode democratic institutions and rights protections. This will have consequences for LBH. Civil society already seems bruised and exhausted by recent battles.[3] It is not hard to understand why. The government essentially got what it wanted – almost all draft laws and policies that activists and students demonstrated against in 2019 have since been passed. LBH staff now talk about fighting for democracy and human rights as a long-term war, comparing their struggle to that of their predecessors under Soeharto. As Isnur told me: "In the 1980s, under the New Order, the situation was incredibly oppressive. [But] LBH was not expecting to change anything in one or two years. We must think about change in a ten to 15-year timeframe."[4]

LBH is intensifying efforts to build a broad-based social movement at the grassroots. Since assuming the YLBHI leadership, Isnur has overseen the establishment of a new branch office in West Kalimantan, and has plans to open several additional LBH offices before his term is finished. Regional LBH offices will be expected to play the kind of role that YLBHI has been playing at the national level, acting as decentralised hubs for oppositional civil society in their respective regions. With the assumption that government repression will continue, LBH offices will be expected to strengthen local social movements, coordinate activism, and maintain networks of survival at the regional level. In a sign of just how bad LBH considers things to have become, it is again prioritising documentation of human rights abuses. YLBHI is encouraging LBH offices in the regions to carefully document rights violations and use this information to dispute the government narrative of events and demand accountability. This was an essential strategy under the New Order, and it is becoming important again now. Similarly, recognising that many students were mobilised online during the 2019 *Reformasi Dikorupsi* protests, LBH is undertaking belated efforts to make better use of digital platforms. Each LBH office is training networks of online activists, who will be tasked with using social media to build greater public awareness of rights abuses.

LBH's plans are ambitious, perhaps overly so. It is too early to assess whether the 2019 and 2020 mass protests were high-water marks for LBH's movement building efforts. There were specific reasons why there was no mass mobilisation against the revised Criminal Code in 2022. Civil society was more divided than in previous years, and was also bruised by the repressive police response to past demonstrations. Yet it would be premature to write off

LBH or broader civil society over the comparatively muted response to the revised Criminal Code in 2022. When the next great threat to Indonesian democracy emerges, as it inevitably will, LBH's movement building skills will again be put to the test. It will quickly become clear the degree to which LBH is able to bring diverse civil society organisations together in a coordinated movement for democratic change.

Final thoughts

When I first began thinking about this book, I imagined I was going to be telling the story of LBH's decline – that despite the institutional advantages and opportunities provided by democratisation, LBH had, paradoxically, struggled. While LBH certainly did struggle for a period, it is not struggling now. LBH remains a vital organisation, as the 2019 and 2020 protests showed. It was vital not only because of its efforts to protect the rights of demonstrating students when they were arrested, but also in helping to bring civil society groups together and continuing to hold the state to account.

One of the most worrying aspects of recent democratic regression has been the lack of any pushback by members of the elite (Power and Warburton 2020, 11). Oppositional civil society – however weak, fragmented and ineffective in the face of oligarchic power – is really the only source of resistance to democratic decline in contemporary Indonesia. And LBH remains one of civil society's most important voices. Back in 1994, Buyung commented that:

> "As long as the LBH and groups like it exist, the people cannot be stripped entirely of an independent voice. If they cannot have a say *in* government, at least they can occasionally talk back to it, announcing their aspirations for freedom, justice, and self-rule."
>
> (Nasution 1994, 117).

While the public have far more channels and opportunities to speak back to the state now than they did under the New Order, Buyung's comments remain relevant, three decades later. In a sense, this book is a positive story for LBH and structural legal aid, in that it has returned to relevance again. But it is a depressing story for Indonesia. LBH and structural legal aid has become so important again because of the bad reasons that led to its creation in the first place.

Notes

1 The National Consortium for Law Reform, the law reform organisation established by YLBHI in 1998.
2 "*Jagalah LBH/YLBHI. Teruskan pemikiran dan perjuangan bagi si miskin dan tertindas*".
3 Interviews, Asfinawati, Meila Nurul Fajriah, March 2023.
4 Interview, Muhamad Isnur, February 2023.

References

Amnesty International Indonesia. 2022. "*Silencing Voices, Suppressing Criticism: The Decline in Indonesia's Civil Liberties.*" Jakarta: Amnesty International Indonesia. https://www.amnesty.org/en/documents/asa21/6013/2022/en/.

Aspinall, Edward. 2004. "Indonesia: Transformation of Civil Society and Democratic Breakthrough." In *Civil Society and Political Change in Asia: Expanding and Contracting Democratic Space*, edited by Muthiah Alagappa, 61–96. Stanford, California: Stanford University Press. https://doi.org/10.1515/9780804767545-008.

Aspinall, Edward. 2005. *Opposing Suharto: Compromise, Resistance, and Regime Change in Indonesia*. East-West Center Series on Contemporary Issues in Asia and the Pacific. Stanford, California: Stanford University Press. https://doi.org/10.1515/9780804767316.

Aspinall, Edward. 2010. "*Assessing Democracy Assistance: Indonesia.*" FRIDE Project Report: Assessing Democracy Assistance. FRIDE.

Baik, Tae-Ung. 2010. "Public Interest Litigation in South Korea." In *Public Interest Litigation in Asia*, edited by Po Jen Yap and Holning Lau, 115–135. London: Routledge. https://doi.org/10.4324/9780203842645.

Banks, Nicola, David Hulme, and Michael Edwards. 2015. "NGOs, States, and Donors Revisited: Still Too Close for Comfort?" *World Development* 66: 707–718. https://doi.org/10.1016/j.worlddev.2014.09.028.

Chang, Wen-Chen. 2010. "Public Interest Litigation in Taiwan: Strategy for Law and Policy Changes in the Course of Democratisation." In *Public Interest Litigation in Asia*, edited by Po Jen Yap and Holning Lau, 136–160. London: Routledge. https://doi.org/10.4324/9780203842645.

Chua, Lynette J. 2014. *Mobilizing Gay Singapore: Rights and Resistance in an Authoritarian State*. Philadelphia: Temple University Press.

Chua, Lynette J. 2022. *The Politics of Rights and Southeast Asia*. Elements in Politics and Society in Southeast Asia. Cambridge: Cambridge University Press. https://doi.org/10.1017/9781108750783.

Cummings, Scott L., and Louise Trubek. 2008. "Globalizing Public Interest Law." *UCLA Journal of International Law and Foreign Affairs* 13 (1): 1–53. https://doi.org/10.2139/ssrn.1338304.

Epp, Charles R. 1998. *The Rights Revolution: Lawyers, Activists, and Supreme Courts in Comparative Perspective*. Chicago: University of Chicago Press. https://doi.org/10.7208/chicago/9780226772424.001.0001.

Fu, Hualing. 2011. "Challenging Authoritarianism Through Law: Potentials and Limit." *National Taiwan University Law Review* 6 (1): 339–365.

Fu, Hualing. 2018. "The July 9th (709) Crackdown on Human Rights Lawyers: Legal Advocacy in an Authoritarian State." *Journal of Contemporary China* 27 (112): 554–568. https://doi.org/10.1080/10670564.2018.1433491.

Fu, Hualing, and Richard Cullen. 2010. "The Development of Public Interest Litigation in China." In *Public Interest Litigation in Asia*, edited by Po Jen Yap and Holning Lau, 9–34. London: Routledge. https://doi.org/10.4324/9780203842645.

Ginsburg, Tom. 2007. "Law and the Liberal Transformation of the Northeast Asian Legal Complex in Korea and Taiwan." In *Fighting for Political Freedom: Comparative Studies of the Legal Complex and Political Liberalism*, edited by

Terence C. Halliday, Lucien Karpik, and Malcolm M. Feeley, 43–63. Oxford: Hart Publishing. https://doi.org/10.5040/9781472560179.ch-002.

Girei, Emanuela. 2016. "NGOs, Management and Development: Harnessing Counter-Hegemonic Possibilities." *Organization Studies* 37 (2): 193–212. https://doi.org/10.1177/0170840615604504.

Gobe, Eric, and Lena Salaymeh. 2016. "Tunisia's 'Revolutionary' Lawyers: From Professional Autonomy to Political Mobilization." *Law & Social Inquiry* 41 (2): 311–345. https://doi.org/10.1111/lsi.12154.

Goedde, Patricia. 2009. "From Dissidents to Institution-Builders: The Transformation of Public Interest Lawyers in South Korea." *East Asia Law Review* 4 (1): 63–89. https://doi.org/10.1017/9781108893947.009.

Hilbink, Thomas M. 2004. "You Know the Type: Categories of Cause Lawyering." *Law & Social Inquiry* 29 (3): 657–698. https://doi.org/10.1086/430155.

Hunt, Alan. 1986. "The Theory of Critical Legal Studies." *Oxford Journal of Legal Studies* 6 (1): 1–45. https://doi.org/10.1093/ojls/6.1.1.

Kelman, Mark G. 1987. *A Guide to Critical Legal Studies.* Cambridge, Massachusetts and London, England: Harvard University Press.

Kuwado, Fabian Januarius. 2015. "Adnan Buyung Menulis Pesan Terakhirnya Sambil Menangis…" *Kompas.com*, 23 September 2015. https://nasional.kompas.com/read/xml/2015/09/23/15443901/Adnan.Buyung.Menulis.Pesan.Terakhirnya.Sambil.Menangis.

Lindsey, Tim. 2004. "Legal Infrastructure and Governance Reform in Post-Crisis Asia: The Case of Indonesia." *Asian-Pacific Economic Literature* 18 (1): 12–40. https://doi.org/10.1111/j.1467-8411.2004.00142.x.

Liu, Sida, and Terence C. Halliday. 2016. *Criminal Defense in China: The Politics of Lawyers at Work.* Cambridge Studies in Law and Society. Cambridge: Cambridge University Press. https://doi.org/10.1017/9781316677230.

Lubis, Todung Mulya. 1985. "Legal Aid in the Future (A Development Strategy for Indonesia)." *Third World Legal Studies* 4 (1): 133–148.

McEvoy, Kieran, and Anna Bryson. 2022. "Boycott, Resistance and the Law: Cause Lawyering in Conflict and Authoritarianism." *The Modern Law Review* 85 (1): 69–104. https://doi.org/10.1111/1468-2230.12671.

Meili, Stephen. 2009. "Staying Alive: Public Interest Law in Contemporary Latin America." *International Review of Constitutionalism* 9 (1): 43–71.

Mietzner, Marcus. 2015. "Dysfunction by Design: Political Finance and Corruption in Indonesia." *Critical Asian Studies* 47 (4): 587–610. https://doi.org/10.1080/14672715.2015.1079991.

Moustafa, Tamir. 2007a. "Mobilising the Law in an Authoritarian State: The Legal Complex in Contemporary Egypt." In *Fighting for Political Freedom: Comparative Studies of the Legal Complex and Political Liberalism*, edited by Terence C. Halliday, Lucien Karpik, and Malcolm M. Feeley, 193–218. Oxford: Hart Publishing. https://doi.org/10.5040/9781472560179.ch-006.

Moustafa, Tamir. 2007b. *The Struggle for Constitutional Power: Law, Politics, and Economic Development in Egypt.* Cambridge: Cambridge University Press. https://doi.org/10.1017/cbo9780511511202.

Munger, Frank, Scott L. Cummings, and Louise Trubek. 2013. "Mobilising Law for Justice in Asia: A Comparative Approach." *Wisconsin International Law Journal* 31 (3): 353–420.

Mustafina, Renata. 2022. "Turning on the Lights? Publicity and Defensive Legal Mobilization in Protest-Related Trials in Russia." *Law & Society Review* 56 (4): 601–622. https://doi.org/10.1111/lasr.12631.

Nasution, Adnan Buyung. 1994. "Defending Human Rights in Indonesia." *Journal of Democracy* 5 (3): 114–123. https://doi.org/10.1353/jod.1994.0048.

Nasution, Adnan Buyung. 2011. "*Towards Constitutional Democracy in Indonesia.*" 1. Papers on Southeast Asian Constitutionalism. Melbourne: Asian Law Centre, Melbourne Law School.

Nesossi, Elisa. 2015. "Political Opportunities in Non-Democracies: The Case of Chinese Weiquan Lawyers." *International Journal of Human Rights* 19 (7): 961–978. https://doi.org/10.1080/13642987.2015.1075305.

O'Brien, Kevin J. 2023. "Neither Withdrawal nor Resistance: Adapting to Increased Repression in China." *Modern China* 49 (1): 3–25. https://doi.org/10.1177/00977004221119082.

Pils, Eva. 2014. *China's Human Rights Lawyers: Advocacy and Resistance.* Routledge Research in Human Rights Law. London and New York: Routledge. https://doi.org/10.4324/9780203769061.

Power, Thomas, and Eve Warburton, eds. 2020. *Democracy in Indonesia: From Stagnation to Regression?* Singapore: ISEAS Publishing. https://doi.org/10.1355/9789814881524.

Robison, Richard, and Vedi R. Hadiz. 2004. *Reorganising Power in Indonesia: The Politics of Oligarchy in an Age of Markets.* London and New York: RoutledgeCurzon. https://doi.org/10.4324/9780203401453.

Sarat, Austin, and Stuart A. Scheingold. 1998. "Cause Lawyering and the Reproduction of Professional Authority: An Introduction." In *Cause Lawyering: Political Commitments and Professional Responsibilities*, edited by Austin Sarat and Stuart A. Scheingold, 3–28. New York; Oxford: Oxford University Press. https://doi.org/10.1093/oso/9780195113198.003.0001.

Sarat, Austin, and Stuart A. Scheingold. 2001. "State Transformation, Globalization, and the Possibilities of Cause Lawyering: An Introduction." In *Cause Lawyering and the State in a Global Era*, edited by Austin Sarat and Stuart A. Scheingold, 3–31. New York: Oxford University Press. https://doi.org/10.1093/0195141172.003.0001.

Scheingold, Stuart A., and Austin Sarat. 2004. *Something to Believe In: Politics, Professionalism, and Cause Lawyering*. Stanford, California: Stanford University Press.

Stern, Rachel E. 2013. *Environmental Litigation in China: A Study in Political Ambivalence.* New York: Cambridge University Press. https://doi.org/10.1017/cbo9781139096614.

Stern, Rachel E., and Kevin J. O'Brien. 2012. "Politics at the Boundary: Mixed Signals and the Chinese State." *Modern China* 38 (2): 174–198. https://doi.org/10.1177/0097700411421463.

Tam, Waikeung. 2013. *Legal Mobilization Under Authoritarianism: The Case of Post-Colonial Hong Kong*. Cambridge: Cambridge University Press. https://doi.org/10.1017/cbo9781139424394.

Vieira, Oscar Vilhena. 2008. "Public Interest Law: A Brazilian Perspective." *UCLA Journal of International Law and Foreign Affairs* 13: 219.

Winters, Jeffrey A. 2011. *Oligarchy.* Cambridge: Cambridge University Press. https://doi.org/10.1017/cbo9780511793806.

Winters, Jeffrey A. 2013. "Oligarchy and Democracy in Indonesia." *Indonesia*, no. 96 (October): 11–33. https://doi.org/10.5728/indonesia.96.0099.

Yap, Po Jen, and Holning Lau. 2010. "Public Interest Litigation in Asia: An Overview." In *Public Interest Litigation in Asia*, edited by Po Jen Yap and Holning Lau, 1–8. London: Routledge. https://doi.org/10.4324/9780203842645.

Zen, A. Patra M., Robertus Robet, Daniel Hutagalung, Daniel Pandjaitan, Ikravany Hilman, and Rita Novella. 2004. "*Public Report: One Year Period Yayasan LBH Indonesia the Board of Director (October 2002-December 2003)*." Annual Report. Jakarta: Yayasan Lembaga Bantuan Hukum Indonesia.

Index